BIOMATERIALS AND NANOTECHNOLOGY FOR TISSUE ENGINEERING

Editor

Emeka Arendse

scitus academics

Biomaterials and Nanotechnology for Tissue Engineering
Edited by **Emeka Arendse**

ISBN: 978-1-68117-720-5
Library of Congress Control Number: 2016958814

SCITUS Academics LLC,
616, Corporate Way, Suite 2, 4766,
Valley Cottage, NY 10989

www.scitusacademics.com

Notice

Printed in United States of America on Acid Free Paper

PREFACE

Nanotechnology and high-end characterization techniques have emphasized the value of tissue engineering material choice. Tissue engineering includes the seeding of cells—allowing adhesive surfaces—on bio-mimicked frameworks. Researchers, though, deal with a number of challenges in tissue generation that can be avoided by using nanotechnology. It supplies cell adhesion substrates and agents and proliferation for cell growth and can also be used to create nanoparticles and nanostructures to help engineer different types of tissue. Due to the tissue and organ shortage, tissue engineering strategies combining engineering principles and biological sciences and methods for the creation of implantable tissues have probably been considered the most promising regeneration technologies for damaged organs and tissues. Since native tissues or organs consist of cell proteins and nanoscale interact directly with nanostructured extracellular matrices, nanobiomaterials such as nanofibres, nanotubes, nanoparticles and other nanofabricated devices smaller than 100 nm can contribute to cell growth and tissue regeneration at least in one dimension.

This book integrates up-to-date information on methods for the design and use of tissue regeneration biomaterials, on the new surface characteristics of nanomaterials, as well as on theranostics, immunology and nano-toxicology aspects. The book also illustrates the manufacturing techniques for the production of frames for a range of tissue-specific applications of frames in tissue engineering for various types and specific tissue biomaterials. Some nanoscale delivery systems are highlighted, which are important milestones in the history of drug delivery. In addition, developments in the delivery of nanodrugs, gene therapy and nanotechnology for cancer are described.

The book helps readers to learn more about the nanotechnology aspects of tissue engineering and will be of great use to those involved in the construction of specific tissue substitutes in order to achieve their goal more efficiently. It shows the latest nanotechnology achievement and its development in the field of tissue engineering. It is intended for researchers and practitioners, laboratory engineers, lecturers and students in tissue engineering or, more generally, regenerative medicine, nanomedicine, medical equipment, nanofabrication, biofabrication, biomaterials and biomedical engineering.

TABLE OF CONTENTS

1

BIOMATERIAL TECHNOLOGY FOR TISSUE ENGINEERING APPLICATIONS

Yasuhiko Tabata

Department of Biomaterials, Field of Tissue Engineering, Institute for Frontier Medical Sciences, Kyoto University, 53 Kawara-cho Shogoin, Sakyo-ku, Kyoto 606-8507, Japan

ABSTRACT

Tissue engineering is a newly emerging biomedical technology and methodology to assist and accelerate the regeneration and repairing of defective and damaged tissues based on the natural healing potentials of patients themselves. For the new therapeutic strategy, it is indispensable to provide cells with a local environment that enhances and regulates their proliferation and differentiation for cell-based tissue regeneration. Biomaterial technology plays an important role in the creation of this cell environment. For example, the biomaterial scaffolds and the drug delivery system (DDS) of biosignalling molecules have been investigated to enhance the proliferation and differentiation of cell potential for tissue regeneration. In addition, the scaffold and DDS technologies contribute to develop the basic research of stem cell biology and medicine as well as obtain a large number of cells with a high quality for cell transplantation therapy. A technology to genetically engineer cells for their functional manipulation is also useful for cell research and therapy. Several examples of tissue engineering applications with the cell scaffold and DDS of growth factors and genes are introduced to emphasize the significance of biomaterial technology in new therapeutic and research fields.

Keywords: biomaterials; drug delivery system; biosignalling molecules; tissue engineering; tissue regeneration

1. SIGNIFICANCE OF BIOMATERIAL TECHNOLOGIES IN TISSUE ENGINEERING APPLICATIONS

Advanced surgical therapies currently available consist of reconstruction surgery and organ transplantation. Although there is no doubt that these therapies have saved and improved countless lives, they have several therapeutic and methodological limitations. In the case of reconstruction surgery, biomedical devices cannot completely substitute the biological functions even for a single tissue or organ, and consequently cannot prevent the progressive deterioration of injured or damaged tissues and organs. One of the biggest issues for organ transplantation is the shortage of donor tissues or organs. Additionally, the continuous and permanent use of immunosuppressive agents to prevent immunological rejection responses often causes side effects, such as the high possibility of bacterial infection, carcinogenesis and virus infection. To resolve these issues in the two advanced therapies, a new therapeutic solution that is clinically mild to patients is required.

In this clinical situation, a new therapeutic trial, in which disease healing can be achieved based on the natural healing potential of patients, has been explored. This trial is termed tissue regeneration therapy where the regeneration of tissues and organs is naturally induced to therapeutically treat diseases by artificially promoting the potential of cell proliferation and differentiation. To realize this cell-induced regeneration therapy, there are two approaches. One is cell transplantation where cells with a high potential of proliferation and differentiation are transplanted to induce tissue regeneration based on their potentials. The other is the therapeutic approach with biomaterials and technologies. In the latter approach, an *in vivo* local environment that enables cells to promote their proliferation and differentiation is created by making use of biomaterials and technologies. If the environment efficiently manipulates the cells inherently present in the body to enhance the biological potentials of tissue regeneration, cell-induced natural healing of tissues and organs will be achieved without cell transplantation. This approach is called tissue engineering. This basic concept of biomaterial-based tissue engineering was originally introduced by Langer & Vacanti (1993). Cell scaffold and biosignalling molecule delivery technologies with biomaterials have been demonstrated to create cell environments suitable for tissue regeneration (Saltzman & Olbricht 2002; Bach *et al.* 2003; Bannasch *et al.* 2003; Chen & Mooney 2003; Hubbell 2004; Langer & Tirrell 2004; Silva & Mooney 2004; Kuo *et al.* 2006; Leo & Grande 2006; Tabata 2008). For the former approach, generally, cells are transplanted into the body by the bolus injection or infusion method. However, few cells are retained at the transplanted site and their grafted rate is very low because of their excretion and death. In addition, the cells infused are hardly accumulated and grafted at the site to be regenerated. This low grafting rate of cells transplanted and their consequent poor functions often cause low therapeutic efficacy of cell transplantation. To overcome these problems, it is necessary to give the cells an environment suitable to their survival and functional achievement. Biomaterials play a key role in creating the environment for cells. The scaffold to promote the proliferation and differentiation is prepared from biomaterials, while the biomaterial is used as the delivery carrier of biosignalling molecules as the cell nutrients to biologically activate cells. Combination with biomaterials enables an angiogenic factor to efficiently induce *in vivo*angiogenesis, which gives nutrients and oxygen to the transplanted cells. Biomaterials need to assist the approach of cell transplantation and enhance the therapeutic efficacy.

This new regeneration therapy cannot always therapeutically substitute the reconstructive surgery and organ transplantation clinically available, and has advantages and disadvantages. However, it is clinically expected as the third therapeutic choice. If this tissue regeneration therapy is realized, it will enable us to create new therapeutic strategies as well as increase the therapeutic choice of clinicians, which consequently brings about large therapeutic benefits for patients who have not been able to receive clinically effective therapies. There are three objectives of regeneration therapy. The first objective is to create a new therapeutic strategy of surgery and internal medicine, which is generally well known. The second objective is to enlarge the clinical application of therapies conventionally available. Conventional surgical therapy is not always effective in treating patients who are aged or suffer from other diseases, such as diabetes and hyperlipaemia, or cannot be applied because the number of key cells is small and their potential for proliferation and differentiation is low. In this case, it is practically possible that combination with the technology and methodology to promote cell-based self-healing potentials improves the therapeutic efficacy even for patients who have been clinically treated. The third objective is to suppress the progressive deterioration of diseases. The deterioration and progress of disease conditions are suppressed by artificially promoting cell potentials to induce tissue regeneration. For example, in chronic fibrosis diseases, the fibrous tissue of excessive collagen fibres and fibroblasts causes the impairment of natural healing processes at the disease site. If the fibrosis can be loosened and digested, and additionally the natural healing potential of the surrounding healthy tissue can be augmented, it is highly expected that the disease deterioration and progression can be suppressed in a physiologically natural manner.

At present, there is no effective medical therapy for chronic fibrosis diseases, such as lung fibrosis, cirrhosis, dilated cardiomyopathy and chronic nephritis. For these diseases, the injured site is normally occupied with fibrous tissue of excessive collagen fibres and the fibroblasts excessively proliferate. It is possible that this tissue occupation causes the physical impairment of healing processes at the disease site. Even in adults, the natural healing potential for tissue regeneration still remains, but cannot operate naturally for certain reasons in disease conditions. For example, therefore, if the fibrosis can be loosened and digested to disappear by any method of drug treatment, it is highly expected that the disease site will be regenerated and repaired based on the natural regeneration potential of the surrounding healthy tissue. This trial is a new and possible therapy for chronic fibrosis diseases and defined as 'tissue regeneration of internal medicine' because of the drug treatment application of internal medicine (figure 1). This therapeutic approach is similar to the surgical regeneration therapy where the cell, the scaffold and the growth factors, or their combination, are surgically applied to the tissue defect for generation therapy, because both approaches are based on the natural healing potential of patients. We have demonstrated that the controlled release of a matrix metalloproteinase (MMP)-1 plasmid DNA at the medulla of chronic renal sclerosis induced the histological regeneration of kidney structure, in contrast to the plasmid DNA solution (Aoyama *et al.* 2003). The intraperitoneal release of hepatocyte growth factor (HGF) histologically cured the liver fibrosis of rats with liver cirrhosis (Oe *et al.* 2003). A biodegradable hydrogel could achieve the controlled release of small interfering RNA (siRNA) for transforming growth factor (TGF)-β1 type II receptor and regenerate and repair the fibrosis of chronic renal sclerosis (Kushibiki *et al.* 2006).

The basic idea of tissue regeneration therapy is to take advantage of the natural healing potentials of patients themselves. Thus, this is applicable for another therapy of internal medicine. For conventional catheter treatment, an aneurysm occlusion with blood clots has been clinically performed. However, sometimes the recurrence of aneurysm due to clot lysis is clinically problematic. As one trial to overcome the problem, aneurysm occlusion with tissue organization has been achieved by using coils incorporating basic fibroblast growth factor (bFGF; Kawakami *et al.* 2005). The bFGF release promoted the cell proliferation inside the aneurysm to allow occlusion by the tissue organized. The tissue regeneration strategy will be applied to the therapy of internal medicine.

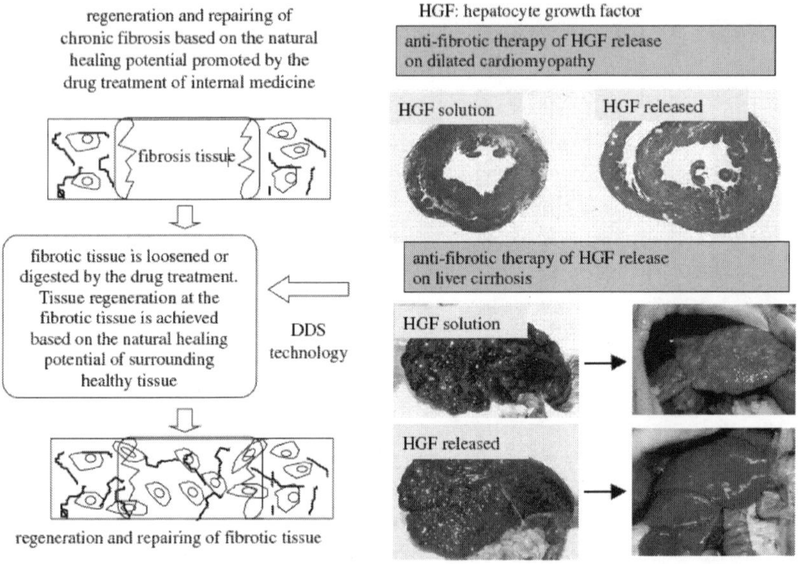

Figure 1. A new therapeutic strategy for chronic fibrotic diseases based on the natural healing potentials of patients themselves. The potential is assisted and promoted for tissue regeneration by DDS technology.

One of the therapeutic advantages is the ability to accelerate the natural healing of body injury through promoted angiogenesis or the infiltration and recruitment of key cells at the injured site. This will enable patients to shorten the healing period and suppress the deterioration process of disease even under inflammation and infection conditions. A disadvantage of this therapy is that, generally, at least a few days are required to induce and activate cell-based tissue regeneration. Consequently, it cannot be expected that tissue regeneration therapy alone will achieve the rapid healing of wounds or diseases. Depending on the clinical situation, it is necessary for better medical treatment to combine the conventional therapies with the tissue regeneration strategy.

2. FUNDAMENTAL BIOMATERIAL TECHNOLOGY AND METHODOLOGY FOR TISSUE ENGINEERING-BASED REGENERATIVE THERAPY

Basically, body tissue is composed of two components: cells and the surrounding environment. The latter includes the extracellular matrix (ECM) for cell proliferation and differentiation (natural scaffold) as the living place of cells and biosignalling molecules as the nutrients of cells. There are some cases where tissue regeneration is achieved by the single or combinational use of the components in an appropriate way. However, since successful tissue regeneration cannot always be expected only by their simple combination, it is necessary to biomedically contrive the way to combine. To this end, proper and positive assistance of biomaterial technology will be practically promising. Biomaterials play a key role in designing and creating substitutes for ECM and the drug delivery system (DDS) of biosignalling molecules to enhance their biological activities. In addition to therapeutic applications, biomaterials are also useful in the progress of research and development of stem cell biology and medicine.

As the biomaterials, various synthetic and natural materials, such as polymers, ceramics and metals or their composites, have been investigated and used in different manners. Among them, biodegradable biomaterials are explained here. From the practical viewpoint, metals and ceramics except for calcium carbonate and tricalcium phosphate are not biodegradable. On the other hand, some polymers are biodegradable materials (table 1). The word 'biodegradation' is defined to be the phenomenon where a material is degraded or water solubilized by any process in the body to disappear from the site implanted. There are two ways of material disappearance. First, the main chain of the material is hydrolysed or enzymatically digested to decrease the molecular weight, and finally disappears. Second, the material is chemically cross-linked to form a hydrogel insoluble in water. When the cross-linking bond is degraded to generate water-soluble fragments, the fragments are washed out from the site implanted, resulting in the disappearance of material. Synthetic polymers are generally degraded by simple hydrolysis while natural polymers are mainly degraded enzymatically. The synthetic polymers can be modified with ease to change their chemical composition and molecular weight, which affect the physicochemical property of the materials. Natural polymers of proteins, polysaccharides and nucleic acids are available. Their degree of freedom for property modification is small when compared with that of synthetic polymers, but they can be chemically modified to produce various derivatives. The natural polymers are normally used in the formulation of hydrogels prepared by chemical cross-linking. Generally, the synthetic polymers are hydrophobic and mechanically strong when compared with natural ones; in other words, their degradation rates are comparatively slow. For the purpose of material application to tissue regeneration, the retention of biomaterials implanted in the body often causes the physical impairment of tissue regeneration. On the other hand, an appropriate mechanical strength of materials is also required. Generally, the mechanical strength of materials weakens as their degradation becomes faster. The two opposite properties should be balanced by material design and combination.

Table 1. Biodegradable polymers used for tissue engineering of cell scaffold and biosignalling molecule release.

synthetic polymers	natural polymers
poly(L-lactic acid) (PLLA)	collagen
poly(glycolic acid) (PGA)	gelatin
poly(e-caprolactone) (PCA)	fibrin
copoly(LL-GA)	
copoly(LL-CA)	hyaluronic acid[a]
copoly(LLA-ethylene glycol (EG))	alginate[a]
copoly(fumarate-EG)	chitosan, chitin

[a]There are no enzymes in the body to directly degrade these polymers. They are washed out by body fluids to disappear from the implanted site.

There are five key technologies or methodologies that are necessary for biomaterial-based tissue regeneration therapy and the basic stem cell research that scientifically supports the future regeneration therapy of cell transplantation. The first key technology is for the preparation of cell scaffolds to promote cell proliferation and differentiation for *in vivo*tissue regeneration (figure 2*a*). ECM is not only a physical support for cells, but also provides a natural environment for cell proliferation and differentiation or morphogenesis, which contributes to cell-based tissue regeneration and organogenesis. Generally, it is difficult to naturally regenerate and repair a large-size tissue defect only by supplying cells to the defective site, because both the cells and the ECM as well as the surrounding environment are lost. Therefore, to induce tissue regeneration at the defective site, one possible way is to artificially build a local environment for cells, which is a three-dimensional scaffold of artificial ECM to initially assist their attachment and subsequent proliferation and differentiation, inducing cell-based tissue regeneration. It is expected that cells residing around the implanted scaffold infiltrate into the scaffold and consequently proliferate and differentiate therein if the artificial ECM is biologically compatible. Biomaterials play an important role in the preparation of cell scaffolds. Basically, the scaffold should be porous and biodegradable. The pore structure is necessary for the infiltration of cells into the scaffold, the supply of oxygen and nutrients to cells proliferated therein and the washout of cell wastes. Once the cell-induced tissue regeneration is naturally initiated, cells eventually produce the ECM of natural scaffold. The scaffold is a temporary platform of cell activities. The long-term retention of cell scaffold sometimes causes physical hindrance against the natural process of tissue regeneration. It is a key for successful tissue regeneration to control the time profile of scaffold biodegradation at the defect as well as the three-dimensional structure. The cell scaffold developed initially is a physical entity that does not have any functions to positively promote the proliferation and differentiation of cells. However, a new type of cell scaffold has been investigated to combine biomaterials with biological ligands reactive for cell receptors, biosignalling molecules and cell-adhesive substances for artificially promoted cell-based tissue regeneration (Lutolf & Hubbell 2005; Chan & Mooney 2008). The mechanical property of biomaterial scaffolds is also important. If the mechanical strength is low, the scaffold readily deforms in the body. As a result, the regeneration of bulky tissue cannot be always expected. An appropriate mechanical design is much required. For example, the incorporation of fibres and ceramic granules enabled biomaterial sponges to mechanically reinforce and enhance *in vivo* tissue regeneration following the implantation of sponges combined with stem cells (Hiraoka *et al.* 2003; Takahashi *et al.* 2005).

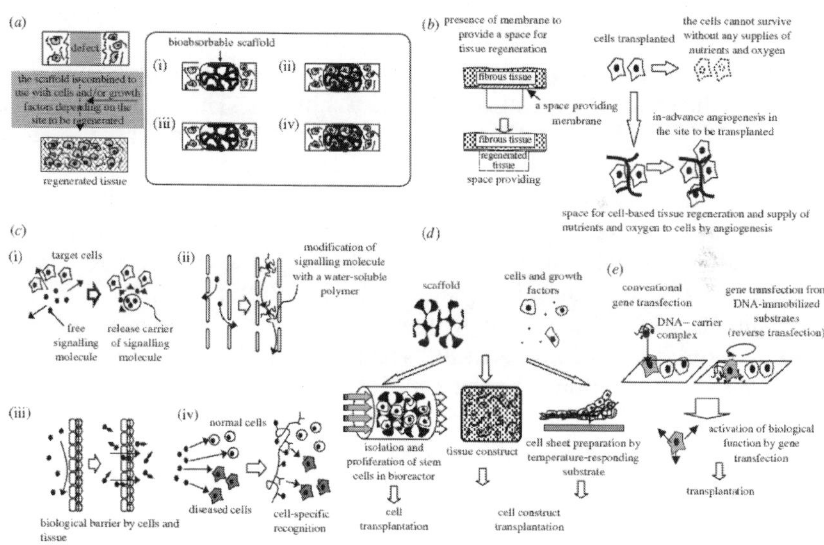

Figure 2. Role of biomaterials in tissue engineering-based regeneration therapy. (*a*) Biomaterials for cell scaffold to induce *in vivo* tissue regeneration. Bioabsorbable scaffold: (i) without cells and growth factors, (ii) with cells, (iii) with growth factors, ...

The second key technology is to provide the space for cell-based tissue regeneration and supply nutrients and oxygen to cells by angiogenesis (figure 2*b*). When a body defect is generated, the defect space is generally occupied rapidly by the fibrous tissue produced by fibroblasts, which are ubiquitously present in the body and can proliferate rapidly. This is one of the typical wound healing processes to temporarily fill and repair a body defect in case of emergency. However, once this ingrowth of fibrous tissue into the space to be regenerated takes place, the regeneration and repairing of a target tissue at the space can no longer be expected. To prevent tissue ingrowth, a barrier membrane to provide a space for tissue regeneration is required. The scaffold (figure 2*a*) sometimes functions to provide the space for tissue regeneration. To expect cell-based tissue regeneration, the site at which the cells are to be transplanted should be physiologically arranged. For the transplantation of cells, if there are no supplies of nutrients and oxygen, it is not practically expected that the cells transplanted would survive and function well to induce tissue regeneration. A similar problem is observed for the transplantation of stem cells although their potential for proliferation and differentiation is superior to that of matured cells. For example, as one possible strategy to tackle the issue, it is promising to induce *in vivo* angiogenesis by biomaterial technology and methodology (Tabata & Ikada 1999; Tabata 2003*a*, 2005*a*,*b*;Tambara *et al.* 2005; Soto-Gutierrez *et al.* 2006; Takehara *et al.* 2008).

When the tissue around a defect does not have the inherent potential to regenerate, tissue regeneration cannot always be expected if only the scaffold or the space-providing membrane is supplied. The cell scaffold and membrane of biomaterials should be used in combination with the cells or/and signalling molecules (e.g. growth factors, cytokines, chemokines) that have the potential to accelerate cell-induced tissue regeneration. Currently, several stem cells with a high potential for proliferation and differentiation are being prepared and applied to a tissue defect to induce tissue regeneration therein (Salgado*et al.* 2006; Fernyhough *et al.* 2008; Mountford 2008). Although there are some cases where a growth factor is required to promote tissue regeneration, the direct injection of growth factor in solution into the site to be regenerated is generally not effective. This is because the growth factor rapidly diffuses from the injected site and is enzymatically digested or deactivated. To enable the growth factor to efficiently exert its biological function, a new technology is required. This comes in the form of the third key technology of tissue engineering: DDS (figure 2*c*). Although every DDS technology is available for tissue engineering, the release technology of growth factors and genes has been mainly applied to induce tissue regeneration. For example, the controlled release of growth factor at the site of action over an extended period of time is achieved by incorporating the factor into an appropriate biomaterial carrier. It is also possible that when incorporated in the release carrier, the growth factor is protected against proteolysis so as to prolong the activity retention *in vivo*. The release carrier should be degraded in the body, because it becomes useless after the release function is completed. In addition to the controlled release, the improvement of signalling molecule stabilization, the prolongation of the *in vivo* half-life and the targeting to the site of action will enhance the factor-induced tissue regeneration.

Over time, DDS has been investigated and developed as the only technology or methodology to enhance the *in vivo* efficacy of therapeutic drugs. Based on fixed ideas and historical background, it has been thought that DDS cannot be scientifically and technologically applied to tissue regeneration therapy. There has been little investigation of DDS-based tissue regeneration. Considering that drugs applicable for regenerative therapy include proteins and genes effective in promoting the proliferation and differentiation of cells to induce tissue and organ regeneration, generally, they are biologically unstable *in vivo*. Therefore, upon *in vivo* administration of biosignalling molecules, it is necessary for enhanced *in vivo* biological activity to make use of the DDS technology and methodology. Many types of biomaterials have been used for DDS applications. The application of DDS to tissue regenerative therapy will be described later.

As mentioned earlier, it is necessary to prepare cells for transplantation therapy. Therefore, basic research of cell biology and medicine is actively ongoing to select candidate cells suitable for cell therapy. Among the candidates, cells that have proliferation and differentiation potentials higher than matured cells, so-called stem cells, are promising from the viewpoint of cell-based tissue regeneration. For example, in addition to haemopoietic stem cells, mesenchymal stem cells (MSC) are present in adult bone marrow. It has been elucidated that the MSC of adult stem cells have a proliferative ability and an inherent potential to differentiate into osteogenic, chondrogenic, adipogenic and myocardial cell lineages. Presently, human MSC are isolated and are commercially available (Pittenger *et al.* 1999) while clinical experiments have been begun. If it is possible to clinically use a differentiated type of a patient's own MSC, immunological rejection will no longer be a question. Many researchers of tissue regeneration therapy with stem cells, especially MSC, have reported their therapeutic feasibility in tissue regeneration (Leo & Grande 2006; Docheva *et al.* 2007; Fibbe *et al.* 2007; Picinich *et al.* 2007; Shanti *et al.* 2007). In addition, neural stem cells (Hsu *et al.* 2007; Kornblum 2007) and stem cells isolatable from fat tissue (Strem & Hedrick 2005; Gimble *et al.* 2007;Schaffler & Buchler 2007) have been extensively investigated. They can be prepared from the foetus and adult and have exhibited plasticity for cell differentiation and are expected to be promising cell candidates for regenerative medical therapy. Embryonic stem (ES;Mountford 2008) and inducible pluripotent stem (iPS) cells (Yamanaka 2007) have been established and expected as cell sources for transplantation therapy and the research and development of drugs.

However, one of the problems is the shortage of cells clinically available. Therefore, it is necessary to develop a technology or methodology for the preparation of a large number of stem cells of high quality. For this purpose, isolation, induction and *in vitro* culture technologies of stem cells are required. The fourth technology is for the efficient preparation and proliferation of cells (figure 2*d*), which are achieved by providing a cell culture substrate as the artificial ECM. The cell scaffold for *in vivo* tissue regeneration mentioned previously can be used for the purpose of culture substrate. The three-dimensional substrate can be designed and prepared from biomaterials of cytocompatibility. From the viewpoint of nutrients and oxygen supplies, research and development of cell culture methods and bioreactors are required (Holtorf *et al.* 2006;Mironov *et al.* 2008).

The fifth technology is to genetically engineer cells for their functional manipulation and basic biology. There are some cases where cells transplanted do not function well to induce cell-based tissue regeneration. As one trial to tackle this issue, cells are genetically engineered with biomaterials to activate the biological functions. It is necessary for the genetic engineering of cells to develop a carrier of gene transfection and cell culture system for efficient gene expression. Generally, viral vectors have been scientifically used for gene transfection because of their high efficiency. However, viruses cannot be used to treat patients; thus, non-viral gene carrier biomaterials are required to be developed from the clinical viewpoint of cell therapy. This technology is also applicable for the basic research of stem cell biology and medicine, which gives important knowledge and results for cell therapy. This combination of cell scaffold, space protection and DDS technologies is practically promising to create an environment that promotes the proliferation and differentiation of cells for cell-induced tissue regeneration. Both the culturing and genetic engineering of cells are key technologies to prepare cells clinically available for cell therapy. Every biomaterial-based technology is important not only to develop the basic research of stem cell biology and medicine, but also to realize cell-based tissue regeneration therapy.

3. CLINICAL ASPECTS OF TISSUE ENGINEERING-BASED TISSUE REGENERATION

Tissue engineering for clinical regeneration therapy can be classified as either *in vitro* or *in vivo* depending on the site where tissue regeneration or organ substitution is performed. *In vitro* tissue engineering involves tissue reconstruction by cell culture methods and organ substitution with functional cells—termed bioartificial hybrid organ. If a tissue can be reconstructed *in vitro* in factories or laboratories on a large scale, it can be supplied to patients when required. For example, human skin fibroblasts are cultured in a collagen sponge to prepare an artificial dermis for skin grafting (Kuroyanagi *et al.* 2001; Kumagai 2002; Ichioka *et al.*

2005; Takemoto *et al.* 2008). To prepare pulmonary vein and bone substitutes for human treatment, the cells of blood vessel and bone marrow-derived stem cells are cultured in the porous scaffolds of tube-shape polylactide (Shin'oka *et al.* 2001) and cubic hydroxyl apatite (Ohgushi & Caplan 1999). However, it is difficult to reproduce the *in vivo* event completely *in vitro* by using the basic knowledge of biology and medicine or cell culture technologies currently available. At present, it is difficult to realize*in vitro* tissue engineering because the artificial arrangement of a biological environment to induce cell-based tissue reconstruction is practically impossible. Even if a three-dimensional tissue-like construct is prepared *in vitro*, it is practically difficult to allow the construct to survive and function *in vivo* after grafting. In addition, the construct does not always connect with surrounding natural tissue biologically. The *in vivo* environment for the construct implanted should be designed. Another application of *in vitro* tissue engineering is the substitution of organ functions by the use of allo- or xenogeneic cells. The cells are combined with an immunoisolation membrane of biomaterials to substitute the physiological functions of liver and pancreas (Falqui *et al.* 1991; Olle *et al.* 1996;Yamashita *et al.* 2002; Kobayashi *et al.* 2003; Ehashi *et al.* 2006). Biomaterials have been investigated to design and create a biological environment that can assist the proliferation and differentiation of cells and maintain their biological functions.

Distinct from the *in vitro* tissue engineering, *in vivo* tissue engineering is advantageous from the viewpoint of the environment to induce tissue regeneration. It is likely that most of the biological components necessary for tissue regeneration, such as growth factors and cytokines, are naturally supplied by the body. Based on this advantage, almost all the approaches of tissue engineering have been performed *in vivo* with or without biodegradable cell scaffolds. There are several examples where *in vivo* tissue regeneration is achieved by use of the scaffold with or without cells (Tabata 2003*b*, 2005*b*, 2008). As described previously, if patients are young and healthy, and the tissue to be repaired has a high potential to induce regeneration, active and immature cells infiltrate the matrix of an implanted biodegradable scaffold from the surrounding healthy tissue, resulting in the formation of new tissue. However, additional means are required if patients are aged and/or suffer from other diseases, such as diabetes and hyperlipaemia, or if the regeneration potential of tissue is low as a result of, for example, a low concentration of cells and growth factors. The simplest method is to supply a growth factor to the site of regeneration for cell differentiation and proliferation in a controllable fashion. As described previously, to allow a growth factor of *in vivo* instability to efficiently function in the body, it is necessary to make use of the DDS technology.

One of the largest problems in the release technology of growth factor protein is the loss of biological activity of protein released from a protein-carrier formulation. It has been demonstrated that this activity loss mainly results from denaturation and deactivation of protein during the preparation process of the formulation (Gombotz & Wee 1998; Tabata 2003*b*). Therefore, a method to prepare the formulation of protein release with biomaterials should be exploited to minimize protein denaturation. From this viewpoint, a polymer hydrogel may be a preferable candidate as a protein release carrier because of its biosafety and its high inertness towards protein drugs. However, it will be practically impossible to achieve the controlled release of protein over a long period of time from hydrogels only when the protein is simply combined in the hydrogels. The protein combined is normally localized in the water phase inside the hydrogel and the release is generally diffusion controlled through the water pathway of hydrogels. The release profile is regulated by modifying the cross-linking density of hydrogel polymers. However, there is a limitation for the regulation of protein diffusion by the cross-linking density. Therefore, it is practically impossible to achieve protein release for more than a few days. Considering protein-induced activation of cell functions, protein release for at least 7 days or longer is required. Thus, it is necessary to contrive a method of protein release. A possible approach is to immobilize a growth factor in a biodegradable hydrogel. It should be noted that chemical immobilization is not suitable for this purpose, because the chemical reaction to growth factor often causes a loss of biological activity. Physical immobilization is preferable. The immobilized factor is not released by simple diffusion, but only by the solubilization of the factor in water accompanied with the generation of water-soluble hydrogel fragments as a result of hydrogel biodegradation. In such a release system, the time profile of growth factor release is governed only by that of the *in vivo*hydrogel degradation. The requirements of a hydrogel polymer for this release system are to have biodegradability and the ability to physically interact with protein for immobilization. In addition, from the viewpoint of therapeutic applications, it is preferable that the polymer has been in previous clinical use.

There have been several reports on the controlled release of proteins (Andrianov & Payne 1998; Fujioka *et al.* 1998; Gombotz & Wee 1998; Tessmar & Gopferich 2007). Some hydrogels were effective in enhancing the *in vivo* biological activity of growth factors to induce tissue regeneration (Lutolf & Hubbell 2005; Coviello *et al.* 2007; Van Tomme & Hennink 2007). Based on the requirement described above, we have selected gelatin to prepare a biodegradable hydrogel protein release carrier, because it has been clinically used in medical and pharmaceutical applications and proven to be biocompatible. As expected, the hydrogel of gelatin immobilized the growth factor through the physicochemical interaction between the gelatin and factor. The growth factor could be released from the hydrogel accompanied with the degradation of carrier hydrogel (Tabataet *al.* 1994, 1999; Tabata & Ikada 1999). Gelatin is easily chemically derivatized to change the molecular nature, which is susceptible to physical interaction with protein. Using the gelatin hydrogel, the controlled release of bioactive growth factors over a time range of 5 days to three months was possible. The controlled release of various growth factors has succeeded in the regeneration therapy of various tissues (figure 3). bFGF is one of the angiogenic factors and has the ability to enhance wound healing through an induction of angiogenesis and induce the regeneration of bone, cartilage, skin, nerve and fat tissues (Tabata 2007, 2008). When human recombinant bFGF (Fibrast spray, Kaken Pharmaceutical Co., Tokyo; http://www.kaken.co.jp) was incorporated into a gelatin hydrogel and subcutaneously implanted into a mouse back, significant angiogenic effect was observed around the implanted site, in marked contrast to the injection of bFGF solution even at higher doses (Ikada & Tabata 1998). Vascular endothelial growth factor (VEGF) is another angiogenic biosignalling molecule and has been extensively investigated to demonstrate the potential to induce angiogenesis in different forms of DDS modifications (Tabata *et al.* 2000*a*; Zisch *et al.* 2003; Patel *et al.* 2008). In general, however, blood vessels regenerated by VEGF are fine when compared with those of bFGF and VEGF often causes tissue oedema. Comparing the points, it is preferable to use bFGF for angiogenic therapy from the viewpoint of the therapeutic necessity of wide blood vessels.

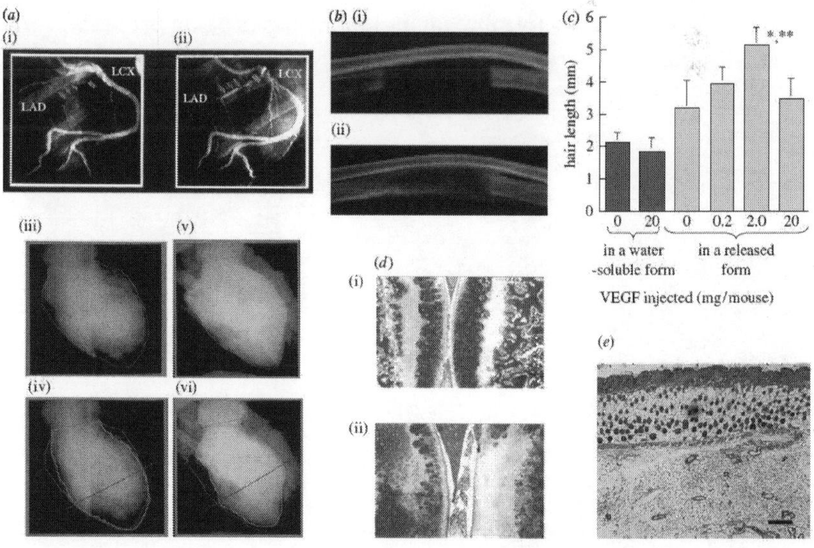

Figure 3. Examples of tissue regeneration with biodegradable hydrogels for growth factor release. (*a*) Regeneration of coronary artery: (i) bFGF

solution, and (iii) diastole, (iv) systole; (ii) gelatin microspheres incorporating bFGF, and (v) diastole, (vi) systole ...

There are two important objectives of angiogenesis in tissue engineering: the therapy of ischaemic disease and 'in-advance angiogenesis' for cell transplantation. As a first example, when injected into the ischaemic site of myocardial infarction (Iwakura *et al.* 2003) or leg ischaemia (Nakajima *et al.* 2004), gelatin microspheres incorporating bFGF induced angiogenesis to a significantly greater extent than the bFGF solution. This angiogenic therapy for leg ischaemia has been clinically shown to demonstrate good therapeutic efficacy (Marui *et al.* 2007; figure 4). This is the first report on clinical angiogenic therapy by using the DDS technology of growth factor without any cell transplantation. The granules of gelatin hydrogels incorporating bFGF were injected into the femoral muscle of the patient's ischaemic leg. bFGF is released locally around the injected site over two weeks and no bFGF in the blood circulation was detected. When compared before and after the bFGF treatment, the pain score, the tissue oxygen and the blood flow were all significantly improved. In addition, the walking distance of patients treated increased by a clinically significant extent. We have performed the bFGF-induced angiogenic clinical treatment for 25 patients in four university hospitals in Japan (ages: 27–73; male/female ratio=15/10). Depending on the patients' conditions, in the case of diabetic foot ulcer, out of seven patients, four have been healed to recover a condition clinically acceptable. In addition, the bFGF-induced angiogenic therapy was also effective in accelerating regeneration healing of the intractable foot ulcer of diabetes and the defect of periodontal tissue. Regeneration therapy of sternum, fat and meniscus is being proceeded with clinically (table 2).

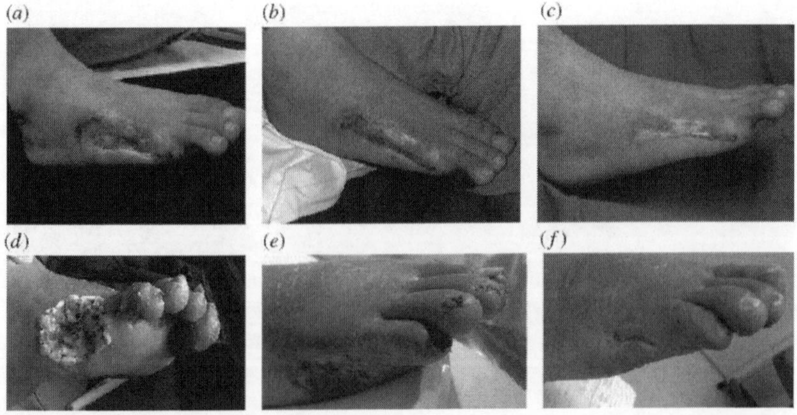

Figure 4. Examples of clinical tissue regeneration for ischaemic ASO and diabetic foot ulcer of intractable disease with biodegradable hydrogels for

bFGF release. Intractable diseases could be repaired only by the intramuscular injection or implantation of hydrogel ...

Table 2. Ongoing clinical experiments of tissue regeneration with the release technology of growth factors.

disease or operation	growth factor	effect	no. of facilities
vascular graft surgery for heart	bFGF	angiogenesis	1
arteriosclerosis obliterans (ASO) and buerger diseases	bFGF	angiogenesis	4
diabetic skin ulcer	bFGF	angiogenesis, dermatogenesis	3
periodontal disease	bFGF	regeneration of alveolar bone	3
hardness of hearing	IGF-1	inhibition of acoustic nerve ageing	2
meniscus injury	PRP	chondrogenesis	1
facial plastic surgery	bFGF	chondrogenesis, soft tissue regeneration	1
chest plastic surgery	bFGF	regeneration of sternum bone, angiogenesis	1
plastic surgery of soft tissues	bFGF	angiogenesis	1
facial nerve paralysis	bFGF	nerve function recovery	1

The sufficient supply of nutrients and oxygen to cells transplanted is indispensable for their survival and maintenance of biological functions *in vivo* (figure 1*b*). For successful cell transplantation, it is beneficial to induce angiogenesis in advance throughout the site where cells are transplanted, by using the bFGF release system. This technology of in-advance angiogenesis efficiently enhanced the grafting rate and improved the biological functions of pancreatic islets (Balamurugan *et al.* 2003), hepatocytes (Ogawa *et al.* 2001), cardiomyocytes (Sakakibara *et al.* 2002) and kidney cells (Saito *et al.* 2003), as well as the engrafting of a bioartificial dermis–epidermis skin-tissue construct (Tsuji-Saso *et al.* 2007). Consequently, the therapeutic efficacy induced by the cells and constructs implanted was significantly augmented compared with that of no angiogenesis. The release system enabled the enhanced activity of various growth factors, such as bFGF, TGF-β1 and bone morphogenetic protein-2 (BMP-2) to induce bone regeneration and bone healing, as well as to synergistically promote bone regeneration induced by MSC of bone marrow (Tabata*et al.* 2000*b*). It is well known that insulin-like growth factor (IGF)-1 suppresses the apoptosis of nerve cells. Controlled release of IGF-1 from the gelatin hydrogel inhibited the ageing of acoustic nerve, resulting in suppressed deterioration of hearing hardness (Iwai *et al.* 2006). This IGF-1 treatment has been started clinically to demonstrate good therapeutic efficacy (table 2). In addition, the hydrogel system can release not only one type of growth factor but also two or more types in different concentrations and release profiles. Upon applying a hydrogel incorporating a low dose of either bFGF or TGF-β1 to a bone defect of rabbit skulls, no bone tissue was regenerated at the defect. However, a synergistic effect on bone regeneration was observed by the simultaneous release of two factors (L. Hong, Y. Tabata, S. Miyamoto, K. Yamada, Y. Ikada 2000, unpublished data).

The synergistic angiogenesis of bFGF and HGF was observed (Marui *et al.* 2005). The platelet contains a cocktail of autologous growth factors. The controlled release of the cocktail by the hydrogel enabled the regeneration of bone (Hokugo *et al.* 2005), knee meniscus (Ishida *et al.* 2007) and intervertebral disc (Nagae *et al.* 2007), in marked contrast with the use of cocktail alone.

4. FURTHER EXPERIMENTAL ASPECTS OF TISSUE ENGINEERING-BASED REGENERATION THERAPY

The cell scaffolding technology can not only be applied to the *in vivo* cell-based tissue regeneration, but can also assist and promote the basic sciences *in vitro* proliferation and differentiation of stem cells. The latter is for the preparation of cells with a good quality applicable for cell therapy and the research development of stem cell biology. To manipulate the proliferation and differentiation of stem cells *in vitro*, there are two scientific and technological approaches: the modification of culture medium and cell substrate. There have been several trials to add various soluble factors in the culture medium to manipulate cell behaviour. Considering that normally most cells cannot survive and biologically function without attachment on to the culture substrate, there is no doubt that the substrate greatly affects the profiles of cell proliferation and differentiation. For example, it has been demonstrated that the direction of cell differentiation can be modified by the softness (Engler *et al.* 2006) and size (Nakamura *et al.* 2005) of cell substrates and the surface modification of biosignalling molecules (Nagaoka *et al.* 2006; Benoit *et al.* 2008). In addition, there have been reports on a three-dimensional scaffold that has a steric gradient of bioactive molecule concentration inside the material (Yamamoto *et al.* 2007). It has been well recognized that a biological niche is the local environment of stem cells to naturally regulate their proliferation and differentiation *in vivo* (Arai & Suda 2007). When the molecular mechanism of stem cell niche is biologically clarified and the key components can be elucidated and used, the combination with the conventional cell scaffold will enable stem cells to artificially manipulate their potentials for tissue regeneration. Future development of stem cell biology and substantial collaboration with biomaterials technology will open a new research field of stem cells and consequently lead to a promising strategy of cell therapy.

DDS technology and methodology play an important role in the basic research of stem cell biology. Biosignalling molecules to regulate the functions and fate of stem cells have been elucidated. In this connection, for the future, the necessity for DDS will be undoubtedly increased to develop the basic research of tissue regeneration and consequently realize stem cell-based regeneration therapy. It is practically impossible to allow biosignalling proteins to function *in vivo* without DDS assistance. In addition to the growth factors, genes have been used for tissue regeneration therapy. There are two future directions of gene therapy. The first direction is the conventional gene therapy where a plasmid DNA and adenovirus are directly injected to give the therapeutic effect. For gene therapy, it is practically important to enhance the efficiency of gene transfection. Generally, the efficacy of viral vectors is higher than that of non-viral carriers. However, the viral vectors are not clinically suitable because of their toxicity and immunogenicity. Therefore, it is necessary to improve the *in vivo* efficacy of gene transfection. One of the possible ways is to use DDS technologies. Several gene carrier biomaterials, such as cationic liposomes and cationic polymers, have been investigated to enhance the level of gene expression (Kushibiki *et al.* 2003; Kushibiki & Tabata 2004; Jo & Tabata 2008). In addition, the release technology is also effective in enhancing gene transfection and expression. The release of plasmid DNA from a biodegradable hydrogel of cationized gelatin derivative enhanced the level of gene expression as well as prolonged the time period of expression (Kushibiki *et al.* 2003; Kushibiki & Tabata 2004). The injection of the cationized gelatin hydrogel incorporating plasmid DNA enhanced the *in vivo*therapeutic effects to a significantly greater extent than plasmid DNA solution alone (Kushibiki *et al.* 2004). The microspheres incorporating plasmid DNA also enhanced the gene expression of cells to genetically activate their biological functions and consequently increased the efficacy of cell therapy. Using the microspheres incorporating plasmid DNA, intracellular controlled release of plasmid DNA was achieved to enhance the efficiency of gene transfection to a higher level than that of adenovirus transfection. The cells genetically engineered functioned well to achieve higher therapeutic efficacy (Yamamoto & Tabata 2006; Jo & Tabata 2008). With the recent development of genomics researches, the DNA sequence of the human genome has been elucidated and disease therapy on the genetic level will develop in the future. As proteins, genes are also unstable *in vivo*. Therefore, it is no doubt that DDS technology will have an important role in gene therapy.

The second direction is to genetically engineer cells for their functional activation, which is applicable for cell transplantation therapy and stem cell biology. There are some cases where the transplantation of stem cells alone does not always induce a clinically acceptable therapeutic effect. A promising and practical way to overcome this problem would be to genetically engineer stem cells by gene transfection for the activation of their biological functions. So far, such cell activation has been tried by using virus vectors (Lai*et al.* 2008). This trial has been of great success, but the good results cannot be applied to clinical therapies because of the virus use. Therefore, it is necessary to develop a system of non-virus gene transfection. The DDS technology and methodology are very effective in developing a non-viral system of gene transfection with the efficiency of gene transfection as high as that of the viral system (Yamamoto & Tabata 2006; Jo & Tabata 2008). In addition to the microspheres for the intracellular release of plasmid DNA, a new non-viral carrier has been prepared from cationized polysaccharides. The carrier of gene transfection enabled plasmid DNA to enhance the level of gene expression for stem cells with less cytotoxicity than commercially available cationized liposomes. Gene transfection with the cationized polysaccharide was effective in enhancing the gene expression of stem cells to genetically engineer the biological function (Jo & Tabata 2008). In addition, the stem cells engineered showed the efficacy of cell therapy greater than the original cells (Jo *et al.* 2007). The level of gene expression by the non-viral carrier was also enhanced by contriving the methodology of gene transfection culture, such as a reverse transfection method and bioreactor combination (Okazaki *et al.* 2007; Nagane *et al.* in press). Thus, the gene engineering technology with biomaterials is effective in enhancing stem cell-based regeneration therapy and also developing the basic research of stem cell biology and medicine. The technology of gene transfection will be available for non-viral induction of iPS cells and also allow several bioactive substances to internalize into cells for an investigation of their biological functions in stem cell biology. The efficient action of siRNA by non-viral carriers is effective in genetically regulating the cell function. The biomaterial carrier enables siRNA to enhance the silencing effect *in vitro* and *in vivo*, resulting in the artificial modification of cell functions, which is applicable for stem cell biology and future cell therapy.

5. CLOSING REMARKS

Tissue regeneration therapy—a new therapy based on the natural induction of tissue regeneration through cell transplantation and tissue engineering—is a third therapy following reconstructive surgery and organ transplantation. To achieve the regeneration therapy by use of tissue engineering technologies, substantial collaborative research between material, pharmaceutical, biological and clinical scientists is needed. Even if superior stem cells can be practically obtained, it is impossible to therapeutically treat patients only by transplanting the cells prepared even combined with the scientific knowledge of cells and related substances, unless a local environment of cells suitable to promote the proliferation and differentiation is created and provided properly. To create the environment, there is no doubt that the biomaterials and the technology are needed.

However, one of the main problems to create the regeneration environment at present is the absolute shortage of biomaterial researchers who investigate the cell scaffold, the DDS, the protective barrier and cell culture, aiming at tissue regeneration and the biological substitution of organ functions. Such researchers must have knowledge in medicine, dentistry, biology and pharmacology, in addition to material sciences. It is indispensable to educate researchers of an interdisciplinary field, who have an engineering background and can also understand basic biology, medicine and clinical medicine. One of the representative interdisciplinary research fields is DDS technology, which is also applicable for producing non-viral vectors in the preparation of genetically engineered cells for regeneration therapy. The development of non-viral vectors with a high efficiency of gene transfection for stem cells is of a high priority.

Tissue engineering technology is not only used surgically for tissue defects, but also applied to develop a therapeutic method for chronic fibrosis diseases by making use of internal medical treatment. Tissue engineering is still in its infancy, although some research projects have already come close to the level of clinical applications. The increasing significance of biomaterials for cell scaffolding and DDS in future will help progress in basic and applied tissue engineering. We will be happy if this review stimulates readers' interest in the idea and research field of tissue engineering to assist in understanding how important biomaterials are to develop the basic research of stem cell biology and medicine as well as realize tissue regeneration therapy.

ACKNOWLEDGMENTS

The ethics committee of Kyoto University and Nippon Medical School approved the study protocol. Patient enrolment began in February 2005.

FOOTNOTES

One contribution of 10 to a Theme Supplement 'Japanese biomaterials'.

REFERENCES

1. Andrianov A.K., Payne L.G. Protein release from polyphosphazene matrices. Adv. Drug Deliv. Rev. 1998;31:185–196.
2. Aoyama T., Yamamoto S., Kanematsu A., Ogawa O., Tabata Y. Local delivery of matrix metalloproteinase gene prevents the onset of renal sclerosis in streptozotocin-induced diabetic mice. Tissue Eng. 2003;9:1289–1299.
3. Arai F., Suda T. Maintenance of quiescent hematopoietic stem cells in the osteoblastic niche. Ann. NY Acad. Sci. 2007;1106:41–53.

Bach A.D., Stem-Straeter J., Beier J.P., Bannasch H., Stark G.B. Engineering of muscle tissue. Clin. Plast. Surg. 2003;30:589–599.

4. Balamurugan A.N., et al. Bioartificial pancreas transplantation at prevascularized intermuscular space: effect of angiogenesis induction on islet survival. Pancreas.2003;26:279–285.

5. Bannasch H., Föhn M., Unterberg T., Bach A.D., Weyand B., Stark G.B. Skin tissue engineering. Clin. Plast. Surg. 2003;30:573–579.

6. Benoit D.S., Schwartz M.P., Durney A.R., Anseth K.S. Small functional groups for controlled differentiation of hydrogel-encapsulated human mesenchymal stem cells.Nat. Mater. 2008;7:816–823.

7. Chan G., Mooney D.J. New materials for tissue engineering: towards greater control over the biological response. Trends Biotechnol. 2008;26:382–392.

8. Chen R.R., Mooney D.J. Polymeric growth factor delivery strategies for tissue engineering. Pharm. Res. 2003;20:1103–1112.

9. Coviello T., Matricardi P., Marianecci C., Alhaique F. Polysaccharide hydrogels for modified release formulations. J. Control. Release. 2007;119:5–24.

10. Docheva D., Popov C., Mutschler W., Schieker M. Human mesenchymal stem cells in contact with their environment: surface characteristics and the integrin system. J. Cell. Mol. Med. 2007;11:21–38.

11. Ehashi T., Ohshima N., Miyoshi H. Three-dimensional culture of porcine fetal liver cells for a bioartificial liver. J. Biomed. Mater. Res. A. 2006;77:90–96.

12. Engler A.J., Sen S., Sweeney H.L., Discher D.E. Matrix elasticity directs stem cell lineage specification. Cell. 2006;126:677–689.

13. Falqui L., Finke E.H., Carel J.C., Scharp D.W., Lacy P.E. Marked prolongation of human islet xenograft survival (human-to-mouse) by low-temperature culture and temporary immunosuppression with human and mouse anti-lymphocyte sera. Transplantation.1991;51:1322–1324.

14. Fernyhough M.E., Hausman G.J., Guan L.L., Okine E., Moore S.S., Dodson M.V. Mature adipocytes may be a source of stem cells for tissue engineering. Biochem. Biophys. Res. Commun. 2008;368:455–457.

15. Fibbe W.E., Nauta A.J., Roelofs H. Modulation of immune responses by mesenchymal stem cells. Ann. NY Acad. Sci. 2007;1106:272–278.

16. Fujioka K., Maeda M., Hojo T., Sano A. Protein release from collagen matrices. Adv. Drug Deliv. Rev. 1998;31:247–266.

17. Gimble J.M., Katz A.J., Bunnell B.A. Adipose-derived stem cells for regenerative medicine. Circ. Res. 2007;100:1249–1260.

18. Gombotz W.R., Wee S. Protein release from alginate matrices. Adv. Drug Deliv. Rev.1998;31:267–285.

19. Hiraoka Y., Kimura Y., Ueda H., Tabata Y. Fabrication and biocompatibility of collagen sponge reinforced with poly(glycolic acid) fiber. Tissue Eng. 2003;9:1101–1112.

20. Hokugo A., Ozeki M., Kawakami O., Sugimoto K., Mushimoto K., Morita S., Tabata Y. Augmented bone regeneration activity of platelet-rich plasma by biodegradable gelatin hydrogel. Tissue Eng. 2005;11:1224–1233.

21. Holtorf H.L., Jansen J.A., Mikos A.G. Modulation of cell differentiation in bone tissue engineering constructs cultured in a bioreactor. Adv. Exp. Med. Biol. 2007;585:225–241.

22. Hsu Y.C., Lee D.C., Chiu I.M. Neural stem cells, neural progenitors, and neurotrophic factors. Cell Transplant. 2007;16:133–150.

23. Hubbell J.A. Tissue and cell engineering. Curr. Opin. Biotechnol. 2004;15:381–382.

24. Ichioka S., Kouraba S., Sekiya N., Ohura N., Nakatsuka T. Bone marrow-impregnated collagen matrix for wound healing: experimental evaluation in a microcirculatory model of angiogenesis, and clinical experience. Br. J. Plast. Surg. 2005;58:1124–1130.

25. Ikada Y., Tabata Y. Protein release from gelatin matrices. Adv. Drug Deliv. Rev.1998;31:287–301.

26. Ishida K., et al. The regenerative effects of platelet-rich plasma on meniscal cells *in vitro*and its *in vivo* application with biodegradable gelatin hydrogel. Tissue Eng.2007;13:1103–1112.

27. Iwai K., Nakagawa T., Endo T., Matsuoka Y., Kita T., Kim T.S., Tabata Y., Ito J. Cochlear protection by local insulin-like growth factor-1 application using biodegradable hydrogel. Laryngoscope. 2006;116:529–533.

28. Iwakura A., Fujita M., Kataoka K., Tambara K., Sakakibara Y., Komeda M., Tabata Y. Intramyocardial sustained delivery of basic fibroblast growth factor improves angiogenesis and ventricular function in a rat infarct model. Heart Vessels.2003;18:93–99.

29. Jo J.I., Tabata Y. Non-viral gene transfection technologies for genetic engineering of stem cells. Eur. J. Pharm. Biopharm. 2008;68:90–104.

30. Jo J., Nagaya N., Miyahara Y., Kataoka M., Harada-Shiba M., Kangawa K., Tabata Y. Transplantation of genetically engineered mesenchymal stem cells improves cardiac functon in rats with myocardial infarction: benefit of a novel nonviral vector, cationized dextran. Tissue Eng. 2007;13:313–322.

31. Kawakami O., Miyamoto S., Hatano T., Yamada K., Hashimoto N., Tabata Y. Accelerated embolization healing of aneurysms by polyethylene terephthalate coils seeded with autologous fibroblasts. Neurosurgery. 2005;56:1075–1080.

32. Kobayashi N., Okitsu T., Nakaji S., Tanaka N. Hybrid bioartificial liver: establishing a reversibly immortalized human hepatocyte line and developing a bioartificial liver for practical use. J. Artif. Organs. 2003;6:236–244.

33. Kornblum H.I. Introduction to neural stem cells. Stroke. 2007;38:810–816.

34. Kumagai N. Clinical application of autologous cultured epithelia for the treatment of burns and disfigurement of skin surfaces. Methods Mol. Biol. 2002;188:185–196.

35. Kuo C.K., Li W.J., Mauck R.L., Tuan R.S. Cartilage tissue engineering: its potential and uses. Curr. Opin. Rheumatol. 2006;18:64–73.

36. Kuroyanagi Y., Yamada N., Yamashita R., Uchinuma E. Tissue-engineered product: allogeneic cultured dermal substitute composed of spongy collagen with fibroblasts.Artif. Organs. 2001;25:180–186.

37. Kushibiki T., Tabata Y. A new gene delivery system based on controlled release technology. Curr. Drug Deliv. 2004;1:153–163.

38. Kushibiki T., Tomoshiqe R., Fukunaka Y., Kalemi M., Tabata Y. *In vivo* release and gene expression of plasmid DNA by hydrogels of gelatin with different cationization extents. J. Control. Release. 2003;90:207–216.

39. Kushibiki T., Matsumoto K., Nakamura T., Tabata Y. Suppression of tumor metastasis by NK4 plasmid DNA released from cationized gelatin. Gene Ther. 2004;11:1205–1214.

40. Kushibiki T., Nagata-Nakajima N., Sugai M., Shimizu A., Tabata Y. Enhanced anti-fibrotic activity of plasmid DNA expressing small interference RNA for TGF-beta type II receptor for a mouse model of obstructive nephropathy by cationized gelatin prepared from different amine compounds. J. Control. Release. 2006;110:610–617.

41. Lai Y., et al. Genetic modification of cells for transplantation. Adv. Drug Deliv. Rev.2008;60:146–159.

42. Langer R., Tirrell D.A. Designing materials for biology and medicine. Nature.2004;428:487–492.

43. Langer R., Vacanti J.P. Tissue engineering. Science. 1993;260:920–926.

44. Leo A.J., Grande D.A. Mesenchymal stem cells in tissue engineering. Cells Tissues Organs. 2006;183:112–122.

45. Lutolf M.P., Hubbell J.A. Synthetic biomaterials as instructive extracellular microenvironments for morphogenesis in tissue engineering. Nat. Biotechnol.2005;23:47–55.

46. Marui A., et al. Simultaneous application of basic fibroblast growth factor and hepatocyte growth factor to enhance the blood vessels formation. J. Vasc. Surg. 2005;41:82–90.

47. Marui A., et al. novel approach to therapeutic angiogenesis for patients with critical limb ischemia by sustained release of basic fibroblast growth factor using biodegradable gelatin hydrogel: an initial report of the phase I-IIa study. Circ. J. 2007;71:1181–1186.

48. Mironov V., Kasyanov V., Markwald R.R. Nanotechnology in vascular tissue engineering: from nanoscaffolding towards rapid vessel biofabrication. Trends Biotechnol. 2008;26:338–344.

49. Mountford J.C. Human embryonic stem cells: origins, characteristics and potential for regenerative therapy. Transfus. Med. 2008;18:1–12.

50. Nagae M., et al. Intervertebral disc regeneration using platelet-rich plasma and biodegradable gelatin hydrogel microspheres. Tissue Eng. 2007;13:147–157.

51. Nagane K. *et al.* In press. Practical induction system for dopamine-producing cells from bone marrow stromal cells by spermine-pullulan-mediated reverse transfection method.*Tissue Eng.*

52. Nagaoka M., Koshimizu U., Yuasa S., Hattori F., Cehn H., Tanaka T., Okabe M., Fukuda K., Akaike T. E-cadherin-coated plates maintain pluripotent ES cells without colony formation. PLoS ONE. 2006;1:e15.

53. Nakajima H., et al. Therapeutic angiogenesis by the controlled release of basic fibroblast growth factor for ischemic limb and heart injury: toward safety and minimal invasiveness. J. Artif. Organs. 2004;7:58–61.

54. Nakamura M., et al. Biocompatible inkjet printing technique for designed seeding of individual living cells. Tissue Eng. 2005;11:1658–1666.

55. Oe S., Fukunaka Y., Hirose T., Yamaoka Y., Tabata Y. A trial on regeneration therapy of rat liver cirrhosis by controlled release of hepatocyte growth factor. J. Control. Release. 2003;88:193–200.

56. Ogawa K., Asonuma K., Inomata Y., Kim I., Ikada Y., Tabata Y., Tanaka K. The efficacy of prevascularization by basic FGF for hepatocyte transplantation using polymer devices in rats. Cell Transplant. 2001;10:723–729.

57. Ohgushi H., Caplan A.I. Stem cell technology and bioceramics: from cell to gene engineering. J. Biomed. Mater. Res. 1999;48:913–927.

58. Okazaki A., Jo J., Tabata Y. A reverse transfection technology to genetically engineer adult stem cells. Tissue Eng. 2007;13:245–251.

59. Olle K., Karlsten R., Sundler F., Jansson L. Blood flow regulation in the transplanted fetal endocrine pancreas. Acquisition of a nitric oxide-dependent glucose-induced increase in blood flow. Transplantation. 1996;61:772–777.

60. Patel Z.S., Ueda H., Yamamoto M., Tabata Y., Mikos A.G. *In vitro* and *in vivo* release of vascular endothelial growth factor from gelatin microparticles and biodegradable composite scaffolds. Pharm. Res. 2008;25:2370–2378.

61. Picinich S.C., Mishra P.J., Glod J., Banerjee D. The therapeutic potential of mesenchymal stem cells. Cell- & tissue-based therapy. Expert Opin. Biol. Ther. 2007;7:965–973.

62. Pittenger M.F., et al. Multilineage potential of adult human mesenchymal stem cells.Science. 1999;284:143–147.

63. Saito A., et al. Bioengineered implantation of megalin-expressing cells: a potential intracorporeal therapeutic model for uremic toxin protein clearance in renal failure. J. Am. Soc. Nephrol. 2003;14:2025–2032.

64. Sakakibara Y., Nishimura K., Tambara K., Yamamoto M., Lu F., Tabata Y., Komeda M. Prevascularization with gelatin microspheres containing basic fibroblast growth factor enhances the benefits of cardiomyocyte transplantation. J. Thorac. Cardiovasc. Surg.2002;124:50–56.

65. Salgado A.J., Oliveira J.T., Pedro A.J., Reis R.L. Adult stem cells in bone and cartilage tissue engineering. Curr. Stem Cell Res. Ther. 2006;1:345–364.

66. Saltzman W.M., Olbricht W.L. Building drug delivery into tissue engineering. Nat. Rev. Drug Discov. 2002;1:177–186.

67. Schaffler A., Buchler C. Concise review: adipose tissue-derived stromal cells—basic and clinical implications for novel cell-based therapies. Stem Cells. 2007;25:818–827.

68. Shanti R.M., Li W.J., Nesti L.J., Wang X., Tuan R.S. Adult mesenchymal stem cells: biological properties, characteristics, and applications in maxillofacial surgery. J. Oral Maxillofac. Surg. 2007;65:1640–1647.

69. Shin'oka T., Imai Y., Ikada Y. Transplantation of a tissue-engineered pulmonary artery. N. Engl. J. Med. 2001;344:532–533.

70. Silva E.A., Mooney D.J. Synthetic extracellular matrices for tissue engineering and regeneration. Curr. Top. Dev. Biol. 2004;64:181–205.

71. Soto-Gutierrez A., et al. Reversal of mouse hepatic failure using an implanted liver-assist device containing ES cell-derived hepatocytes. Nat. Biotechnol. 2006;24:1412–1419.

72. Strem B.M., Hedrick M.H. The growing importance of fat in regenerative medicine.Trends Biotechnol. 2005;23:64–66.

73. Tabata Y. Princeton International Publishing Co. Ltd; Lawrenceville, NJ: 2003a. Tissue engieering by release technology based on biodegradable polymer hydrogel.

74. Tabata Y. Tissue regeneration based on growth factor release. Tissue Eng. 2003b;9:S5–S15.

75. Tabata Y. Nanomaterials of drug delivery systems for tissue regeneration. Methods Mol. Biol. 2005a;300:81–100.

76. Tabata Y. Significance of release technology in tissue engineering. Drug Discov. Today.2005b;10:1639–1646.

77. Tabata Y. A new concept of biomaterials to induce tissue regeneration. Mater. Sci. Forum.2007;561–565:1467–1470.

78. Tabata Y. Current status of regenerative medical therapy based on drug delivery technology. Reprod. BioMed. Online. 2008;16:70–80.

79. Tabata Y., Ikada Y. Vascularization effect of basic fibroblast growth factor released from gelatin hydrogels with different biodegradabilities. Biomaterials. 1999;20:2169–2175.

80. Tabata Y., Hijikata S., Ikada Y. Enhanced vascularization and tissue granulation by basic fibroblast growth factor impregnated in gelatin hydrogels. J. Control. Release.1994;31:189–199.

81. Tabata Y., Nagano A., Ikada Y. Biodegradation of hydrogel carrier incorporating fibroblast growth factor. Tissue Eng. 1999;5:127–138.

82. Tabata Y., Miyao M., Ozeki M., Ikada Y. Controlled release of vascular endothelial growth factor released by use of collagen hydrogels. J. Biomater. Sci. Polym. Ed.2000a;11:915–930.

83. Tabata Y., et al. Bone formation at a rabbit skull defect by autologous bone marrow cells combined with gelatin microspheres containing TGF-beta1. J. Biomater. Sci. Polym. Ed. 2000b;11:891–901.

84. Takahashi Y., Yamamoto M., Tabata Y. Osteogenic differentiation of mesenchymal stem cells in biodegradable sponges composed of gelatin and beta-tricalcium phosphate.Biomaterials. 2005;26:3587–3596.

85. Takehara N., et al. Controlled delivery of basic fibroblast growth factor promotes human cardiosphere-derived cell engraftment to enhance cardiac repair for chronic myocardial infarction. J. Am. Coll. Cardiol. 2008;52:1858–1865.

86. Takemoto S., Morimoto N., Kimura Y., Taira T., Kitagawa T., Tomihata K., Tabata Y., Suzuki S. Preparation of collagen/gelatin sponge scaffold for sustained release of bFGF. Tissue Eng. Part A. 2008;14:1629–1638.

87. Tambara K., et al. Administration of control-released hepatocyte growth factor enhances the efficacy of skeletal myoblast transplantation in rat infarcted hearts by greatly increasing both quantity and quality of the graft. Circulation. 2005;112:I129–I134.

88. Tessmar J.K., Gopferich A.M. Matrices and scaffolds for protein delivery in tissue engineering. Adv. Drug Deliv. Rev. 2007;59:274–291.

89. Tsuji-Saso Y., Kawazoe T., Morimoto N., Tabata Y., Taira T., Tomihata K., Utani A., Suzuki S. Incorporation of basic fibroblast growth factor into preconfluent cultured skin substitute to accelerate neovascularisation and skin reconstruction after transplantation. Scand. J. Plast. Reconstr. Surg. Hand Surg. 2007;41:228–235.

90. Van Tomme S.R., Hennink W.E. Biodegradable dextran hydrogels for protein delivery applications. Expert Rev. Med. Devices. 2007;4:147–164.

91. Yamamoto M., Tabata Y. Tissue engineering by modulated gene delivery. Adv. Drug Deliv. Rev. 2006;58:535–554.

92. Yamamoto M., Yanase K., Tabata Y. Fabrication of three dimensional cell scaffolds with spatial gradients of biomolecules. Inflamm. Regen. 2007;27:102–106.

93. Yamanaka S. Strategies and new developments in the generation of patient-specific pluripotent stem cells. Cell Stem Cell. 2007;1:39–49.

94. Yamashita Y., Shimada M., Ijima H., Nakazawa K., Funatsu K., Sugimachi K. Hybrid-artificial liver support system. Surgery. 2002;131:S334–S340.

95. Zisch A.H., Lutolf M.P., Hubbell J.A. Biopolymeric delivery matrices for angiogenic growth factors. Cardiovasc. Pathol. 2003;12:295–310.

2

CARBON NANOTUBES PLAY AN IMPORTANT ROLE IN THE SPATIAL ARRANGEMENT OF CALCIUM DEPOSITS IN HYDROGELS FOR BONE REGENERATION

Giulia Cancian[1], Gianluca Tozzi[2], Amirul Ashraf Bin Hussain[1], Arianna De Mori[1], Marta Roldo[1]

[1]School of Pharmacy and Biomedical Science, University of Portsmouth, St Michael's Building, White Swan Road, PO1 2DT Portsmouth, UK

[2]School of Engineering, University of Portsmouth, Anglesea Building, Anglesea Road, PO1 3DJ Portsmouth, UK

ABSTRACT

Age related bone diseases such as osteoporosis are considered among the main causes of reduced bone mechanical stability and bone fractures. In order to restore both biological and mechanical function of diseased/fractured bones, novel bioactive scaffolds that mimic the bone structure are constantly under development in tissue engineering applications. Among the possible candidates, chitosan-based thermosensitive hydrogel scaffolds represent ideal systems due to their biocompatibility, biodegradability, enhanced antibacterial properties, promotion of osteoblast formation and ease of injection, which makes them suitable for less invasive surgical procedures. As a main drawback, these chitosan systems present poor mechanical performance that could not support load-bearing applications. In order to produce more mechanically-competent biomaterials, the combined addition of hydroxyapatite and carbon nanotubes (CNTs) is proposed in this study. Specifically, the aim of this work is to develop thermosensitive chitosan hydrogels containing stabilised single-walled and multi-walled CNTs, where their effect on the mechanical/physiochemical properties, calcium deposition patterns and ability to provide a platform for the controlled release of protein drugs was investigated. It was found that the addition of CNTs had a significant effect on the sol–gel transition time and significantly increased the resistance to compression for the hydrogels. Moreover, in vitro calcification studies revealed that CNTs played a major role in the spatial arrangements of newly formed calcium deposits in the composite materials studied, suggesting that they may have a role in the way the repair of fragile and/or fractured bones occurs in vivo.

1 INTRODUCTION

Bone is a dynamic, mineralised and highly vascularised tissue whose major functions are body support, motion and organ protection. Major alterations in its structure, due to injuries or metabolic diseases, can cause long-term pain, infection, inflammation and loss of mobility that can lead to a general decrease in functionality and quality of life [1]. Current therapies for complex bone fractures are mainly based on auto- and allo-bone grafts, or on the use of inert materials such as metals and ceramics that replace bone. These invasive strategies have, however, many serious limitations such as an inadequate revascularisation and mineralisation of the bone grafts, risk of infection, rejection and immunoreaction [2].

Bone tissue engineering (BTE) provides an interdisciplinary approach to promote the physiological regeneration of bone tissues, overcoming the downfalls of current methods. BTE research focuses on developing bioactive scaffolds that, mimicking the bone structure, work as temporary matrices for cell proliferation, extracellular matrix deposition and vascularization of newly formed tissue [3, 4]. In order to reach these targets, an ideal scaffold for BTE applications should possess the following properties: have a porous structure; be biocompatible; have good mechanical properties to support tissue ingrowth; be fully biodegradable; resorb while bone formation occurs, eliminating the need for a revision surgery; not form toxic products; and finally deliver bioactive molecules or drugs in a controlled manner where needed [5].

Among the possible scaffolds for tissue engineering, hydrogels represent ideal systems due to their biocompatibility and biodegradability [6, 7]. These gels can be formulated from different natural polymers such as collagen, chitosan, fibrin, hyaluronic acid and alginate [8, 9]. Chitosan is a linear polysaccharide derived from *N*-deacetylation of chitin, which is commonly found in the exoskeletons of crustaceans and insects [10]; its biocompatibility, antibacterial properties and structure amenable to simple chemical modifications, make chitosan a very interesting polymer for biomedical applications [11]. Chitosan only hydrogels present poor bioactive and mechanical properties; however, we have previously demonstrated that the combined addition of hydroxyapatite (HA) and carbon nanotubes (CNTs) can improve gels properties [12]. HA is a natural, nontoxic and bioactive ceramic, shown to enhance osteoconduction, bone bonding and stiffness in scaffolds. Moreover, HA favours deposition of calcium phosphate, improving the bone-matrix interface strength [13, 14]. CNTs are cylindrical molecules of carbon atoms in sp^2 conformation with outstanding chemical, electrical, mechanical and thermal properties [15]. They have recently been shown to have a role in the differentiation of osteoblasts suggesting that they might have a double function in scaffolds for BTE as structural reinforcement and osteiniductive materials [16]. However, their application is dependent on their successful dispersion in physiological conditions. In fact, CNTs have a tendency to form large bundles in aqueous media, resulting from Van der Waals interaction and π–π stacking [17–19]. Such bundles may cause slippage between nanotubes, become stress concentrators or initiate cracks under applied forces. Many covalent and non-covalent modifications of CNTs have been proposed in order to increase their water dispersability and hence their biocompatibility. Non-covalent modifications, in particular, are based on the absorption or the wrapping of various functional molecules, such as biomolecules, polymers, surfactants and phospholipids around the CNTs [20]. In a previous study, we have demonstrated the possibility of efficiently stabilising single-walled carbon nanotubes with the amphiphilic and self-assembling chitosan derivative *N*-octyl-*O*-sulfate chitosan (NOSC) [21].

In the present study, we developed thermosensitive chitosan hydrogels containing NOSC stabilised CNTs (SWNTs or MWNTs). Our aim was to develop a more flexible formulation strategy by using non-covalent CNTs stabilisation instead of chemical grafting and studying the effects of this strategy on the mechanical and physicochemical properties of the formulations. Image analysis was used to qualitatively and quantitatively assess the effect of CNTs on salts deposition within the formulated hydrogels. The ability of the gel to provide a platform for the controlled release of protein drugs was also evaluated.

2 MATERIALS AND METHODS

2.1 Materials

All salts and solvents, unless otherwise stated, were obtained from Fisher Scientific (UK). Sodium phosphate monobasic monohydrate, low viscosity chitosan from shrimp, glycerol phosphate disodium salt hydrate, tris(hydroxymethyl)amino-methane [$(HOCH_2)_3CNH_2$], albumin from bovine serum (BSA), albumin from chicken egg white (OVA) and QuantiPro™ Bicinchoninic Acid Assay Kit were obtained from Sigma-Aldrich (UK). L(+)-lactic acid 90 % solution in water was obtained from Acros Organics (USA). Dialysis membranes (size 10, MWCO 12–14 kDa) were obtained from Medicell International Ltd (UK). MWCTs and SWCTs were purchased from Cheap Tubes Inc. (USA). The MWCTs used in the study were 8–15 nm in diameter and 10–50 μm in length. The SWCTs used were 1–4 nm in diameter and 5–30 μm in length. Di-potassium hydrogen orthophosphate anhydrous (K_2HPO_4) was acquired from British Drug Houses (UK) and troclosene sodium dehydrate from Guest Medical (UK). Water used in all experiments was purified water. NOSC and chitosan modified hydroxyapatite (HACS) were prepared as described previously [12, 21].

2.2 Formulation of hydrogels

Clear solutions of chitosan were prepared by dissolving chitosan (200 mg) in lactic acid (0.1 M, 7 ml), HACS (86 mg) was then added and dispersed by vigorous mixing and sonication. CNTs loaded formulations [CS-HACS-(NOSC-SWNTs) and CS-HACS-(NOSC-MWNTs)] were prepared by addition of stable CNTs suspensions in NOSC (2 ml) to the HACS containing chitosan solutions. NOSC suspensions of CNTs were obtained by adding CNTs (2.5 mg) to a solution of NOSC (1.25 mg/ml) sonicating for 2 h, followed by overnight rest and further sonication for 1 h. At this point all suspensions were brought to a final volume of 9 ml by addition of lactic acid 0.1 M (if needed) and the thermosenstive crosslinker glycerol phosphate was added (1 ml, 1.12 g/ml) to all samples, which were then stirred for 10 min. Gelation was obtained by incubating the formulations at 37 °C. The gels prepared were labelled CS (chitosan control gel), CS-HACS (chitosan gel containing HACS),CS-HACS-(NOSC-SWNTs) and CS-HACS-(NOSC-MWNTs) (gels containing HACS and SWNTs or MWNTs, respectively). All gels were prepared in triplicate; gelation time was evaluated visually by the inversion method [22].

2.3 Texture analysis of the gels

A Texture Analyzer (XT *plus*, Stable Micro Systems Ltd, UK) was used to determine syringeability of the suspensions before gelation as well as gel compressibility. The methods employed have been described previously by Yasmeen et al. [12]. Briefly, for syringeability the liquid formulations were loaded onto 5 ml plastic syringes fitted with 19 G, 25 mm long needles. The vertically clamped syringe was actioned by the instrument probe (10 mm diameter) that compressed the barrel (1 mm/s) to a distance of 40 mm and the initial glide force, dynamic glide force and maximum force were measured. Compressibility tests were performed by depressing a polycarbonate probe (10 mm dia) into the gels at 1 mm/s up to a depth of 5 mm, six measurements were taken before and after the sol–gel transition took place. Gels were formed and tested in 14 ml glass vials. Force/displacement curves were obtained and used to calculate the gel compressibility (N·mm). All measurements were performed at room temperature.

2.4 Characterisation of the freeze dried hydrogels

Freeze dried hydrogels were prepared from the suspension described above and then transferred into 24 well-plates (2 ml per well), before incubation at 37 °C. Once formed, the gels were flash-freezed and freeze-dried (Edwards Micro Modulyo RV3, UK). The samples were characterized by FT-IR (Varian Spectrometer 640, UK), DVS (Surface Measurement Systems DVS Advantage, UK) and SEM (JEOL JSM-6060 LV Scanning Electron Microscope, UK). In the DVS analysis the mass change of the gels subjected to a changing water vapour partial pressure at 25 °C was recorded. The partial pressure was increased from 0 to 90 % at incremental steps of 10 %, where the next step was reached either after equilibrium or after a maximum time of 360 min. A full adsorption/desorption cycle was performed and the data collected were used to calculate the adsorption/desorption isotherms.

2.5 In vitro calcification assay

Calcification studies were performed to investigate the ability of the composite hydrogels to induce calcium salts deposition. Simulated body fluid (SBF), prepared according to Kokubo and Takadama [23], was used to mimic the in vivo environment, as it contains ions at very similar concentrations to those of human blood plasma. The freeze dried samples were inserted in a 50 ml tube to which 15 ml of SBF were added. Samples were incubated at 37 °C for 7 or 14 days. The SBF was changed every 4 days. Finally, samples were dried under vacuum at 40 °C overnight.

Controls were hydrated in deionised water for 5 h and dried as above. All samples were gold-coated with sputter coater (Polaron e500, Quoram Technologies, UK) and analysed by SEM, coupled with EDS [Silicon Drift Detector (SDD)—X-MaxN, Oxford Instruments, UK] and microCT (XTH225, Nikon Systems, UK). The microCT scanner was set to a voltage of 55 kV and a current of 75 μA. With an isotropic voxel size of 10.6 μm and exposure of 2 s, the image acquisition was performed at a rotational step of 0.23° over 360° for 90 min approximately. The images were then analysed using ImageJ (1.48v, NIH, USA) as described below. To quantify calcium deposition, image thresholding (global) based on the grey scale histogram was carried out to distinguish between the different material densities (different attenuation). Total volume of the sample and total volume of the denser material in the sample (i.e. the deposits of calcium salts) were obtained and the percentage of calcium deposition was calculated for each sample (n = 5). EDS images for calcium (Ca) and phosphorus (P) were also analysed after being overlaid to corresponding SEM images. The total area of overlapping for Ca and P onto the total sample surface area was used to calculate percentage of Ca and P salt deposition (n = 3).

2.6 Release of model drugs

Release studies were performed including bovine serum albumin (BSA, 1 mg/ml) in the hydrogels (5 ml). After the gelation of the samples, prepared in a 15 ml tubes (n = 4), PBS (5 ml) was added. The samples were then incubated at 37 °C and at set time points 2 ml of PBS were sampled and replaced with 2 ml of fresh buffer. Release studies were also performed including ovalbumin (OVA, 1 mg/ml) in the hydrogels. The suspensions were transferred (1 g per sample) into a Flot-A-Lyzer G2 dialysis device (cut off 50 kDa) and PBS (15 ml) was added to them. The samples were subsequently incubated at 37 °C and at set time points the sampled buffer was replaced with fresh buffer. Determination of both BSA and OVA concentrations was carried out with a QuantiPro™ Bicinchoninic Acid Assay Kit (Sigma, UK) and samples prepared according to manufacturer's instructions, reading the UV absorbance at 570 nm. Unloaded gels were used as a control to eliminate any false response due to chitosan interaction with the assay kit. The two model drugs were tested with slightly different methods due to limitations imposed by their molecular weight.

3 RESULTS

3.1 Texture analysis of the thermosentive hydrogels

A chitosan only formulation was used as a control to evaluate the effect of the addition of HACS and CNTs suspensions on the physical properties of gels before and after formation. The chitosan only formulation underwent sol–gel transition at 37 °C in 60 min, whereas the addition of HACS and NOSC suspensions of SWNTs and MWNTs reduced this time to 7 and 9 min, respectively. The syringeability of the suspensions was also studied. A force of 30 N was considered as the maximum acceptable force for an injection to be performed without particular patient discomfort [24]; none of the samples tested presented values above 10 N (Table 1) [25]. Furthermore, texture analysis was performed to determine gel compressibility (Fig. 1a).

Table 1. Syringeability of composite formulations

	Stiction (N)	Plateau force (N)	End capture (N)
Chitosan	8.36 ± 0.33	8.56 ± 0.29	8.92 ± 0.13
Chitosan-HACS	8.86 ± 0.39	8.56 ± 0.34	9.09 ± 0.34
Chitosan-HACS-(NOSC-SWNTs)	6.69 ± 0.25***	6.40 ± 0.24***	6.56 ± 0.21***
Chitosan-HACS-(NOSC-MWNTs)	7.03 ± 0.15***	6.92 ± 0.15***	7.15 ± 0.22***

Stiction, plateau force and end capture are expressed as mean ± SD (n = 3). One way ANOVA, $P < 0.001$; Tukey–Kramer multiple comparison test

*** $P < 0.001$, compared to the chitosan control

All suspensions tested showed significant difference in strength before and after gelation at 37 °C (at least $P = 0.0004$, two tailed, *t* test with Welch's correction) apart from the chitosan control. In particular, after gelation the formulation containing SWNTs showed significantly higher values compared to all other formulations (0.273 ± 0.001 N mm).

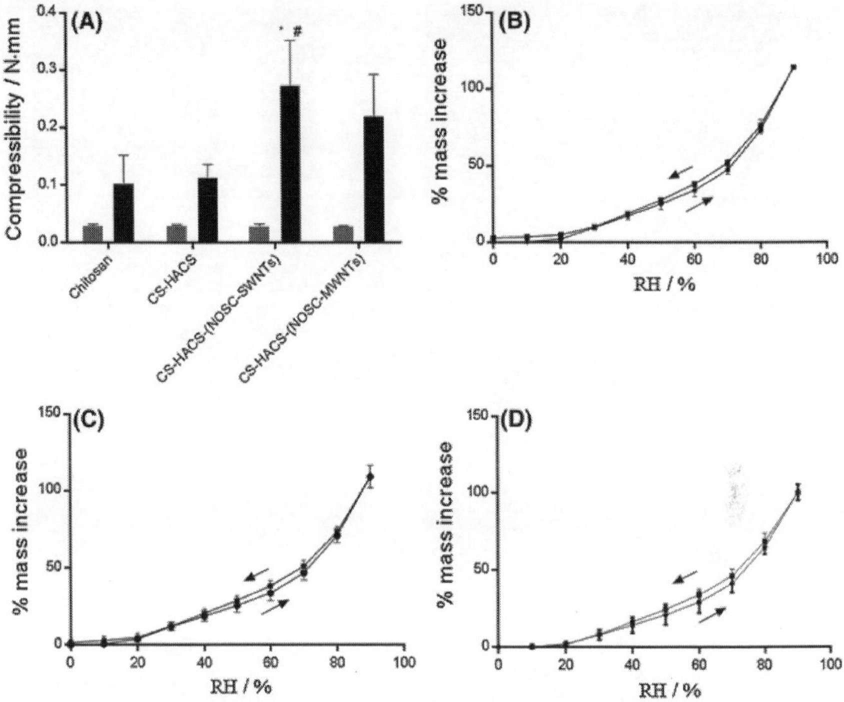

Figure1. a Compressibility of the different formulations before (*grey*) and after (*black*) sol–gel transition. Results are reported as mean ± SD (n = 3). One way ANOVA ($\alpha = 0.05$) on the values after gelation returned $P = 0.0131$; Tukey's multiple comparisons test: $^{*}P < 0.05$, compared to chitosan; $^{\#}P < 0.05$ compared to CS-HACS. Water sorption and desorption isotherms of the freeze dried hydrogels: **b**chitosan; **c** CS-HACS-(NOSC-SWNTs); **d** CS-HACS-(NOSC-MWNTs). Data are presented as mean ± SD (n = 3)

3.2 Characterisation of the freeze dried hydrogels

The morphology of the freeze-dried hydrogels was evaluated by SEM (Fig. 2). The chitosan only gel appeared to be scaly and rough while the gels containing CNTs showed a more defined porous structure with evidence of coated CNTs filaments bridging the structure; this was previously observed by Ormsby et al. who observed CNTs bridging sub-micron voids and preventing pore coalescence [26].

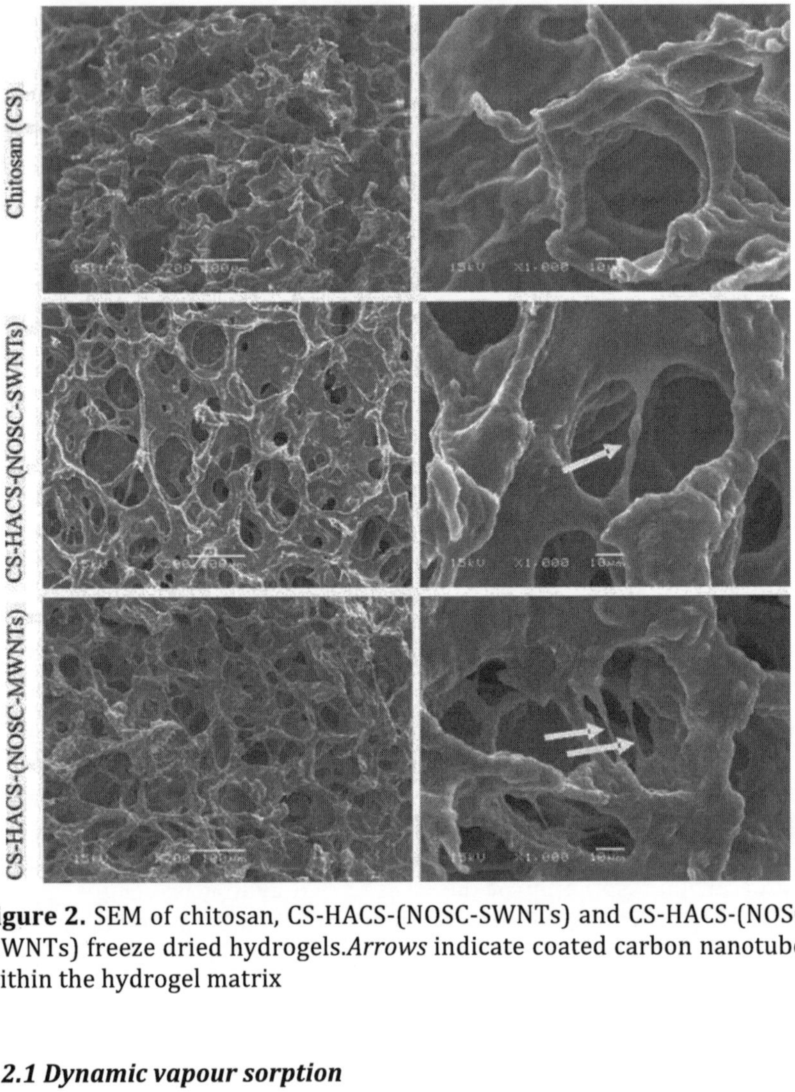

Figure 2. SEM of chitosan, CS-HACS-(NOSC-SWNTs) and CS-HACS-(NOSC-MWNTs) freeze dried hydrogels.*Arrows* indicate coated carbon nanotubes within the hydrogel matrix

3.2.1 Dynamic vapour sorption

DVS analysis was performed to evaluate the amount and kinetics of water sorption of the hydrogels. Previous work showed that the addition of chitosan grafted CNTs to chitosan hydrogel significantly decreased the hydrophilicity of the formulation [12]; this was not the case when the added CNTs were non-covalently coated with NOSC (Fig. 1b–d). Applying Korsmeyer–Peppas model to the absorption versus time curves, the mechanism of water penetration described by the value of $n_p > 0.89$ was determined to be super case-II transport (Table 2). Hence, it seems that the dynamic swelling behaviour of hydrogels is dependent on the contribution of both the penetrant (water) diffusion and polymer relaxation [27].

Table 2. Kinetic parameter values for water penetration in freeze-dried hydrogels

	n_p	K_p	R^2
CS control	1.67 ± 0.15	0.000369 ± 0.000314	0.968 ± 0.015
CS-HACS-(NOSC-SWNTs)	1.79 ± 0.23	0.000260 ± 0.000335	0.973 ± 0.016
CS-HACS-(NOSC-MWNTs)	1.65 ± 0.16	0.000461 ± 0.000298	0.983 ± 0.007

Data are reported as mean ± SD (n = 3)

3.2.2 In vitro calcification studies

Bioactivity of a scaffold proposed for bone regeneration can be evaluated by studying the calcium salts deposition on its surface and internal structure [23, 28]. Samples were soaked in SBF for 7 and 14 days, at 37 °C, and superficial Ca/P co-deposition was evaluated by EDS coupled SEM (Fig. 3); deposition, in the whole volume, of a denser material was also visualised and quantified by microCT. The quantitative analysis of the EDS data (Fig. 4a) revealed that no calcium-phosphate deposits were found in the gels containing only chitosan, both at 7 and 14 days. Samples containing CNTs also contained HACS, this accounted for 3.06 ± 0.09 and 3.52 ± 0.18 % co-localised Ca/P present at time zero, for gels containing SWNTs and MWNTs respectively. Furthermore, the SEM images (Fig. 3) showed that HACS was more homogeneously distributed in the samples containing SWNTs. In both samples an increase in the concentration of co-localised Ca and P was observed, with significantly higher values for the gel containing SWNTs (7.8 ± 0.1 %). Further analysis of areas characterised by Ca/P co-localisation revealed that the Ca/P ratio of the newly formed deposits was 1.55 (Table 3) for gels containing MWNTs, this corresponds to calcium deficient HA. Gels containing SWNTs presented deposits with Ca/P ratio <1.50, suggesting a combination of crystal phases of lower Ca/P values such as monetite, brushite and octacalcium phosphate.

Salts deposition was also observed by micro CT (Fig. 4b), were volumes of denser material were identified and quantified, these may represent the deposition of other salts as well as calcium phosphate salts that possess similar x-ray attenuations. The micro CT analysis evidenced differences in the microarchitecture of the hydrogel, were chitosan alone gels presented a less homogenous and more porous structure (Fig. 5). This may be attributed to the drying process, during which the migration of water from the core to the periphery of the gel might have dragged the chitosan to condense on the edge leaving a porous structure in the core. This phenomenon is much less noticeable in the gels containing CNTs, confirming how CNTs have an important role in the formation and preservation of the 3D gel structure. As a consequence, a marked difference in the distribution of the salts deposition was observed and resulted in a much more homogeneous distribution in gels containing CNTs. Quantitative analysis (Fig. 4b) also revealed a higher deposition of salts in CNTs containing gels, with no significant difference between SWNTs and MWNTs (34.6 ± 2.6 and 36.0 ± 1.1 % salt deposition over total sample volume, respectively) [29]. These data are in good agreement with the EDS/SEM data and confirm that calcification is notably more pronounced in samples containing HACS and CNTs.

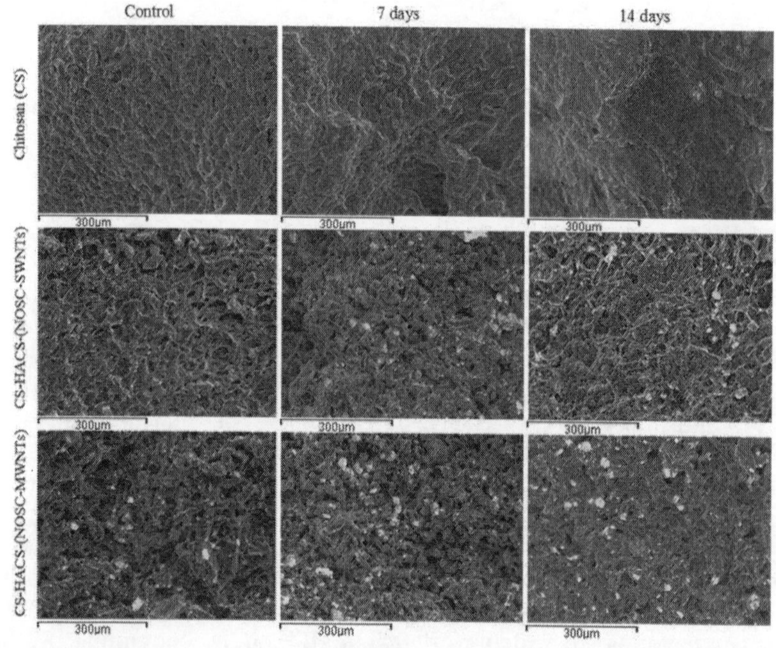

Figure 3. Overlay images of SEM pictures with EDS data: Ca (*green*), P (*red*) and co-localisation of Ca/P (*yellow*) (Color figure online)

Table 3. Ca/P ratio of salt deposits on hydrogels incubated with SBF for 7 and 14 days, *t*-test (paired, two-tailed)

	7 days	14 days
CS-HACS-(NOSC-SWNTs)	1.45 ± 0.54	1.32 ± 0.04
CS-HACS-(NOSC-MWNTs)	1.55 ± 0.30	1.55 ± 0.15*

* $P < 0.05$, compared to the HACS included in the gels which had a Ca/P ratio of 1.34 ± 0.06

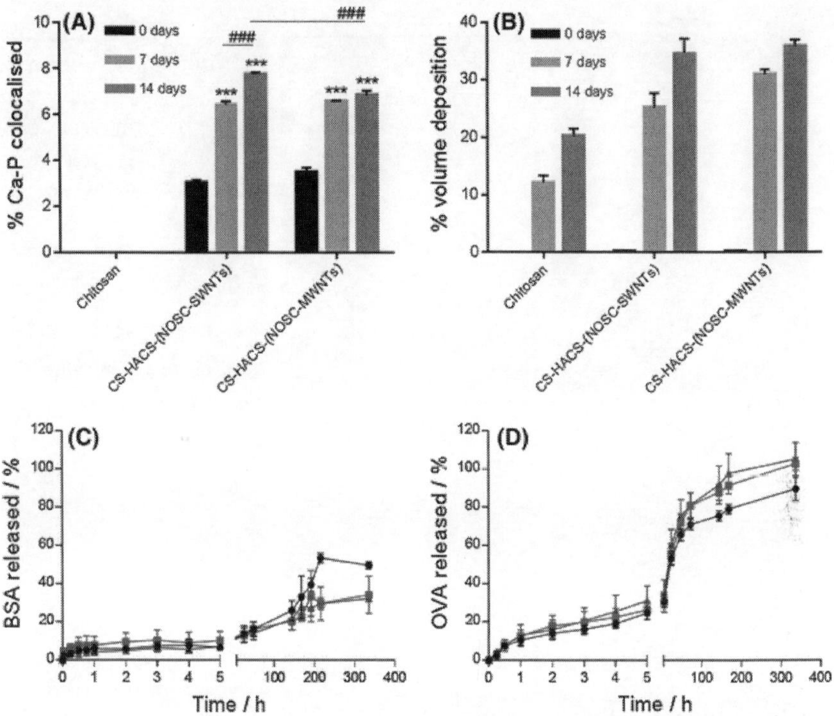

Figure 4. a Quantitative analysis of surface calcium deposition presenting % of surface covered by co-localised Ca and P (from SEM). Data are reported as mean ± SD (n = 3). One-way ANOVA ($\alpha = 0.05$) $P < 0.0001$ for both gels containing CNTs; Tukey's multicomparison test, $^{***}P < 0.001$ compared to control at time zero, $^{\#\#\#}P < 0.001$ compared to a specific sample as indicated in the figure. **b** Average % salt deposition on different types of hydrogel samples (from microCT). Data are reported as mean ± SD (n = 5). One-way ANOVA ($\alpha = 0.05$) $P < 0.0001$ for both gels containing CNTs, Tukey's multicomparison test, all samples presented $P < 0.001$ compared to control, and $P < 0.001$ was obtained for all comparisons between 7 and 14 days. Release profile of **c** BSA **d** OVA from (*black closed circle*) chitosan; (*red closed square*) CS-HACS-(NOSC-SWNTs); (*blue closed triangle*) CS-HACS-(NOSC-MWNTs) gels. Data are reported as mean ± SD (n = 3) (Color figure online)

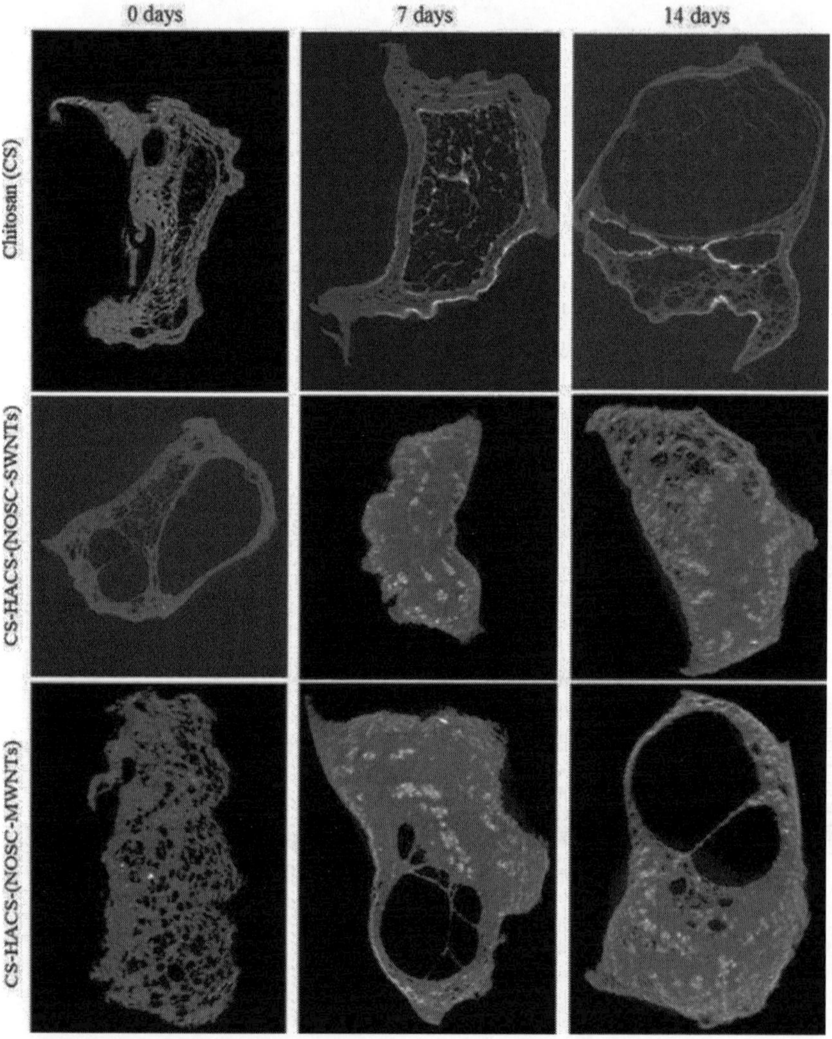

Figure 5. Representative micro CT slices of different gel samples incubated in SBF for 0, 7 and 14 days

3.3 Release of model drugs

The release of two macromolecular model drugs, albumin form chicken egg (OVA, Mw 45 kDa, Fig. 4c) and bovine serum albumin (BSA, Mw 66 kDa, Fig. 4d), were studied for 14 days. All the hydrogels tested showed the potential to provide controlled release within the 2 weeks of the experiment. Analysis of the release during the first 5 h revealed very similar release profiles for all gels, indicating that the initial release is governed by the behavior of the chitosan component of the gel. The addition of HACS and CNTs decreased the amount of BSA release over 14 days, but this did not have a significant effect on the release of OVA. This suggests that the mechanism of release for OVA is mainly dependent on the interaction of the protein with the chitosan network. Also gel degradation has an effect, indeed a marked change in the gel structure was noticed after 14 days of BSA release study. Hydrogels before the release study presented a typical porous structure, with well-defined pores of different size; while hydrogels after the release presented a scaly appearance.

4 DISCUSSION

Composite thermosensitive chitosan hydrogels were formulated. The addition of CNTs was confirmed to have a significant effect on the sol–gel transition time; these results are in accordance to those obtained by addition of chitosan grafted CNTs [12], indicating that both covalently and non-covalently coated CNTs present similar thermal behaviour and lead to similar effects on the gelation time of the formulations [30, 31]. Unexpectedly, the formulations containing NOSC suspensions of CNTs had better syringeability performance than chitosan only or chitosan with HACS formulation, while previous studies showed that the addition of chitosan grafted CNTs increased the force needed to expel the suspension from the syringe [12]. These data show that covalent and non-covalent coating of CNTs can have very different effect on the viscosity of the suspensions and their syringeability. In our previous work, we demonstrated that NOSC tends to stretch and align along the surface of CNTs [32], reducing the opportunity for formation of physical entanglements between NOSC and the free chitosan chains present in the gel. It is postulated that under the shear stress applied during injection non-covalently coated CNTs are able to align within the flowing liquid formulation therefore present a reduced resistance compared to a crosslinked system. Conversely, CNTs afforded a significant increase in resistance to compression for the composite hydrogels. In comparison with published data, where chitosan grafted CNTs were added instead of non-covalently coated ones, no statistical difference was noticed in the hydrogels containing MWNTs, while SWNTs produced statistically stronger gels ($P < 0.05$) when suspended in NOSC solutions [12]. The use of NOSC for the stabilisation of CNTs reduced the hydrophobicity of the nanotubes when compared to chitosan grafted CNTs. This phenomenon can be explained by considering the shielding effect that NOSC has on the surface of the nanotubes. In fact, five chains of NOSC arrange themselves along the entire surface of CNTs, thus resulting in a net decrease of surface hydrophobicity [21]. Reduced hydrophobicity has been linked with better in vivo outcomes [26, 33, 34]; furthermore, previous studies demonstrated the cytocompatibility of CNTs stabilised in NOSC (data not shown). The in vitro calcification studies revealed that CNTs played a major role in the spatial arrangements of newly formed calcium deposits in the composite materials studied, suggesting that they may have a role in the way the repair of fragile and/or fractured bones occurs in vivo. Finally, in model drug delivery studies, the composite gels demonstrated potential as delivery platforms for macromolecular drugs with minimum interference by CNTs in the determination of the mechanism of drug release.

In conclusion the present study reports of a simple non-covalent method for the inclusion of CNTs into hydrophilic chitosan hydrogels. The non-covalent stabilisation of the nanotubes is advantageous as it facilitates the injectability of the formulation while affording gels with higher compressibility. CNTs also proved to be key in guaranteeing a homogeneous deposition of salts suggesting that they can be used for guided mineralisation of scaffold for bone regeneration. Finally the gels have potential also as platforms for prolonged drug delivery.

ACKNOWLEDGMENTS

This research was co-funded by the Institute of Biomedical and Biomolecular Science Quality Research fund and the Research Development Fund of the University of Portsmouth, UK.

REFERENCES

1. Barrère F, Mahmood TA, de Groot K, van Blitterswijk CA. Advanced biomaterials for skeletal tissue regeneration: instructive and smart functions. Mater Sci Eng R. 2008;59(1–6):38–71.

2. Lu XY, Qiu T, Wang XF, Zhang M, Gao XL, Li RX, et al. Preparation of foam-like carbon nanotubes/hydroxyapatite composite scaffolds with superparamagnetic properties. Appl Surf Sci. 2012;262:227–30.

3. Porter JR, Ruckh TT, Popat KC. Bone tissue engineering: a review in bone biomimetics and drug delivery strategies. Biotechnol Prog. 2009;25(6):1539–60.

4. Bose S, Roy M, Bandyopadhyay A. Recent advances in bone tissue engineering scaffolds. Trends Biotechnol. 2012;30(10):546–54.

5. Amini AR, Laurencin CT, Nukavarapu SP. Bone tissue engineering: recent advances and challenges. Crit Rev Biomed Eng. 2012;40(5):363–408.

6. Klouda L, Mikos AG. Thermoresponsive hydrogels in biomedical applications—a review. Eur j pharm biopharm. 2008;68(1):34–45.

7. Zhu J, Marchant RE. Design properties of hydrogel tissue-engineering scaffolds. Expert Rev Med Devices. 2011;8(5):607–26.

8. Im O, Li J, Wang M, Zhang LG, Keidar M. Biomimetic three-dimensional nanocrystalline hydroxyapatite and magnetically synthesized single-walled carbon nanotube chitosan nanocomposite for bone regeneration. Int J Nanomed. 2012;7:2087–99.

9. Zhang L, Hu J, Athanasiou KA. The Role of tissue engineering in articular cartilage repair and regeneration. Crit Rev Biomed Eng. 2009;37(1–2):1–57.

10. Croisier F, Jérôme C. Chitosan-based biomaterials for tissue engineering. Eur Polym J. 2013;49(4):780–92.

11. Roldo M, Fatouros GD. Chitosan derivative based hydrogels as drug delivery platforms. In: Zilberman M, editor. Biomaterials and nanostructures for active implants. studies of mechanobiology, tissue engineering and biomaterials. Newyork: Springer; 2011.

12. Yasmeen S, Lo MK, Bajracharya S, Roldo M. Injectable scaffolds for bone regeneration. Langmuir. 2014;30(43):12977–85.

13. Wang T, Yang X, Qi X, Jiang C. Osteoinduction and proliferation of bone-marrow stromal cells in three-dimensional poly (ε-caprolactone)/hydroxyapatite/collagen scaffolds. J Transl Med. 2015;13:152.

14. Verron E, Khairoun I, Guicheux J, Bouler J-M. Calcium phosphate biomaterials as bone drug delivery systems: a review. Drug Disc Today. 2010;15(13–14):547–52.

15. Roldo M, Fatouros DG. Biomedical applications of carbon nanotubes. Ann Rep Sec C. 2013;109:10–35.

16. Zancanela DC, de Faria AN, Simao AM, Goncalves RR, Ramos AP, Ciancaglini P. Multi and single walled carbon nanotubes: effects on cell responses and biomineralization of osteoblasts cultures. J Mater Sci Mater Med. 2016;27(3):62.

17. Bianco A, Kostarelos K, Partidos CD, Prato M. Biomedical applications of functionalised carbon nanotubes. Chem Commun. 2005;5:571–7.

18. Shi X, Hudson JL, Spicer PP, Tour JM, Krishnamoorti R, Mikos AG. Injectable nanocomposites of single-walled carbon nanotubes and biodegradable polymers for bone tissue engineering. Biomacromolecules. 2006;7(7):2237–42.

19. Barabás R, Katona G, Bogya ES, Diudea MV, Szentes A, Zsirka B, et al. Preparation and characterization of carboxyl functionalized multiwall carbon nanotubes–hydroxyapatite composites. Ceramics International. 2015;41(10, Part A):12717–27.

20. Fatouros GD, Roldo M, Van der Merwe SM. Stabilisation of carbon nanotube suspensions. In: Pastorin G, editor. Carbon nanotubes: from bench chemistry to promising biomedical applications. Singapore: Pan Stanford Publishing Pte. Ltd.; 2011. p. 1–22

21. Roldo M, Power K, Smith JR, Cox PA, Papagelis K, Bouropoulos N, et al. N-Octyl-O-sulfate chitosan stabilises single wall carbon nanotubes in aqueous media and bestows biocompatibility. Nanoscale. 2009;1(3):266–373.CrossRef

22. Nazar H, Fatouros DG, van der Merwe SM, Bouropoulos N, Avgouropoulos G, Tsibouklis J, et al. Thermosensitive hydrogels for nasal drug delivery: the formulation and characterisation of systems based on N-trimethyl chitosan chloride. Eur J Pharm Biopharm. 2011;77(2):225–32.

23. Kokubo T, Takadama H. How useful is SBF in predicting in vivo bone bioactivity? Biomaterials. 2006;27(15):2907–15.

24. Burckbuchler V, Mekhloufi G, Giteau AP, Grossiord JL, Huille S, Agnely F. Rheological and syringeability properties of highly concentrated human polyclonal immunoglobulin solutions. Eur J Pharm Biopharm. 2010;76(3):351–6.

25. Cilurzo F, Selmin F, Minghetti P, Adami M, Bertoni E, Lauria S, et al. Injectability evaluation: an open issue. AAPS Pharmscitech. 2011;12(2):604–9.

26. Ormsby R, McNally T, O'Hare P, Burke G, Mitchell C, Dunne N. Fatigue and biocompatibility properties of a poly(methyl methacrylate) bone cement with multi-walled carbon nanotubes. Acta Biomater. 2012;8(3):1201–12.

27. Gierszewska-Drużyńska M, Ostrowska-Czubenko J. Mechanism of water diffusion into noncrosslinked and ionically crosslinked chitosan membranes. Prog Chem Appl Chitin Deriv. 2012;17:59–66.

28. Chavan PN, Bahir MM, Mene RU, Mahabole MP, Khairnar RS. Study of nanobiomaterial hydroxyapatite in simulated body fluid: formation and growth of apatite. Mater Sci Eng B. 2010;168(1–3):224–30.

29. Shin SR, Bae H, Cha JM, Mun JY, Chen Y-C, Tekin H, et al. CNT reinforced hybrid microgels as scaffold materials for cell encapsulation. ACS Nano. 2012;6(1):362–72.

30. Du F, Fischer JE, Winey KI. Coagulation method for preparing single-walled carbon nanotube/poly(methyl methacrylate) composites and their modulus, electrical conductivity, and thermal stability. J Polym Sci B. 2003;41(24):3333–8.

31. Hone J, Llaguno MC, Biercuk MJ, Johnson AT, Batlogg B, Benes Z, et al. Thermal properties of carbon nanotubes and nanotube-based materials. Appl Phys A. 2002;74(3):339–43.

32. Piovesan S, Cox PA, Smith JR, Fatouros DG, Roldo M. Novel biocompatible chitosan decorated single-walled carbon nanotubes (SWNTs) for biomedical applications: theoretical and experimental investigations. Phy Chem Chem Phys. 2010;12(48):15636–43.

33. Ciapetti G, Granchi D, Devescovi V, Baglio SR, Leonardi E, Martini D, et al. Enhancing osteoconduction of PLLA-based nanocomposite scaffolds for bone regeneration using different biomimetic signals to MSCs. Int J Molecular Sci. 2012;13(2):2439–58.

34. Pan L, Pei X, He R, Wan Q, Wang J. Multiwall carbon nanotubes/polycaprolactone composites for bone tissue engineering application. Colloids Surf B. 2012;93:226–34.

3

NANOHYDROXYAPATITE-REINFORCED CHITOSAN COMPOSITE HYDROGEL FOR BONE TISSUE REPAIR IN VITRO AND IN VIVO

S Dhivya[1] , S Saravanan[1] , T P Sastry[2] and N Selvamurugan[1]

[1] Department of Biotechnology, School of Bioengineering, SRM University, Kattankulathur 603203, Tamil Nadu, India.

[2] Bioproducts Laboratory, Central Leather Research Institute, Chennai 600 020, Tamil Nadu, India.

ABSTRACT

Background

Bone loss during trauma, surgeries, and tumor resection often results in critical-sized bone defects that need to be filled with substitutionary materials. Complications associated with conventional grafting techniques have led to the development of bioactive tissue-engineered bone scaffolds. The potential application of hydrogels as three-dimensional (3D) matrices in tissue engineering has gained attention in recent years because of the superior sensitivity, injectability, and minimal invasive properties of hydrogels. Improvements in the bioactivity and mechanical strength of hydrogels can be achieved with the addition of ceramics. Based on the features required for bone regeneration, an injectable thermosensitive hydrogel containing zinc-doped chitosan/nanohydroxyapatite/beta-glycerophosphate (Zn-CS/nHAp/β-GP) was prepared and characterized, and the effect of nHAp on the hydrogel was examined.

Methods

Hydrogels (Zn-CS/β-GP, Zn-CS/nHAp/β-GP) were prepared using the sol–gel method. Characterization was carried out by scanning electron microscopy (SEM), energy dispersive spectroscopy (EDX), Fourier transform infrared spectroscopy (FTIR), and X-ray diffraction (XRD) as well as swelling, protein adsorption, and exogenous biomineralization studies. Expression of osteoblast marker genes was determined by real-time reverse transcriptase polymerase chain reaction (RT-PCR) and western blot analyses. In vivo bone formation was studied using a rat bone defect model system.

Results

The hydrogels exhibited sol–gel transition at 37°C. The presence of nHAp in the Zn-CS/nHAp/β-GP hydrogel enhanced swelling, protein adsorption, and exogenous biomineralization. The hydrogel was found to be non-toxic to mesenchymal stem cells. The addition of nHAp to the hydrogel also enhanced osteoblast differentiation under osteogenic conditions in vitro and accelerated bone formation in vivo as seen from the depositions of apatite and collagen.

Conclusions

The synthesized injectable hydrogel (Zn-CS/nHAp/β-GP) showed its potential toward bone formation at molecular and cellular levels in vitro and in vivo. The current findings demonstrate the importance of adding nHAp to the hydrogel, thereby accelerating potential clinical application toward bone regeneration.

Keywords: Chitosan Zinc nHAp β-Glycerophosphate Runx2 Bone

BACKGROUND

The rapid shift in the treatment regime from bone grafting to bone tissue engineering is due to surgical complications and biocompatibility issues in terms of immunogenic responses associated with tissue grafting [1, 2]. The use of polymeric scaffolds and ceramic particles has shown promising improvements, extending their application as bone graft materials. Hydrogels have mechanical and structural properties that are very similar to those of natural tissues and extracellular matrices (ECMs) [3]. Hydrogels are exemplified as bone grafts in non-load-bearing regions because of their high degree of flexibility, low toxicity, biocompatibility, biodegradability, sensitivity to external stimuli, injectability, and easily modifiable properties and because they can be used for timed release of growth factors [4, 5, 6, 7, 8].

Chitosan (CS), which is a deacetylated derivative of chitin, is a linear aminopolysaccharide that has been widely used in biomedical applications [9, 10, 11, 12, 13, 14]. CS possesses both structural and compositional similarity to glycosaminoglycans (GAGs) and hence elicits minimal immune response [15]. It demonstrates efficient chelating abilities with metal ions [16] and exhibits swelling properties [17]. CS-based hydrogels have the potential to swell and dehydrate depending on the surrounding environment [18]. Zinc (Zn) is an important trace element found in bone and has structural and regulatory cellular functions [19]. It stimulates osteoblastogenesis and suppresses osteoclastogenesis [20, 21]. In addition, Zn possesses excellent antimicrobial properties [22], which are widely exploited in bone implant materials to tackle implant-associated microbial infections.

Hydroxyapatite (HAp) is a mineral component of natural bone and exhibits osteoconductive properties, osteoinductive properties, bone bonding abilities, and slow degradation in situ [23]. A significant increase in protein adsorption and osteoblast adhesion was observed on nanosized ceramic materials, compared with that on traditional micron-sized ceramic materials [24]. The addition of HAp improves stiffness, interconnectivity, and osteogenic potential in collagen-based scaffolds for bone tissue engineering applications [25]. Organic phosphates, in particular β-glycerophosphate (β-GP), have been used to induce mineralization in cell culture systems [26]. β-GP serves as a source of inorganic phosphates when hydrolyzed by alkaline phosphatase (ALP), and it has been used as an osteogenic supplement in cultures of human bone marrow stromal cells (hBMSCs) [27, 28]. By adding β-GP, which is a weak base, to aqueous chitosan solutions, the polymer remains in solution at neutral pH and room temperature, while homogeneous gelation of this system can be triggered by heat [29]. Controlled hydrogel formation with temperature increase takes place because of the thermally induced transfer of protons from the amine groups of chitosan to the $(PO_4)^{2-}$ groups of β-GP [30, 31].

Based on the biomaterial properties mentioned above, in this study, we aimed to synthesize and characterize an injectable thermosensitive hydrogel containing zinc-doped chitosan/nanohydroxyapatite/β-glycerophosphate (Zn-CS/nHAp/β-GP). Most importantly, the impact of nHAp on the osteogenic potential of the synthesized hydrogel was determined in vitro and in vivo. The molecular mechanism behind the bone formation potential of the hydrogel was also examined. In vitro bone regeneration promoted by the hydrogel was confirmed in vivo in critical-sized bone defects in rats.

RESULTS AND DISCUSSION

Hydrogel preparation

Zn-CS/β-GP and Zn-CS/nHAp/β-GP solutions were prepared by drop-wise addition of an optimized molar concentration of pre-cooled β-GP to pre-cooled solutions of Zn-CS and Zn-CS/nHAp under continuous mixing. The sol form of the hydrogel formed at 4°C was transferred to 37°C for gelation. The gelation time was found to be >10 min at pH 7. The acidic pH of the CS solution changed to 7 with the addition of β-GP.

Physicochemical characterization of hydrogel

Scanning electron microscopy (SEM) analysis of the Zn-CS/β-GP hydrogel (Figure 1a, b) and the Zn-CS/nHAp/β-GP hydrogel (Figure 1c, d) revealed porous architectures indicative of the presence of both micro- and macro-sized pores. The pores in the hydrogels were uniformly present and highly interconnected. In both Zn-CS/β-GP and Zn-CS/nHAp/β-GP hydrogels, the diameter of the pores was in the range of 100–150 μm. This was a suitable pore size for cell penetration into the hydrogels and aided in nutrient transport and metabolic waste disposal. The magnified image of the hydrogel with nHAp shown in Figure 1d indicates that the addition of nHAp resulted in the deposition of sharp needle-shaped nanoparticles along the pore walls and struts, which acted as nucleating sites for mineral deposition, protein adsorption, and cell attachment because of their nanoarchitecture [32, 33, 34].

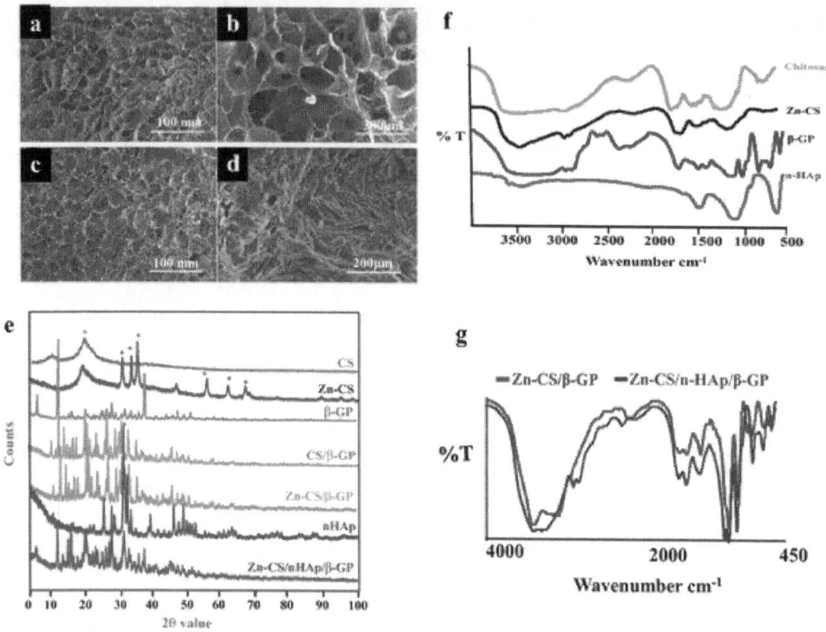

Figure 1. Physicochemical characterization of the hydrogels. Scanning electron microscopic images of **a** Zn-CS/β-GP and **c** Zn-CS/nHAp/β-GP; **b** and **d** are magnified images of the above, respectively. Pores greater than 150 μm in diameter were seen in both hydrogels. The presence of nHAp along the matrix of chitosan is clearly visible under high magnification. **e** XRD spectra of the individual components and of the hydrogels. FTIR spectra of **f** the individual components and of **g** the Zn-CS/β-GP and Zn-CS/nHAp/β-GP hydrogels.

X-ray diffraction (XRD) analyses of the individual components employed in the preparation of the hydrogels and the final resulting hydrogels are shown in Figure 1e. The appearance of broad peaks between 20° and 30° indicated the semi-crystalline nature of chitosan. The sharper peaks observed at 31.66°, 34.2°, 36.16°, 47.86°, 56.52°, 62.66°, and 67.80° corresponded to the (100), (002), (101), (102), (110), (103), and (112) lattice planes, respectively, of the wurtzite phase of ZnO. The values are in strong agreement with those of standard ZnO in JCPDS card No 36-1451 [35]. The occurrence of phases corresponding to CS, ZnO, nHAp, and β-GP was clearly evident in the final hydrogel because of its crystalline nature.

The Fourier transform infrared spectroscopy (FTIR) spectra of the individual components and of the hydrogels are shown in Figure 1f and g. The absorption band at 1,600 cm^{-1}, representing the "free amine" group of chitosan, disappeared in the Zn-CS complex and was replaced by a new peak at 1,657 cm^{-1}, corresponding to the Zn-CS complex [36]. Chitosan showed an absorption peak at 3,425 cm^{-1}, which was attributed to the combined peaks of the $-NH_2$ and $-OH$ stretching vibration. This broader and stronger peak of chitosan was shifted to a higher wavenumber at 3,430 cm^{-1} in Zn-CS, indicating a strong interaction between these groups and ZnO. The absorption peak at 2,921 cm^{-1} was attributed to the asymmetric stretching of–CH_3 in chitosan, and the absorption peaks at 1,657 and 1,093 cm^{-1} were ascribed to the bending vibration of the $-NH_2$ group and the C–O stretching group, respectively. All of these interactions clearly described the attachment of zinc oxide to chitosan chains. Standard FTIR spectra of nHAp and β-GP were also recorded to study the interactions among the gel-forming components. The spectrum of β-GP showed the bending and stretching vibration modes of PO_4, which were identified by the peaks at 548 and 653 cm^{-1}, respectively. The individual spectrum of nHAp showed the Υ_1 and Υ_4 absorption bands of PO_4 at 1,050 and 577 cm^{-1}, respectively [37]. In addition, the peak at 3,567 cm^{-1} in the nHAp spectrum corresponded to the –OH stretching mode of stoichiometric HAp [38].

The FTIR spectrum of the Zn-CS/β-GP hydrogel was compared with that of the Zn-CS/nHAp/β-GP hydrogel (Figure 1g) to identify important interactions between Zn-CS/β-GP and the bioactive component nHAp. The Zn-CS/nHAp/β-GP hydrogel revealed peaks at 1,410 and 1,573 cm^{-1}attributing to C=O stretching, strong peaks at 2,848 and 2,936 cm^{-1} belonging to the C–H stretching region, and small peaks between 2,160 and 2,255 cm^{-1} corresponding to C=C. The peaks at 1,074 and 2,848 cm^{-1} indicated amide II bands, and the peaks at 1,664, 1,573, and 1,550 cm^{-1} were assigned to amide I (C=O), amino ($-NH_2$), and amide II (–NH), respectively [39]. The peak at 960 cm^{-1} in the Zn-CS/β-GP hydrogel, representing the Υ_1 absorption band of PO_4, was intensified and shifted to a higher wavenumber of 971 cm^{-1} with the addition of nHAp. This result indicated a strong interaction between nHAp and Zn-CS in the hydrogel.

Hydrogel characteristics with the addition of nHAp

Swelling behavior and structural integrity are two critical design variables in constructing hydrogels and other scaffolding materials. CS has the ability to readily swell up when exposed in a biological environment. Figure 2a indicates increased fluid uptake associated with the addition of nHAp particles to the hydrogel. A drop in the swelling behavior was noticed with an increase in the incubation period to greater than 12 h. The higher rate of swelling associated with hydrogels containing nHAp may be due to the hydrophilicity offered by the free –OH groups of nHAp. Fluid retention results in the relaxation of mechanically coiled chains of chitosan, leading to an increase in surface area. In turn, the increase in internal surface area caused by swelling facilitates the infiltration of cells into the scaffolds. Moreover, increased swelling enables efficient transport of nutrients from the culture media and ensures better probability of cell infusion in a 3D pattern, mimicking cell growth under physiological conditions in vivo [40, 41]. The persistence and rate of degradation of any implanted material in vivo are the key factors that affect the rate of bone formation. Glycosidic bonds in the CS polymeric network are a major target for hydrolytic scission by lysozymes. The chains of *N*-acetylglucosamine in CS are linked by glycosidic bonds, and their degradation will lead to the release of amino sugars, which can be utilized for several metabolic pathways [42]. In addition, the degradation of scaffolds is essential for the release of bioactive ingredients and determines the mechanical strength of any biomaterial. The degradation rate of both hydrogels remained the same at 48 h. From Figure 2b, it is clear that the addition of nHAp had no effect on the degradation of the hydrogel.

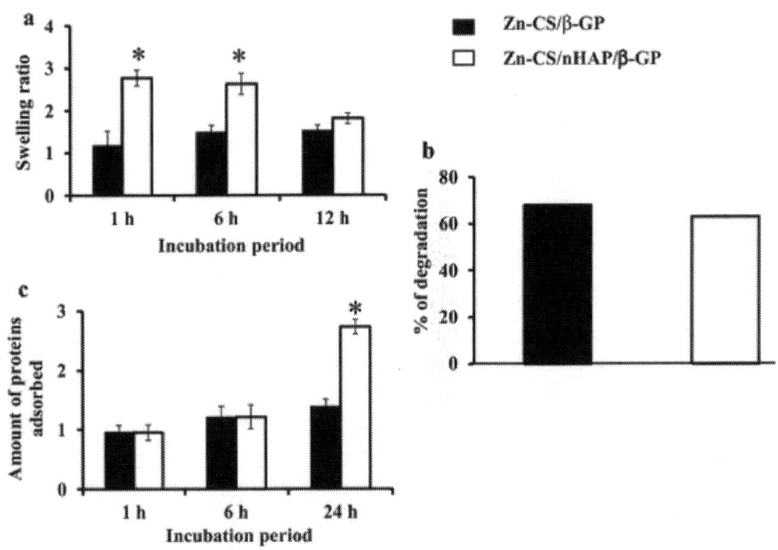

Figure 2. **a** Swelling studies of Zn-CS/β-GP and Zn-CS/nHAp/β-GP hydrogels. Addition of nHAp significantly increased the water retention capacity of the hydrogel. **b** Degradation studies of Zn-CS/β-GP and Zn-CS/nHAp/β-GP hydrogels. Addition of nHAP had no effect on the degradation characteristics of the hydrogel. *Asterisk* indicates significant changes compared with Zn-CS/β-GP ($p < 0.05$) (n = 6). **c**Protein adsorption studies on Zn-CS/β-GP and Zn-CS/nHAp/β-GP hydrogels.

Protein adsorption is a pre-requisite for cell-biomaterial interaction, which can be strengthened by the presence of nanoparticles in the composites. The protein adsorption ability of the hydrogels was studied at different times (Figure 2c). Initial protein adsorption is a vital property of a successful tissue-engineered implant. The addition of nHAp significantly increased the amount of protein adsorbed onto the hydrogel at 24 h. This effect could have been due to the increased hydrophilicity of the hydrogel, which mediated enhanced protein adsorption. In addition, the nanostructures of hydroxyapatite in the hydrogel offered increased surface area for the adsorption of proteins [43].

The in vitro exogenous biomineralization of the hydrogels prepared was examined by immersing them in simulated body fluid (SBF) solutions for 21 days. The presence of apatite crystals on the surface of the hydrogels was examined and confirmed by SEM analyses (Figure 3a–d). Figure 3a and b represent the SEM images of exogenous biomineralized Zn-CS/β-GP and Zn-CS/nHAp/β-GP hydrogels, respectively. Figure 3c and d represent the magnified images of the above. The pattern of apatite formation was intense on the surface of the hydrogels containing nHAp (Figure 3d). This could be due to the availability of negatively charged hydroxyl groups on hydroxyapatite that acted as nucleation sites to initiate crystal deposition. Furthermore, the presence of calcium and phosphate ions in the biomineralized Zn-CS/β-GP (Figure 3e) and Zn-CS/nHAp/β-GP (Figure 3f) hydrogels was confirmed in the energy dispersive spectroscopy (EDX) spectra. The characteristic diffraction peaks of hydroxyapatite at values of 2θ from 31.8° to 52.5°, in accordance to JCPDS-09-0432, were observed in all tested samples (Figure 3g). The observed sharper peaks implied that the degree of crystallinity of the hydrogel samples had increased after biomineralization.

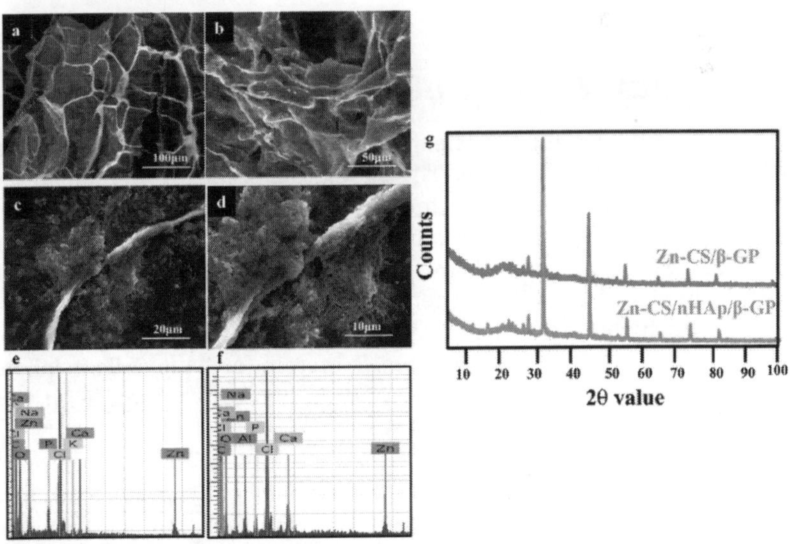

Figure 3. SEM images of exogenous biomineralized **a** Zn-CS/β-GP and **b** Zn-CS/nHAp/β-GP hydrogels in SBF for 21 days. **c** and **d** are *magnified images* of the above, respectively. EDX spectra of **e** Zn-CS/β-GP and **f** Zn-CS/nHAp/β-GP hydrogels. **g** XRD spectra of the mineralized hydrogels.

The antimicrobial nature of any implanted graft material aids in the persistence of the implant by preventing microbial attacks associated with surgeries. The hydrogels exhibited a bactericidal effect against Gram-negative and Gram-positive bacteria, such as *Escherichia coli* and*Streptococcus pyogenes*, respectively. The antimicrobial activity of the Zn-CS/nHAp/β-GP hydrogel was not significantly different compared with that of the Zn-CS/β-GP hydrogel (Table 1). Overall, these results (Figures 2, 3) clearly indicated the superiority of the nHAp-incorporated hydrogel over the native hydrogel (without nHAp).

Table 1. Zone of inhibition observed against Gram-negative and Gram-positive organisms

Hydrogel	Zone of inhibition (mm)	
	Escherichia coli	*Streptococcus pyogenes*
Zn-CS/β-GP	8.0 ± 0.013	9.3 ± 0.305
Zn-CS/n-HAp/β-GP	8.0 ± 0.057	9.6 ± 0.252

Addition of nHAp induced no significant change in the antimicrobial behavior of the hydrogel.

Biocompatible nature of the synthesized hydrogel

Biomaterials intended for tissue engineering applications should not exert any toxicity on mammalian cells. The dissolution of ions from biomaterials plays a vital role in regulating cytotoxicity. Impedance of cell growth and alterations in membrane permeability are signs of cytotoxicity. Hence, the synthesized Zn-CS/nHAp/β-GP hydrogel was subjected to an indirect 3-(4,5-dimethythiazol-2-yl)-2,5-diphenyltetrazolium bromide (MTT) assay. Mouse mesenchymal stem cells (mMSCs) were incubated with medium that was conditioned by immersing hydrogels therein for 24 h. Cells incubated in normal media served as negative controls, and cells treated with 0.1% Triton X-100 served as positive controls. From Figure 4a, it is obvious that the cells experienced cytotoxicity when cultured in 100 μL of conditioned medium, which was the amount equivalent to 5 mg of hydrogel. Since there was no discernible toxicity associated with 50 μL (2.5 mg) of conditioned medium, this condition was chosen for further studies. Direct measurement of cell viability was also performed with fluorescein diacetate (FDA) staining of control mMSCs (Figure 4bI, II) and of the mMSCs in the Zn-CS/nHAp/β-GP hydrogel (Figure 4bIII, IV). The results showed the non-toxic nature of the hydrogel, as indicated by the distinct cellular morphology and highly spread cellular extensions.

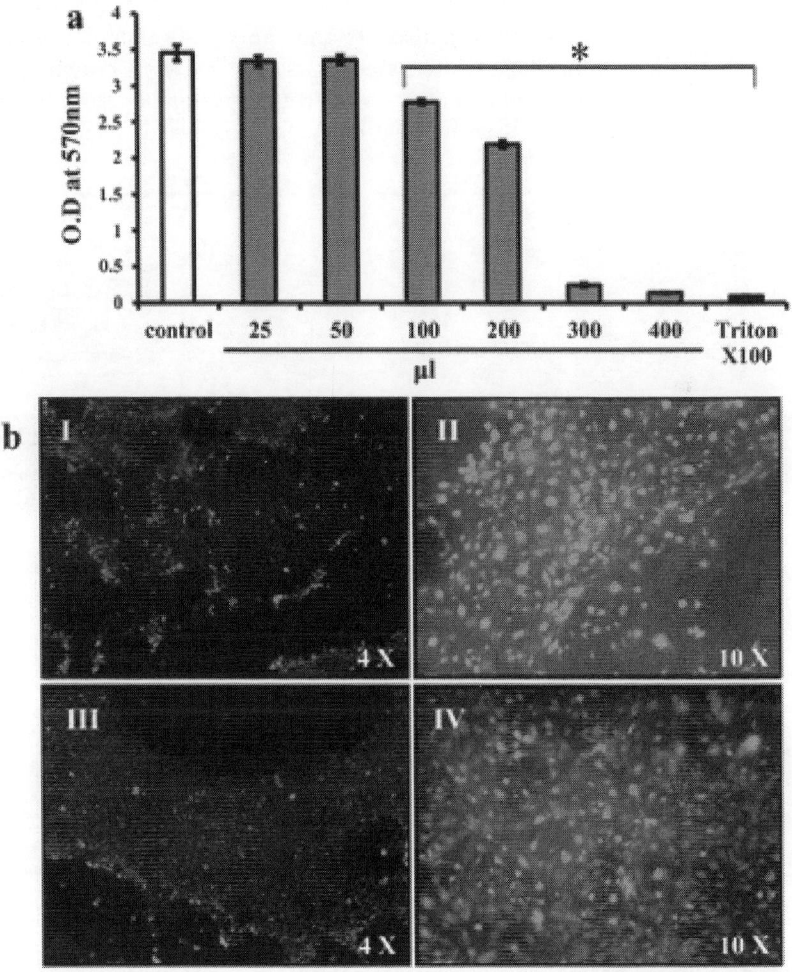

Figure 4. In vitro cytotoxicity analysis of Zn-CS/nHAp/β-GP hydrogel. **a** MTT assay of C3H10T1/2 cells grown for 24 h in medium conditioned with Zn-CS/nHAp/β-GP hydrogel. Wells treated with 0.1% Triton X-100 served as positive controls. *Asterisk* indicates significant decrease compared with control ($p < 0.05$) ($n = 5$). **b** FDA staining of C3H10T1/2 cells grown in Zn-CS/nHAp/β-GP hydrogel for 5 days and assessed with ×4 and ×10 objectives for cellular morphology. Control cells (*I*) and cells grown in hydrogel (*III*) showed highly spread cellular extensions. *Magnified images* of control and hydrogel-treated cells are depicted in (*II*) and (*IV*), respectively.

mMSC differentiation to osteoblasts at the cellular level

To determine the capability of hydrogels and the significance of nHAp in promoting MSC differentiation into osteoblasts, mMSCs were grown in the conditioned media obtained from the Zn-CS/β-GP hydrogel or from the Zn-CS/nHAp/β-GP hydrogel, either in the presence or absence of osteogenic

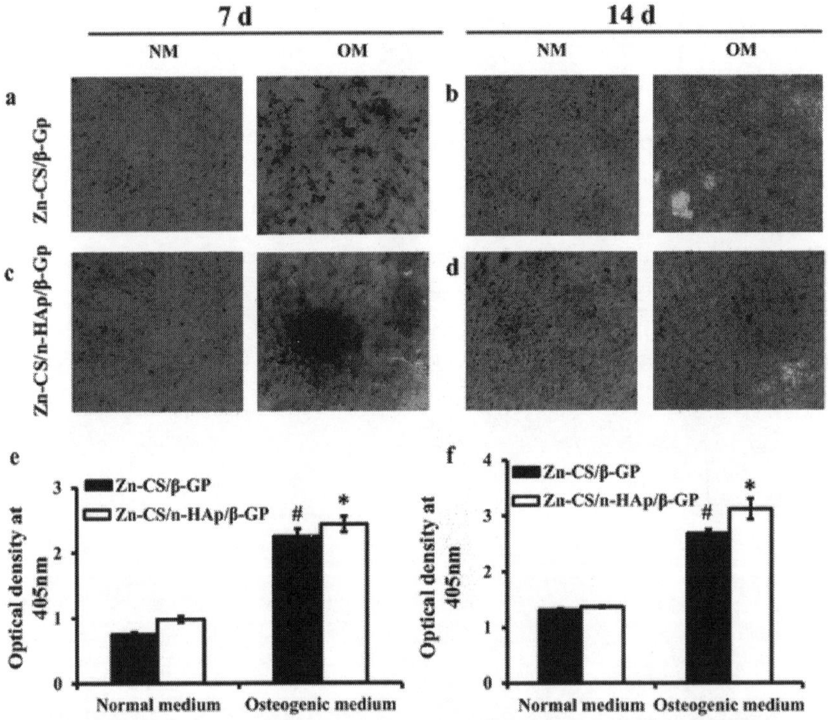

Figure 5. Effect of Zn-CS/nHAp/β-GP hydrogel on mMSC differentiation and mineralization by alizarin staining. Cells were grown in the presence of hydrogels under normal and osteogenic conditions for periods of 7 and 14 days. Fresh medium was replenished every 3 days. Cells were fixed and stained with alizarin red to visualize the calcium deposits. Representative microscopic images of the fixed cells incubated on **a** Zn-CS/β-GP hydrogel for 7 days, **b** Zn-CS/β-GP hydrogel for 14 days, **c** Zn-CS/nHAp/β-GP hydrogel for 7 days, and **d** Zn-CS/nHAp/β-GP hydrogel for 14 days. **e** and **f** are quantification plots of calcium deposits of the above. *NM* normal medium, *OM* osteogenic medium. *Asterisk*indicates significant increase compared with the respective control in normal medium. *Hash*indicates significant increase compared with the respective control in normal medium ($p < 0.05$).

stimulants for 7 and 14 days (Figure 5). Calcium deposition was assessed by alizarin red staining. Microscopic images of the biomineralized deposits indicated that, in the presence of osteogenic stimulation, the addition of nHAp in the Zn-CS/β-GP hydrogel promoted better osteoblast differentiation at 7 days (Figure 5c) and 14 days (Figure 5d) compared with the Zn-CS/β-GP hydrogel (Figure 5a, b). Quantification data for the alizarin red staining also indicated increased mineralization in the Zn-CS/nHAp/β-GP hydrogel at 7 days (Figure 5e) and 14 days (Figure 5f). Hence, the hydrogel exerted maximal differentiation potential under osteogenic conditions and in the presence of nHAp, indicating its osteoconductive nature.

mMSC differentiation into osteoblasts at the molecular level

Real-time reverse transcriptase polymerase chain reaction (RT-PCR) was carried out to determine the expression of osteoblastic differentiation marker genes such as alkaline phosphatase (ALP), type-I collagen (COL-I), osteocalcin (OC), and bone transcription factor Runx2 at the molecular level in mMSCs incubated with the hydrogels (Figure 6). Increased mRNA levels of ALP, COL-I, and Runx2 were observed in cells incubated with the hydrogel containing nHAp (Zn-CS/nHAp/β-GP) for 7 days in osteogenic medium compared with those observed in the Zn-CS/β-GP hydrogel (Figure 6a). A significant increase in the mRNA level of OC was found in mMSCs after 14 days of incubation with the Zn-CS/nHAp/β-GP hydrogel in the presence of osteogenic medium (Figure 6b). Although the mRNA expression of Runx2 was stimulated by nHAp in the Zn-CS/nHAp/β-GP hydrogel even in normal medium, maximal Runx2 expression was observed in osteogenic medium. Runx2 is an essential transcription factor for the expression of various osteoblast-specific marker genes, namely, ALP, COL-I, and OC [44]. ALP and COL-1 are early differentiation marker genes, whereas OC is a late marker gene of osteoblast differentiation. ALP plays a role in the conversion of inorganic pyrophosphate into inorganic phosphate, a process that enhances mineralization process [45]. COL-I aids in the deposition of collagen, the most abundant ECM protein in bone. The voids in the collagen molecules act as the sites of deposition for hydroxyapatite crystals, facilitating subsequent bone mineralization [46]. OC is a major non-collagenous protein component of the bone ECM, acting as a regulator of bone crystal growth by mediating its high affinity binding to hydroxyapatite crystals [47].

The expression of the Runx2 protein was also determined. At 7 and 14 days of cell incubation with Zn-CS/β-GP and Zn-CS/nHAp/β-GP hydrogels in osteogenic medium, the expression of Runx2 was higher than that in normal medium (Figure 7a, c). The addition of nHAp in the Zn-CS/β-GP hydrogel stimulated higher Runx2 protein expression than that seen in the Zn-CS/β-GP hydrogel at an incubation period of 7 days (Figure 7a). The densitometry scanning of western blotting is shown in Figure 7b and d. The consolidation of gene and protein expression results (Figures 6, 7) suggested that the presence of nHAp in the hydrogel enhanced the expression of Runx2 and osteoblast differentiation marker genes under osteogenic conditions, depicting the osteoconductive nature of the hydrogel.

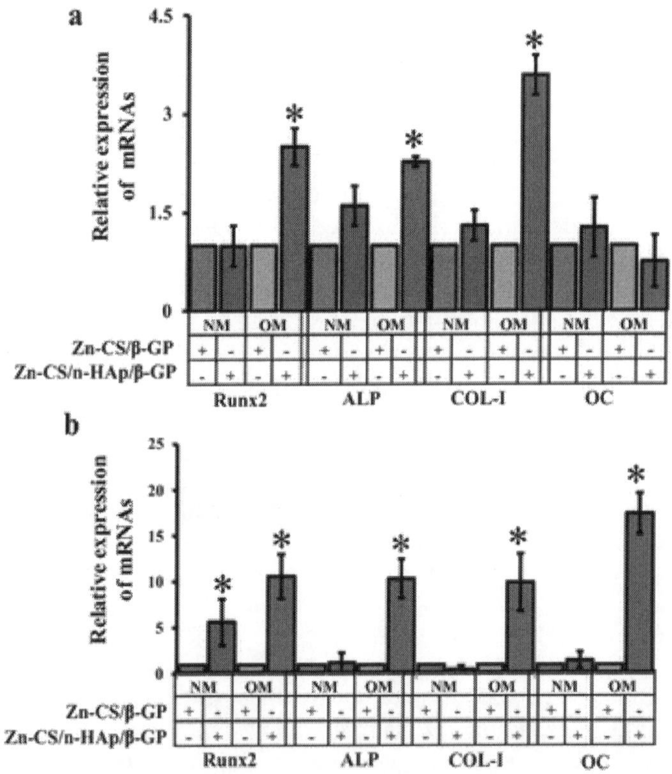

Figure 6. Effect of Zn-CS/nHAp/β-GP hydrogel on the mRNA expression of osteoblast differentiation marker genes in mMSCs. Cells were grown in the presence of Zn-CS/β-GP hydrogel or Zn-CS/nHAp/β-GP hydrogel under normal and osteogenic conditions for **a** 7 days and **b** 14 days. At the end of the incubation periods, total RNA was isolated and real-time RT-PCR was performed using primers specific for Runx2, ALP, COL-I, and OC genes. The fold change in mRNA expression was calculated over the control samples. GAPDH was used as an internal control. *NM* normal medium, *OM*osteogenic medium. *Asterisk* indicates significant increase compared with the respective control, i.e., Zn-CS/β-GP ($p < 0.05$).

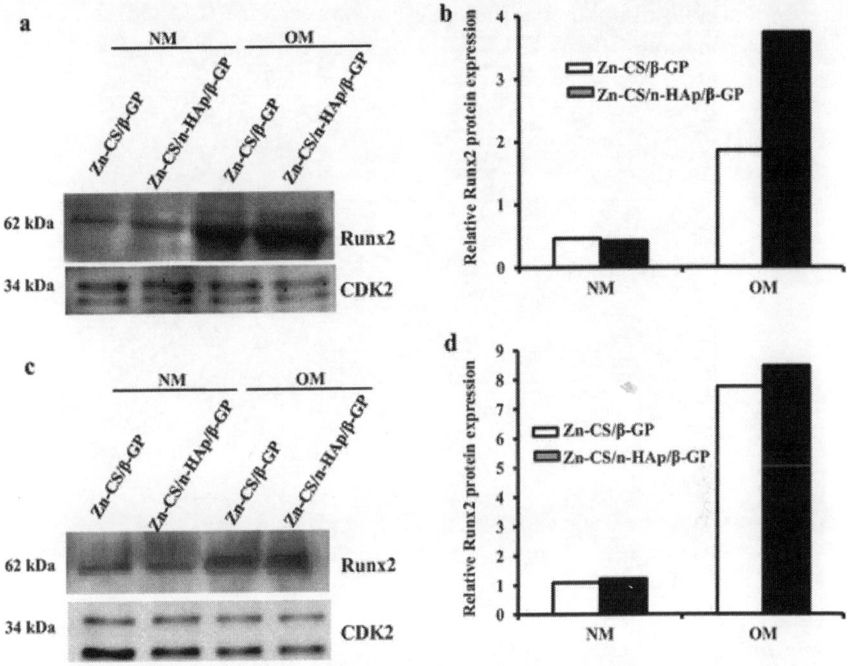

Figure 7. Effect of Zn-CS/nHAp/β-GP hydrogel on the expression of Runx2 protein. mMSCs were grown in the presence of Zn-CS/β-GP hydrogel or Zn-CS/nHAp/β-GP hydrogel under normal and osteogenic conditions for **a** 7 days and **c** 14 days. At the end of the incubation periods, whole cells lysates were prepared and separated onto 12% SDS-PAGE. The proteins were electrophoretically transferred to PVDF membranes and subjected to western blot analysis using Runx2 antibodies. Cdk2 antibodies were used as an internal loading control. **b** and **d** represent the quantification of Runx2 protein blots as mentioned above using Image Lab software version 4.1 (Bio-Rad). *NM* normal medium, *OM*osteogenic medium.

Healing of critical-sized bone defects in rat tibia in vivo

We next assessed the bone-healing property of the hydrogels in rats with tibial defects as well as the significance of nHAp in the hydrogels. From radiographic images taken 2 weeks post-surgery, better tissue organization was observed in tibiae treated with Zn-CS/nHAp/β-Gp hydrogels than in tibiae treated with Zn-CS/β-GP hydrogels and in control tibiae (Figure 8a). The radiographs showed that, in the control group, the tibial bone defect was devoid of any implanted hydrogel, resulting in a radiolucent gap with no signs of the drill hole being filled. The presence of nHAp in the Zn-CS/nHAp/β-Gp hydrogel resulted in better wound closure and bone

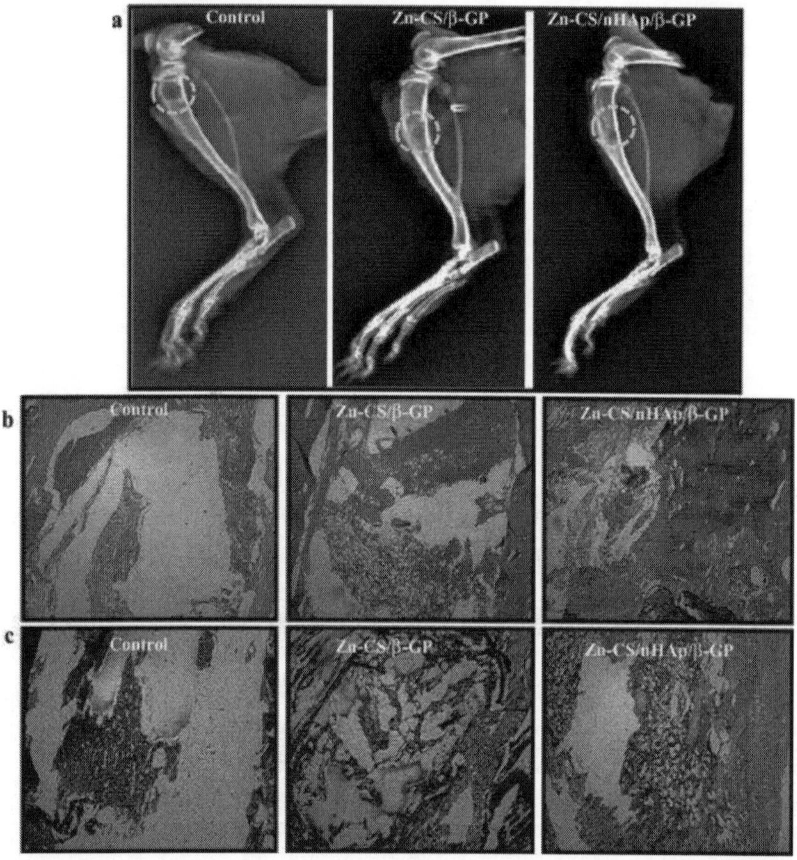

Figure 8. Effect of Zn-CS/nHAp/β-GP hydrogel on bone healing in vivo. **a** Radiographic images of the rat tibial defects obtained 2 weeks after they were filled with Zn-CS/β-GP hydrogel or Zn-CS/nHAp/β-GP hydrogel. **b** and **c** represent hematoxylin and eosin staining and collagen staining of the implants, respectively. Closure of the drill hole, formation of new bone, and deposition of collagen were seen in Zn-CS/nHAp/β-GP-treated animals. *Dotted yellow circle* indicates the drill hole defect and subsequent closure in hydrogel-treated animals. *Purple color-stained areas* indicate the formation of bone, and *blue-colored areas* depict the deposited collagen.

formation, in contrast with the findings for hydrogel without nHAp and the control. The radiographs depicted minimal periosteal reaction and smoothening of cortical bone defect edges in the control and Zn-CS/β-GP-treated groups. The shape of the bone defect changed to an oval morphology, suggesting the initiation of bone healing. Histological analysis involving hematoxylin and eosin (H&E) staining (Figure 8b) and collagen staining (Figure 8c) was performed. Initiation of bone formation was found to be more prominent (purple color) in Zn-CS/nHAp/β-GP hydrogel-treated animals, whereas few areas with purple color were observed in Zn-CS/β-Gp hydrogel-treated animals. No apparent bone formation was observed in control animals (Figure 8b). The presence of nHAp in the hydrogel was a key factor in bone crystal deposition and acted as a nucleating site for bone formation. Collagen is the main component of the ECM involved in the bone repair mechanism. It is clear that collagen deposition (blue-colored deposits) was much more intense in Zn-CS/nHAp/β-GP hydrogel-treated animals than in the control or Zn-CS/β-GP hydrogel-treated animals (Figure 8c). Thus, the addition of nHAp in the Zn-CS/β-GP hydrogel improved bone formation and closure of rat tibial bone defects in vivo.

CONCLUSIONS

The current investigations implicate the importance of incorporating nHAp in a thermosensitive CS-based hydrogel to improve its physical and biological characteristics. Increased protein adsorption, controlled swelling, and decreased susceptibility to lysozyme degradation were exhibited with the addition of nHAp. The hydrogel was also found to be osteoconductive in nature, promoting the differentiation of mMSCs into osteoblasts through the upregulation of Runx2 expression. Moreover, the use of Zn-CS/nHAp/β-GP hydrogels promoted bone healing in critical-sized rat tibial defects. The current findings demonstrate the advantages of adding nHAp to hydrogels, implying their potential clinical application in bone regeneration.

METHODS

Preparation of Zn-doped CS

Zn-doped CS was prepared under refluxing conditions by using an acidic solution of chitosan and a precursor solution for zinc. The procedure has been previously described in detail [48].

Preparation of Zn-CS/nHAp/β-GP

Zn-doped (2% w/v) CS was weighed and dissolved in 0.1 M acetic acid to obtain a clear solution. β-glycerophosphate (β-GP) disodium salt hydrate (Sigma Aldrich) solutions of appropriate molar concentration were prepared by dissolving calculated amounts of β-GP in distilled water. Both solutions were pre-cooled to 4°C before hydrogel preparation. nHAp (Sigma Aldrich) was sonicated in distilled water at 0.1% (w/v) to obtain a homogeneous dispersion and pre-cooled to 4°C prior to hydrogel formation.

To the solution containing Zn-CS/nHAp, β-GP solution was added drop-wise under continuous mixing. The pre-cooled solutions were mixed together at a ratio of 8:1:1 (Zn-CS:β-GP:nHAp) with the final concentration of β-GP in the sol being 0.35 M. The sol form of the hydrogel was poured into 24-well culture plates and incubated at 37°C to allow gelation to take place. Gelation was observed within 10 min at pH 7. The hydrogel was then stored at −20°C overnight, followed by lyophilization at −40°C for 16 h.

Characterization of the hydrogels

The external surface morphology and porous structure of the synthesized Zn-doped CS and the freeze-dried Zn-CS/nHAp/β-GP hydrogel were analyzed by SEM. Thin sections of scaffolds were sliced using a scalpel and the samples were fixed on adhesive carbon tapes, gold-coated, and subjected to SEM analysis using the HR SEM Quanta 200FEG Instrument (Netherlands). The freeze-dried hydrogel samples were ground into a powder form in liquid nitrogen. The powdered scaffolds were analyzed by XRD with a 2θ angle of 5°–80° at a speed of 2° min^{-1} using an analytical XPERT PRO powder diffractometer operating at a voltage of 40 kV (Cu Kα radiation). The intermolecular chemical interactions between the various functional groups within the components present in the Zn-CS/β-GP hydrogel and the Zn-CS/nHAp/β-GP hydrogel were analyzed by FTIR (American Perkin Elmer Spectrum). The samples were scanned in the range from 4,000 to 450 cm^{-1}.

Swelling studies

Equal amounts of freeze-dried Zn-CS/β-GP and Zn-CS/nHAp/β-GP hydrogels were measured and their dry weights were noted as W_i. They were then immersed in 1 × PBS for different time periods (1, 6, 12, and 24 h) at 37°C. At the end of each incubation period, the hydrogels were washed with deionized water to remove any ions adsorbed onto the surfaces and blot-dried using filter paper. The wet weights were recorded as W_f. The swelling ratio of the hydrogel was calculated using the following formula:

$$\text{Swelling ratio}=(Wi-Wt)/(Wi)\text{Swelling ratio}=(Wi-Wt)/(Wi)$$

Biodegradation studies

Equal amounts of hydrogel samples were weighed and their initial weights were noted as W_0. They were then immersed in medium containing lysozymes (Sigma Aldrich) at a concentration similar to that in circulating levels of blood (10,000 U/L) and incubated for 48 h at 37°C [49]. At the end of the incubation period, the supernatant was removed and the hydrogels were freeze-dried. The final weight of the freeze-dried hydrogels was noted as W_t. The degradation percentage of the hydrogels was calculated using the following formula:

Biodegradation %=(Wo−Wt)/Wt×100Biodegradation %=(Wo−Wt)/Wt×10
0

Protein adsorption studies

Equal amounts of hydrogels were weighed and were subjected to pre-wetting in 100% ethanol for 1 h, followed by incubation in PBS for 30 min to reach equilibrium hydration. They were then incubated from 1 to 24 h in bovine serum albumin solutions (BSA, 5 mg/mL) (Sigma Aldrich) at 37°C. After incubation, the hydrogels were removed and the supernatant solution was used for quantification of unabsorbed proteins by Bradford's assay.

Exogenous biomineralization

Equal amount of hydrogels were weighed and tested for their biomineralization abilities by immersing them in simulated body fluid (SBF) for a period of 7 , 14 , and 21 days at 37°C. Equal volumes of fresh SBF were replaced once every 2 days. The SBF solution was prepared according to a previous report [50]. After the desired time intervals, the scaffolds were removed, washed thrice with deionized water, and lyophilized. The freeze-dried samples were then subjected to FTIR and XRD analyses.

Antimicrobial activity

Antibacterial activity was determined by the zone of inhibition test using *S. pyogenes* (Gram-positive bacteria) and *E. coli* (Gram-negative bacteria). Overnight cultures were prepared. A 100-μL aliquot of the overnight culture of the bacteria (1×10^8 CFU/mL) was spread onto Luria–Bertani agar plates, and 20 mg of the lyophilized hydrogel was placed on the plates, followed by incubation for 24 h at 37°C. The clear area (zone of inhibition) was measured and the results were tabulated.

Cytotoxicity assessment

Indirect MTT assay was carried out to test the cytotoxicity of the hydrogels. mMSCs (4×10^4 per cm^2) were seeded in 24-well plates. The freeze-dried Zn-CS/nHAp/β-GP hydrogel (800 mg) was weighed and soaked in 16 mL of DMEM for 24 h. The supernatant, termed as "conditioned medium", was taken and added at different volumes (25, 50, 100, 200, 300, and 400 μL) to the wells containing cells. Triton X-100 (0.1%) served as a positive control. Cells seeded into the wells with normal medium and incubated for 24 h served as negative controls. The medium was carefully aspirated from the wells, and 200 μL of 0.05% MTT solution was added to each well and incubated for another hour at 37°C. DMSO was used to dissolve the formazan crystals formed, and the optical densities (OD) were measured using a spectrophotometer (ULTRA SPEC 2100 pro; Amersham Life Sciences, USA) at 570 nm.

Cell morphology evaluation

Zn-CS/β-GP and Zn-CS/β-GP/nHAp hydrogels were UV-treated for 2 h. Approximately 2×10^5mMSCs were seeded onto the hydrogels and incubated for 4 days. The medium was changed once every 2 days. After 4 days of incubation, the medium was removed and the cells were washed with ice-cold 1 × PBS. Ten milligrams of fluorescein diacetate (FDA) (Sigma Aldrich) was weighed and dissolved in 10 mL of acetone, which served as the stock solution. A working solution was prepared freshly by mixing 30 μL of the stock solution with 10 mL of PBS. One milliliter of this working solution was added onto the films in each well and incubated for 15 min at 37°C in the dark [51]. Cells were observed under a fluorescent microscope with 4× and 10× objectives.

Western blot analysis

mMSCs were grown in normal or osteogenic medium with Zn-CS/β-GP or Zn-CS/β-GP/nHAp hydrogels for different time periods. The cells were then washed with cold phosphate-buffered saline (PBS), and whole cell lysates were prepared. Twenty micrograms of proteins was resolved using 12% sodium dodecyl sulfate–polyacrylamide gel electrophoresis (SDS-PAGE) and transferred to polyvinylidene difluoride (PVDF) membranes by electroblotting. The membranes were blocked with 5% non-fat milk powder (BioRad) and incubated overnight with a primary antibody at 4°C. The membranes were probed with secondary antibodies conjugated with horseradish peroxidase (HRP). Finally, the bands were visualized by adding Super Signal West Dura Extended Duration Substrate (Thermo Scientific) according to the manufacturer's instructions. The images obtained were used for quantification with the ImageJ software, as described previously [52]. Mouse monoclonal Runx2 (1:1,000) antibodies, Cdk2 (1:1,000) antibodies, and secondary antibodies conjugated with HRP were obtained from Santa Cruz Biotechnology. Cdk2 antibodies served as internal loading controls.

Real-time RT-PCR analysis

mMSCs were grown for different time periods in normal or osteogenic medium conditioned with Zn-CS/β-GP or Zn-CS/β-GP/nHAp hydrogels. Total RNA was isolated using the Trizol method (Invitrogen) according to the manufacturer's instructions. cDNA synthesis was carried out with 1 μg of total RNA from each sample using the GeNei AMV RT-PCR kit according to the manufacturer's protocol. SYBR reagent (BioRad) was used for real-time RT-PCR with mouse-specific oligonucleotide primers, as shown in Table 2. Relative mRNA expression was calculated by using the ΔΔCt method of relative quantification [34].

Table 2. Primer sequences used for real-time RT-PCR analysis (F: forward; R: reverse)

Gene	Primer sequence (forward/reverse)
Runx2	5′ CGCCTCACAAACAACCACAG 3′ (F) 5′ TCACTGTGCTGAAGAGGCTG 3′ (R)
ALP	5′ TTGTGCCAGAGAAAGAGAGAGA 3′ (F) 5′ GTTTCAGGGCATTTTTCAAGGT 3′ (R)
COL-I	5′ TAACCCCCTCCCCAGCCACAAA 3′ (F) 5′ TTCCTCTTGGCCGTGCGTCA 3′ (R)
OC	5′ ATGGCTTGAAGACCGCCTAC 3′ (F) 5′ AGGGCAGAGAGAGAGGACAG 3′ (R)
GAPDH	5′ GAGAGACCCCACTTGCTGCCA 3′(F) 5′ CTCACACTGCCCCTCCCTGGT 3′ (R)

In vivo examination of bone formation in drill hole defects in rat tibia

Male Wistar rats (3 months old) weighing 200–250 g were kept in a room with light (12 h light–dark cycle) at a controlled temperature. The procedure for creating drill hole defects was approved by the animal ethical committee, Kovai Medical College and Hospital, Coimbatore, India. Rats were divided into three groups (n = 6 in each group): (1) control, (2) Zn-CS/β-GP, and (3) Zn-CS/β-GP/nHAp. The rats were anesthetized with 10% ketamine and 2% xylazine (1:1, 0.1 mL/100 g body weight, i.m.) and subjected to perforation of the right tibia by using a dental drill (diameter, 3 mm) under constant saline irrigation (0.9% NaCl). The defects were entirely filled with each of the hydrogels in the respective experimental groups. The defects in control animals were left unfilled. After 14 days, the animals were sacrificed under anesthesia and the tibiae were removed and radiographed. The tibiae were then fixed in neutral 10% buffered formalin for 48 h at room temperature. The specimens were embedded in paraffin after decalcification, sectioned at a thickness of 6 μm, and examined with H&E staining and collagen (Masson's trichrome) staining.

Statistical analysis

All experiments were performed in triplicates and the results have been expressed as the mean ± standard deviation (S.D.) Statistical significance was calculated by ANOVA and Student's*t* test analyses. A *p* value of lower than 0.05 was considered to be statistically significant.

DECLARATIONS

Authors' contributions

SD performed all experiments except for the animal studies. SS carried out the animal experiments. SD and SS wrote the manuscript. NS planned all experiments and provided the facilities necessary for the completion of this work. SD, SS, TPS, and NS analyzed and verified the experimental data. All authors have read and approved the final manuscript.

Acknowledgements

We thank Dr. M. Michael Aruldhas from the University of Madras for providing the fluorescent microscope facility. We also thank the Nanotechnology Research Centre at SRM University for providing the SEM facility. This work was supported in part by SRM University and Council for Science and Industrial Research, India [Grant No. 37 (1574)/12/EMR-II to N. S].

Compliance with ethical guidelines

Competing interests The authors declared that they have no competing interests.

REFERENCES

1. Arrington ED, Smith WJ, Chambers HG, Bucknell AL, Davino NA (1996) Complications of iliac crest bone graft harvesting. Clin Orthop Relat Res 329:300–309
2. Schaaf H, Lendeckel S, Howaldt HP, Streckbein P (2010) Donor site morbidity after bone harvesting from the anterior iliac crest. Oral Surg Oral Med Oral Pathol Oral Radiol Endod 109:52–58
3. Drury JL, Mooney DJ (2003) Hydrogels for tissue engineering: scaffold design variables and applications. Biomaterials 24:4337–4351
4. Lee KY, Mooney DJ (2001) Hydrogels for tissue engineering. Chem Rev 101:1869–1880
5. Fedorovich NE, Alblas J, De Wijn JR, Hennink WE, Verbout AJ, Dhert WJ (2007) Hydrogels as extracellular matrices for skeletal tissue engineering: state-of-the-art and novel application in organ printing. Tissue Eng 13:1905–1925
6. Mano JF (2008) Stimuli-responsive polymeric systems for biomedical applications. Adv Eng Mater 10:515–527
7. Huang Z, Tian J, Yu B, Xu Y, Feng Q (2009) A bone-like nano-hydroxyapatite/collagen loaded injectable scaffold. Biomed Mater.
8. Martins AM, Alves CM, Kasper FK, Mikos AG, Reis RL (2010) Responsive and in situ-forming chitosan scaffolds for bone tissue engineering applications: an overview of the last decade. J Mater Chem 20:1638–1645
9. Jayakumar R, Prabaharan M, Nair SV, Tokura S, Tamura H, Selvamurugan N (2010) Novel carboxymethyl derivatives of chitin and chitosan materials and their biomedical applications. Prog Mater Sci 55:675–709
10. Knapczyk J, Krowczynski L, Krzck J, Brzeski M, Nirnberg E, Schenk D et al (1989) Requirements of chitosan for pharmaceutical and biomedical applications. In: Skjåk-Braek G, Anthonsen T, Sandford PA (eds) Chitin and Chitosan: sources, chemistry, biochemistry, physical properties and applications. Elsevier, London, pp 657–663

11. Hirano S, Seino H, Akiyama I, Nonaka I (1990) Chitosan: a biocompatible material for oral and intravenous administration. In: Gebelein CG, Dunn RL (eds) Progress in biomedical polymers. Plenum Press, New York, pp 283–90

12. Muzzarelli RAA (1997) Human enzymatic activities related to the therapeutic administration of chitin derivatives. Cell Mol Life Sci 53:131–140

13. Raafat D, Sahl HG (2009) Chitosan and its antimicrobial potential—a critical literature survey. Microb Biotechnol 2:186–201

14. Freier T, Koh HS, Kazazian K, Shoichet MS (2005) Controlling cell adhesion and degradation of chitosan films by N-acetylation. Biomaterials 26:5872–5878

15. Madihally SV, Matthew HWT (1999) Porous chitosan scaffolds for tissue engineering. Biomaterials 20:1133–1142

16. Varma AJ, Deshpande SV, Kennedy JF (2004) Metal complexation by chitosan and its derivatives: a review. Carbohydr Pol. 55:77–93

17. Rohindra R, Nand AV, Khurma JR (2004) Swelling properties of chitosan hydrogels. South Pac J Nat App Sci. 22:32–35

18. Qiu Y, Park K (2001) Environment-sensitive hydrogels for drug delivery. Adv Drug Deliv Rev 53:321–339

19. Chvapil M (1973) New aspects in the biological role of zinc: a stabilizer of macromolecules and biological membranes. Life Sci 13:1041–1049

20. Yamaguchi M, Weitzmann MN (2011) Zinc stimulates osteoblastogenesis and suppresses osteoclastogenesis by antagonizing NF-κB activation. Mol Cell Biochem 355:179–186

21. Hershfinkel M, Moran A, Grossman N, Sekler J (2001) A zinc-sensing receptor triggers the release of intracellular Ca2+ and regulates ion transport. Proc Natl Acad Sci USA 20:11749–11754

22. Tripathi A, Saravanan S, Pattnaik S, Moorthi A, Partridge NC, Selvamurugan N (2012) Bio-composite scaffolds containing chitosan/nano-hydroxyapatite/nano-copper–zinc for bone tissue engineering. Int J Biol Macromol 50:294–299

23. Zhou H, Lee J (2011) Nanoscale hydroxyapatite particles for bone tissue engineering. Acta Biomater 7:2769–2781

24. Webster TJ, Ergun C, Doremus RH, Siegel RW, Bizios R (2000) Specific proteins mediate enhanced osteoblast adhesion on nanophase ceramics. J Biomed Mat Res 51:475–483

25. Gleeson JP, Plunkett NA (2010) Addition of hydroxyapatite improves stiffness, interconnectivity and osteogenic potential of a highly porous collagen-based scaffold for bone tissue regeneration. Eur Cell Mater 20:218–230

26. Chung H, Golub EE, Forbes E, Tokuoka T, Shapiro IM (1992) Mechanism of action of beta-glycerophosphate on bone cell mineralization. Calcif Tissue Int 51:305–311

27. Chang YL, Stanford CM, Keller JC (2000) Calcium and phosphate supplementation promotes bone cell mineralization: implications for hydroxyapatite (HA)-enhanced bone formation. J Biomed Mat Res 52:270–278

28. Wang L, Stegemann JP (2010) Thermogelling chitosan and collagen composite hydrogels initiated with β-glycerophosphate for bone tissue engineering. Biomaterials 31:3976–3985

29. Cho J, Heuzey MC, Begin A, Carreau PJ (2005) Physical gelation of chitosan in the presence of β-glycerophosphate: the effect of temperature. Biomacromolecules 6:3267–3275

30. Lavertu M, Filion D, Buschmann MD (2008) Heat-induced transfer of protons from chitosan to glycerol phosphate produces chitosan precipitation and gelation. Biomacromolecules 79:640–650

31. Chenite M, Buschmann D, Wang D, Chaput C, Kandani N (2001) Rheological characterization of thermogelling chitosan/glycerol phosphate solutions. Carbohydr Polym 46:39–47

32. Huang J, Best SM, Bonfield W, Brooks RA, Rushton N, Jayasinghe SN et al (2004) In vitro assessment of the biological response to nano-sized hydroxyapatite. J Mater Sci Mater Med 15:441–445

33. Cerroni L, Filocamo R, Fabbri M, Piconi C, Caropreso S, Condo SG (2002) Growth of osteoblast-like cells on porous hydroxyapatite ceramics: an in vitro study. Biomol Eng 19:119–124

34. Zhou H, Lee J (2011) Nanoscale hydroxyapatite particles for bone tissue engineering. Acta Biomater 7:2762–2781

35. Salehi R, Arami M, Mahmoodi NM, Bahrami H, Khorramfar S (2010) Novel biocompatible composite (chitosan–zinc oxide nanoparticle): preparation, characterization and dye adsorption properties. Colloids Surf B Biointerfaces 80:86–93

36. Wang X, Du Y, Fan L, Liu H, Hu Y (2005) Chitosan-metal complexes as antimicrobial agent: synthesis, characterization and structure-activity study. Polym Bull 55:105–113

37. Han JK, Song HY, Saito F, Lee BT (2006) Synthesis of high purity nano-sized hydroxyapatite powder by microwave-hydrothermal method. Mater Chem Phy 99:235–239

38. Balamurugan A, Michel J, Faure J, Benhayoune H, Wortham L, Sockalingum G et al (2006) Synthesis and structural analysis of sol gel derived stoichiometric monophasic hydroxyapatite. Ceram Silik 50:27–3

39. Pawlak A, Mucha M (2003) Thermogravimetric and FTIR studies of chitosan blends. Thermochim Acta 396:153–166

40. Peter M, Ganesh N, Selvamurugan N, Nair SV, Furuike T, Tamura H et al (2010) Preparation and characterization of chitosan-gelatin/nanohydroxyapatite composite scaffolds for tissue engineering applications. Carbohydr Poly 80:687–694

41. Ji C, Khademhosseini A, Dehghani F (2011) Enhancing cell penetration and proliferation in chitosan hydrogels for tissue engineering applications. Biomaterials 32:9719–9729

42. Ren D, Yi H, Wang W, Ma X (2005) The enzymatic degradation and swelling properties of chitosan matrices with different degrees of N-acetylation. Carbohydr Res 340:2403–2410

43. Lord MS, Foss M, Besenbacher F (2010) Influence of nanoscale surface topography on protein adsorption and cellular response. Nano Today 5:66–78

44. Harada H, Tagashira S, Fujiwara M, Ogawa S, Katsumata T, Yamaguchi A et al (1999) Cbfa1 isoforms exert functional differences in osteoblast differentiation. J Biol Chem 274:6972–6978

45. Bellows CG, Aubin JE, Heersche JNM (1991) Initiation and progression of mineralization of bone nodules formed in vitro: the role of alkaline phosphatase and organic phosphate. Bone Min 14:27–40

46. Nudelman F, Pieterse K, George A, Bomans PH, Friedrich H, Brylka LJ et al (2010) The role of collagen in bone apatite formation in the presence of hydroxyapatite nucleation inhibitors. Nat Mater 9:1004–1009

47. Boskey AL, Gadaleta S, Gundberg C, Doty SB, Ducy P, Karsenty G (1998) Fourier transform infrared microspectroscopic analysis of bones of osteocalcin-deficient mice provides insight into the function of osteocalcin. Bone 23:187–196

48. Niranjan R, Koushik C, Saravanan S, Moorthi A, Vairamani M, Selvamurugan N (2012) A novel injectable temperature-sensitive

zinc doped chitosan/β-glycerophosphate hydrogel for bone tissue engineering. Int J Biol Macromol 54:24–29

49. Pattnaik S, Nethala S, Tripathi A, Saravanan S, Moorthi A, Selvamurugan N (2011) Chitosan scaffolds containing silicon dioxide and zirconia nano particles for bone tissue engineering. Int J Biol Macromol 49:1167–1172

50. Jalota S, Bhaduri SB, Tas AC (2008) Using a synthetic body fluid (SBF) solution of 27 mM HCO_3^- to make bone substitutes more osteointegrative. Mater Sci Eng C 28:129–140

51. Moller B, Terheyden H, Açil Y, Purcz NM, Hertrampf K, Tabakov A et al (2012) A comparison of biocompatibility and osseointegration of ceramic and titanium implants: an in vivo and in vitro study. Int J Oral Maxillofac Surg 41:638–64

52. Vimalraj S, Partridge NC, Selvamurugan N (2014) A positive role of microRNA-15b on regulation of osteoblast differentiation. J Cell Physiol 229:1236–1244

4

DEGRADATION OF ZINC CONTAINING PHOSPHATE-BASED GLASS AS A MATERIAL FOR ORTHOPEDIC TISSUE ENGINEERING

Mustafa Al Qaysi[1], Aviva Petrie[2], Rishma Shah[3], Jonathan C. Knowles[1,4]

[1]Division of Biomaterial and Tissue Engineering, UCL Eastman Dental Institute, London, UK

[2]Department of Statistics, UCL Eastman Dental Institute, London, UK

[3]Unit of Orthodontics, Department of Craniofacial Growth and Development UCL Eastman Dental Institute, London, UK

[4]Department of Nanobiomedical Science and BK21 Plus NBM Global Research Center for Regenerative Medicine, Dankook University, Cheonan, Republic of Korea

ABSTRACT

Phosphate-based glasses have been examined in many studies as a potential biomaterial for bone repair because of its degradation properties, which can be controlled and allow the release of various elements to promote osteogenic tissue growth. However most of these experiments studied either tertiary or quaternary glass systems. This study investigated a qinternary system that included titanium dioxide for degradation rate control and zinc that is considered to have a role in bone formation. Zinc and titanium phosphate glass discs of different compositions were melt synthesized and samples of each composition was tested for different physical, chemical and biological characteristics via density measurement, X-ray diffraction, differential thermal analysis, mass loss, ion release, scanning electron microscopy, biocompatibility studies via live/dead assays at three time points (day 1, 4, and 7). The results showed that the glass was amorphous and that the all thermal variables decreased as zinc oxide amount raised, mass loss as well as ion release increased as zinc oxide increased, and the maximum rise was with ZnO15. The cellular studies showed that all the formulation showed similar cytocompatibility properties with MG63 except ZnO15, which displayed cytotoxic properties and this was confirmed also by the scanning electron microscope images. In conclusion, replacing calcium oxide with zinc oxide in proportion less than 10 % can have a positive effect on bone forming cells.

INTRODUCTION

Since the first synthesis of Bioglass® by L.L.Hench in 1971, various glass compositions have been developed to be examined for their suitability as biomaterials for clinical application [1]. The concept behind bioactive glass was their chemical reactivity and ability to form hydroxyl crystalline apatite on the surface of the implanted material [2]. Silicate-based glass has been used widely for hard and soft tissue repair and many studies were done to explore the use of silicate based glass for bone tissue engineering purposes and found the ability of the silicate glass scaffolds to enhance the cellular attachment and enhance both MC3T3-E1 and osteoblast cell proliferation [3, 4]. However, whilst melt derived Bioglass undergoes some initial degradation, it is relatively insoluble [5, 6], which worked as a driver to develop phosphate glass with its compositionally dependant degradation rate. In addition the components making up a phosphate-based glass may be biologically beneficial to the human body [3, 4]. The initial studies were concerned mainly with simple ternary compositions composed of phosphate, calcium oxide, and sodium oxide and that the calcium played a major role in controlling degradation, as it leads to the formation of non- bridging oxygen (P–O–Ca–O–P) bonds that lead to structural and thermal stability of the developed glass, concluding that the increase in calcium oxide percentages may reduce the degradation rate and enhance the glass chemical durability [5, 6]. However, this was followed by many studies that investigated the addition of other various elements that can play a potential role in hard tissue growth as well as enhancing glass properties such as zinc and titanium.

Zinc is not only an essential element for cell growth, proliferation, and differentiation, but it has an important role in DNA replication, growth hormones, growth factors, and enzyme production [7]. The human skeleton is usually considered the main reservoir for zinc, which is mainly stored in the osteoid layer [8, 9] It also has an influential role on bone growth and mineralization through its ability to activate aminoacyl-tRNA synthetase in osteoblastic cells, as well as preventing resorption processes due to its inhibitory action on osteoclasts [10, 11]. It was also found that zinc containing bioactive glass has the potential property to form hydroxy carbonate apatite layers on their surface after soaking in simulating body fluid [12]. Zinc phosphate-based glass was originally developed to examine its physical and biological effect on bioactive glass and was shown to enhance osteoblast like cells (HOB) cell proliferation and cellular attachment when it was added to ternary phosphate glass at different oxide mole percentages which did not exceed 5 mol % of glass composition [13]. Controversially, it displayed poor biocompatibility on MG63 when it was added to pyrophosphate glass in percentages between 10–20 % as this amount of addition may lead to increased degradation and decline in glass stability [13, 14]. Other work concerned with studying the replacement of calcium with zinc in silicate based glasses in different ratios (0.25, 0.5, 1, 1.5, 2.5, 5) mol % and revealed that zinc incorporation resulted in a lower degradation rate and enhance glass durability, which does not show significant results on cellular physiological activities [15]. The addition of zinc containing silicate glass to biphasic calcium phosphates, which is composed of hydroxyapatite and β-tri calcium phosphate showed a positive impact on rat calvarium-derived osteoblasts cell proliferation and behavior [16].

Many studies have been performed to investigate the effect of titanium addition to phosphate glass, and it was found that titanium dioxide incorporation can aid in glass stability and decrease the degradation rate and ion release [17, 18]. It was found that an optimal percentage of titanium dioxide is around 5 mol %, which has shown a biocompatible effect on MG63 cell and the addition of titanium dioxide above that amount does not have a clear and significant effect on glass degradation. The reason behind this is that titanium may tend to make resistant covalent P–O–Ti bonds instead of P–O–P that are more vulnerable to hydration [19, 20], but because of the high valency of titanium, there is a finite number of these bonds that can form.

In the current study, different compositions of quaternary and quinternary titanium stabilized zinc phosphate-based glasses were developed by the melt quenching technique , with the titanium dioxide content fixed at 5 mol % in all the compositions and zinc oxide was varied (0, 5, 10, and 15 mol %) in place of calcium oxide content to investigate the impact of adding zinc oxide on the physical and biocompatibility properties of the glass and whether the effect of titanium dioxide incorporation will decrease the degradation rate to offset the increase seen when adding zinc oxide and therefore overcome the deleterious effects of zinc incorporation that was seen in the previous study [14].

2 MATERIALS AND METHODS

2.1 Glass preparation

Four different glass compositions of phosphate glass were made by using phosphorus pentoxide (P_2O_5 98 %, VWR, Lutterworth, UK), sodium dihydrogen phosphate (NaH_2PO_4, 99 %, VWR), titanium dioxide (TiO2, 99 %, VWR), calcium carbonate ($CaCO_3$, 98.5 %, VWR) and zinc oxide (ZnO, 99.95 %, Sigma-Aldrich Gillingham, UK) as precursors. The glass compositions were $50P_2O_5$—$10Na_2O$—$5TiO_2$—(35-x) CaO—x ZnO (mol %) where x (zinc oxide) was 0, 5, 10, or 15mol%). The required amounts of precursor powders were weighted according to proportions by electronic balance (Sartorius) (Table 1) and blended for 1 min (Stomacher 400 blender/Seward). Once the mixing has been finished, the precursor mixture was then placed into a 200 ml volume Pt/10 %Rh crucible type 71040 (Johnson Matthey, Royston, UK) that was subsequently placed in a furnace (Carbolite) at 1350 °C for 4 h. Following this, the molten glass was then cast into a 420 °C preheated graphite mold, which was placed again in a furnace for one hour at 420 °C then left in the furnace to cool gradually to room temperature overnight to remove the residual stress, the 15 mm diameter glass rods were removed from the mould and cut into 2 mm thick glass discs using a diamond saw (Struers/ Accutom 50) and was checked it was amorphous using X-ray powder diffraction (XRD) analysis. Glass was milled to make a powder of each of the compositions for XRD and Fourier transform infrared spectroscopy analyses (MM 301 Mixer Mill, Retsch GmbH, Hope, UK).

Table 1. Zinc phosphate glass composition.

Zinc phosphate glass composition						
Glass name and composition		Amount (g)				
		P_2O_5	Na_2O	TiO_2	CaO	ZnO
ZnO 0 %	$P_{50}Na_{10}Ti_5Ca_{35}$	56.8	24	4	35	0
ZnO 5 %	$P_{50}Na_{10}Ti_5Ca_{30}Zn_5$	56.8	24	4	30	4
ZnO 10 %	$P_{50}Na_{10}Ti_5Ca_{25}Zn_{10}$	56.8	24	4	25	8.14
ZnO 15 %	$P_{50}Na_{10}Ti_5Ca_{20}Zn_{15}$	56.8	24	4	20	12.21

2.2 Materials characterization

2.2.1 Density determination

The density of each composition was measured using an analytical balance and density determination apparatus according to Archimedes' principal (AG 204 & MS-DNY-43, Mettler Toledo, Beaumont Leys, UK). Ethanol was used as the immersion fluid since the phosphate glasses are water-soluble. The density of the glass discs was calculated using Eq. 1:

ρglass=MairMair−Methanol×ρethanolρglass=MairMair−Methanol×ρethanol

where ρ_{glass} is the density of the specimen (g/cm^3), $\rho_{ethanol}$ is the density of ethanol at the temperature at which the measurement was performed, M_{air} is the mass (g) of the specimen in air and $M_{ethanol}$ is the mass of the specimen under ethanol.

2.2.2 2.2.2 Thermal analysis

Differential thermal analysis was performed for the four compositions by using powder of each glass disc.

Three thermal parameters were measured: (1) T_g the glass transition thermal temperature, (2) T_c the glass crystallization temperature and (3) T_m the melting temperature. A Setaram differential thermal analyzer was used in this experiment using a nitrogen environment and heating temperature extended from room temperature to 1400 °C at 20 °C min^{-1}.

2.2.3 2.2.3 Material degradation

Triplicates of each composition were initially weighed then stored in plastic vials (Sterilin tube) with 25 mm pre adjusted pH deionized water to pH7 by (HCl or NH_4OH) and incubated at 37 °C. After 1, 4, 7, and 14 days, the solution was removed and kept for ion release measurements. While the glass discs were dried and weighed at each time point to determine mass loss. Following weighing, the discs were stored again in new fresh deionized water solution.

2.2.4 2.2.4 Ion release measurements

Ion release measurement was performed for each stored sample from the degradation study, this was done for the anions (PO_4^{3-}, $P_2O_7^{4-}$, $P_3O_9^{3-}$, $P_3O_{10}^{5-}$) and the cations (Na^+, Ca^{2+}) and the transition metal (Zn^{2+}) by using the ion chromatography systems (ICS1000, ICS 2500, Dionex, Thermo scientific, Hemel Hempstead, UK).

For the cation measurements all the samples were filtered prior to measurement to remove the anions (OnGuard II, Dionex).

2.2.5 2.2.5 Cell studies

Initially, preparation of human osteoblast-like osteosarcoma cell line (MG-63, European Collection of Cell Cultures, Porton Down, UK) was done using standard conditions (37 °C, 95 % air, 5 % CO_2, 95 % relative humidity) in Dulbecco's modified Eagle medium (DMEM, Gibco, Life Technologies, Paisley, UK) supplemented with 10 % fetal bovine serum (Gibco) and 1 % penicillin/streptomycin (PAA Laboratories, GE Healthcare, Chalfont St. Giles, UK).

Cell number

Three discs of each composition were sterilized by exposure to ultra-violet light for 10 min on each side before being transferred to 24 well plates on which cells were seeded evenly over the glass surface at a density of 10000 cells per disc for 15 min to allow them to properly attach to the discs then culture media was added to end up with 1 ml of culture media for each glass discs. After that all the samples were incubated at 37 °C in an atmosphere of 5 %CO_2 for 7 days, Standard glass cover slides were used as controls. Cell culture medium was replaced at 3 day periods. Live and dead assay was performed for all the compositions by using calcien AM (4 mM) as an indicator for live cells and Ethidium homodimer-1 (2 mM) for dead cells. This was performed by initial calibration for both of the former reagents to gain the most appropriate molarity that result in the optimal fluorescent color, which was done by seeding MG63 cells on glass slides discs and preparation of various concentration of reagents as recommended by the protocol, the best results were found with 1 µM calcien and 2 µM ethidium. At each time point, culture media was removed and the cells were washed gently with Dulbecco's phosphate buffered saline. After the washing procedure, 3 mL of calcien and ethidium mixed solution was added to each well plate and incubated for about 45 min. Following that, the solution was removed and the samples were ready for the cell counting under confocal (Bio-Rad, USA). The cell counting was done on different 13 pre identified 1 mm squares on each glass discs, these points were positioned on 15 mm circle template, which was made by transparent paper as is shown on (Fig. 1). Live and dead cells counting was performed on three times points (day 1, day 4, and day7) using the Image J software program.

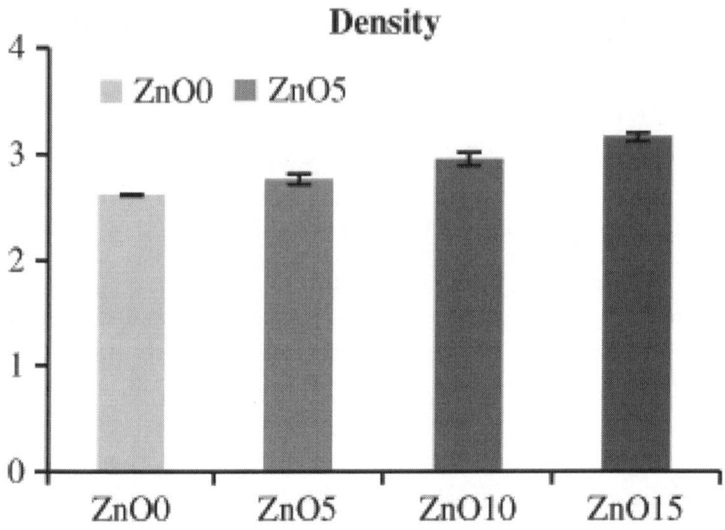

Figure 1. Density of phosphate-based glasses with different CaO and ZnO content. Cumulative zinc content resulted in increased glass density from 2.6 ± 0.004 g cm^{-3} for ZnO0 to 3.16 ± 0.03 g cm^{-3} for ZnO15, Error bars are SD (n = 3)

Metabolic activity of cells

Again triplicate discs of each composition were sterilized as previously mentioned, before being transferred to 24 well plates on which cells were seeded at a density of 10,000 cells per disc in 1 ml of culture media followed by incubation at 37 °C in an atmosphere of 5 % CO_2 for 7 days. Cell culture medium was replaced at 3 day periods and the aspirated culture was kept in another 24 cell culture plate. cytocompatibility assay was done at three times points (days 1, 4, and 7) in which a 10 % solution of water soluble tetrazolium salt-8 (alamar blue, ABD Serotec) was added to the aspirated culture then stored in the incubator for four hours at 5 % CO_2. Alamar blue is a cell viability assay reagent which contains the cell permeable, non-toxic, and weakly fluorescent blue indicator dye called Resazurin, which is used as an oxidation-reduction (Redox) indicator that undergoes colorimetric change in response to cellular metabolic reduction. The reduced form Resorufin is pink and highly fluorescent, and the fluorescence produced is proportional to the metabolic activity of cells and was detected at 570 nm by a fluorimeter (Infinite M200, Tecan, Männedorf, Switzerland).

Imaging of cells

Scanning electron microscopy (SEM) images were done at the same time points for each sample which were fixed initially in 3 % glutaraldehyde followed by dehydration through a graded ethanol (50, 70, 90, 100 %) then drying by hexamethyldisilazane (Aldrich, UK).

2.3 Statistical analysis

Cell work data for both cell counting and metabolic activity studies were statistically assessed by hierarchal ANOVA to find the statistical significance between individual groups and the differences were considered statistically significant at $p < 0.05$.

3 RESULTS

3.1 Density measurement

Fig. 1 shows glass density (g cm-3) in relation to different replacements of calcium oxides with zinc oxides and revealed that density cumulate as the content of incorporated metal oxide increase, with densities ranging from 2.6 ± 0.004 g cm^{-3} for zinc free quaternary glass to stand at 3.16 ± 0.03 g cm^{-3} for Zn15.

3.2 Thermal analysis

In contrast to the density measurements, all the thermal parameters showed a decrease with increasing zinc oxide content (Fig. 2). For the glass transition temperature T_g which is considered an indicator for bulk glass properties, it ranged from about 471 °C for the ZnO 0 % to 407 °C for the Zn 15 %, likewise the crystallization temperature was at 764 °C for ZnO 0 % and decreased with increasing ZnO glass to 720 °C for 15 mol % ZnO. The same pattern was seen for the melting point in which it decreased with the incorporation of zinc oxide, it extended between the wide endothermic peak at 850 °C for the zinc oxide free glass to about 793 °C for ZnO 15 %.

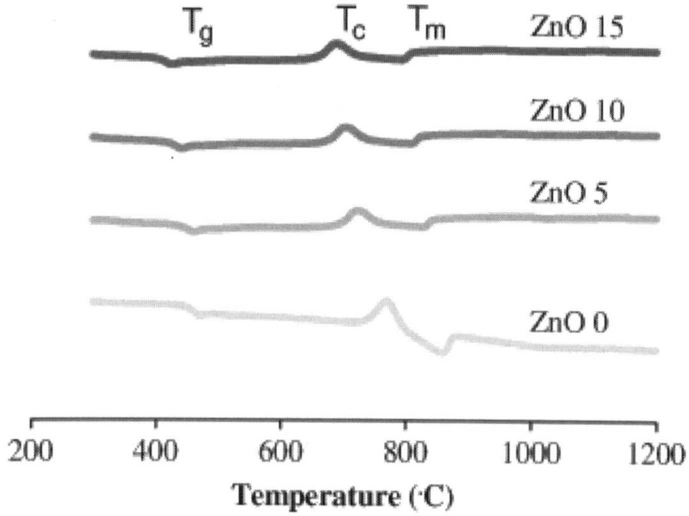

Figure 2. Differential thermal analysis for the whole compositions in which T_g represent glass transition temperature, T_c temperature of crystallization and T_m is melting temperature. Showing that as zinc incorporation increase all the thermal parameters decrease especially the temperature of glass transition, which decreased gradually from 471 °C for Zn0 % to 407 °C for Zn 15 %

3.3 Degradation study

Fig. 3 shows the cumulative mass loss per unit area measured (mg cm^{-2}) over a 1 week period. There was continuous mass loss over time for all the compositions but the highest rate was in the compositions with higher zinc oxide contents. In other words there was clear correlation between Zinc percentage and mass loss rate. For all the time points, the mass loss for ZnO 15 % was about three times that of Zn0 % and was about 0.46 ± 0.038 mg cm^{-2} for Zn 15 % in day 7 and about 0.16 ± 0.33 mg cm^{-2} for Zn 0 % at the same time point. Both Zn5 % and Zn 10 % had similar degradation rates at all the time points and was less than that of Zn15 %.

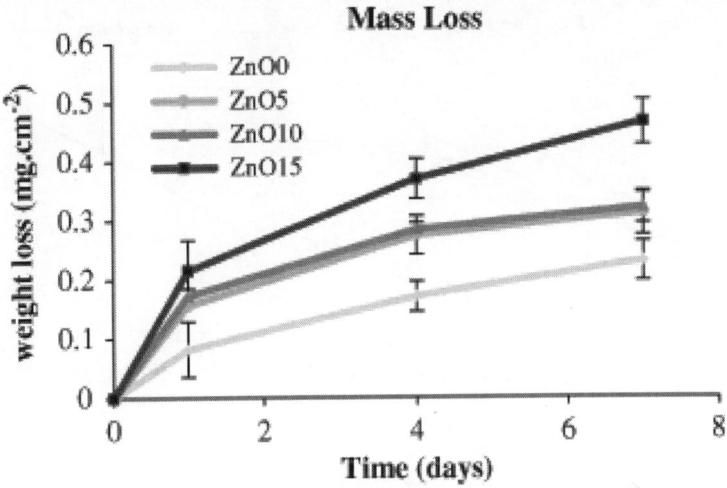

Figure 3. Mass loss (mg cm^{-2}) for the various four zinc phosphate based glass compositions. It shows clear correlation between zinc incorporation and mass loss within time. Which was for ZnO15 about three times that of ZnO0 % in all time points. Error bars are SD ($n = 3$)

3.4 Ion release measurement

Ion release results showed that anion and cation release increased with zinc composition increment as it shown in (Figs. 4, 5), respectively. Regarding the anions, ZnO 15 % displayed the highest release for all the anions, which was significantly higher than the other compositions. For the other formulations (ZnO0, ZnO5, and ZnO10), although the release was higher with the increased zinc composition, this difference was not significantly different. The cation release trend was similar to that of anions and this was clear for Zn^{2+}, Ca^{2+}, and Na^{+} in which cations release was significantly highest in ZnO15 and subsequently decreased as the zinc ion proportion goes down. The only exception was in Ca^{2+} release in which it was highest from the glass with the highest calcium content which is ZnO0.

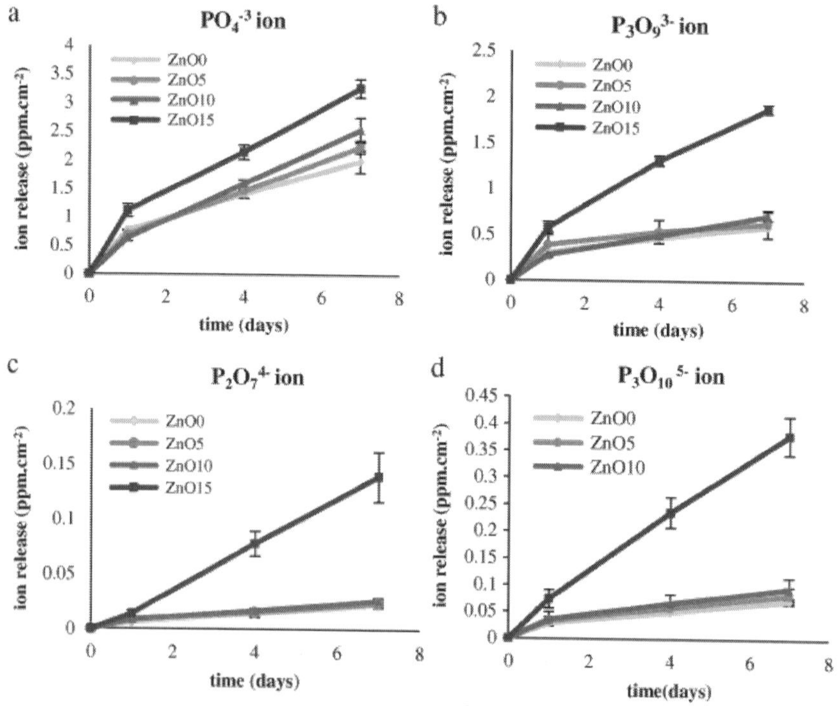

Figure 4. Accumulative release for anion (PPm cm^{-2}) for the whole four zinc phosphate based glass compositions. **a** PO_4^{3-}, **b** $P_3O_9^{3-}$, **c** $P_2O_7^{4-}$, **d** $P_3O_{10}^{5-}$. It shows clear correlation between zinc incorporation and anion within time which was obtained after incubation in deionized water for (1, 4, 7 days). Error bars are SD ($n = 3$)

3.5 Cytocompatibility and metabolic activity

All glass compositions showed good cytocompatibility except for ZnO15, and this was determined by the results of the cell counting assay, which involved determination of both live and dead cells from an initial seeding of around 10,000 cells/ disc. Concerning the live cells (Fig. 6a), between seeding day and day 1, cells number approximately doubled for control, ZnO0 and ZnO5 to reach about 17,500 ~ 23,000 cells/disc this was significantly more than that of ZnO10 and ZnO15, which showed a lower rise in cell density in both of them. By day 4, all the compositions displayed a further doubling to stand at about 40,000 ~ 55,000 cells/disc for control, ZnO0 and ZnO5, respectively, which was again more than that of ZnO10 and ZnO 15 that were about 30,000 cells/disc. For the last time point, cell numbers were much higher, to reach around 100,000 ~ 125,000 for the control and ZnO0 this was clearly significant for the other zinc containing glass ($p < 0.05$).

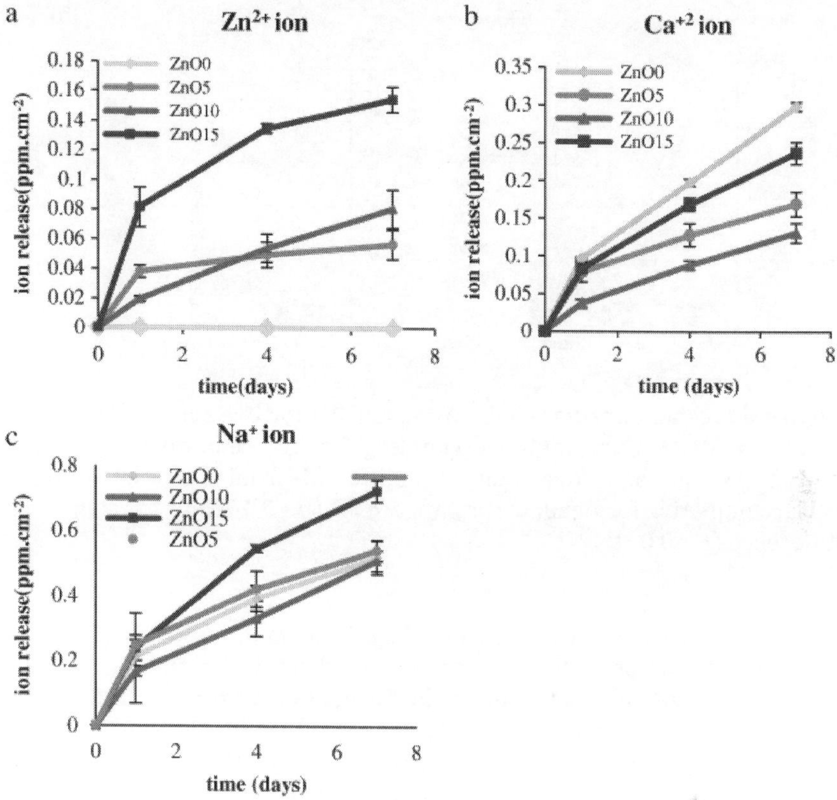

Figure 5. Cumulative release for anion (ppm cm^{-2}) for the whole four zinc phosphate-based glass compositions.**a** Zn^{2+}, **b** Ca^{2+}, **c** Na^{+}. It shows clear correlation between zinc incorporation and cations against time, which was obtained after incubation in deionized water for (1, 4, 7 days). Error bars are SD ($n = 3$)

Conversely and as expected, dead cells follow the opposite pattern to that of live cells, in that dead cell numbers rise as zinc percentage increases and was maximum for ZnO15 for all the days (Fig. 6b). By day 1, dead cells number were about 160 ~ 400 cells/disc just half that of ZnO15 that was about 800 cells/disc, which was significantly higher than the other compositions. By day 4, dead cells number increased by five times for all glass types and was again significantly different in ZnO15 to the other types ($p < 0.05$). By day 7 dead cells number continued to increase especially in the zinc containing glass with maximum number of non-viable cells was in ZnO15 samples which was three times that of control samples.

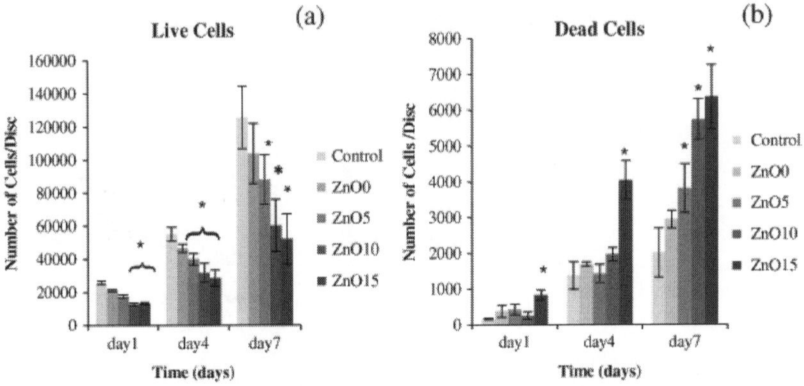

Figure 6. Live and dead assay by using Calcien for live cells and ethidium iodine for dead. **a** Accumulative counting for live cells displaying ZnO0 and ZnO5 has an effect that is indifferent from TCP, while there is **b** accumulative for dead. Error bars are SD ($n = 3$), While * is significant difference ($p < 0.05$)

Cell metabolic activity was determined by the alamar blue assay (Fig. 7) in which the reduction of fluorescence is an indicator of metabolic activity. Metabolic activity increased for all the formulations except ZnO15. At day 1, Zn0 and Zn5 showed comparable values to that for the control which was about 6 ± 1.2 % for control just slightly higher than ZnO0 and ZnO5 activity , while both ZnO10 and ZnO15 showed low reduction ability, which was significantly different from control ($p < 0.05$). By day 4, reduction rate for control increased three times to stand at 17 ± 0.4 %, which was significantly different from ZnO0, ZnO5, and ZnO10, However this statistical difference cannot neglect the existence of increase metabolic activity that happened in these glass formulations. Whereas ZnO15 show sign of significancy with the other remaining groups ($p < 0.05$).

The continuity in reduction percentage persists to day 7, which was about 42 ± 3.3 % for control and was not as high as that for ZnO0. Despite the presence of significant difference between ZnO5 and Zn10 with the control, the mean difference in reduction was little low from the control about (6.5–9 %). Interestingly, reduction rate for ZnO15 showed a clear decline in day 7 to end at 2.5 ± 0.7 % slightly higher than that at day 1 and displaying higher significant difference with the whole batches.

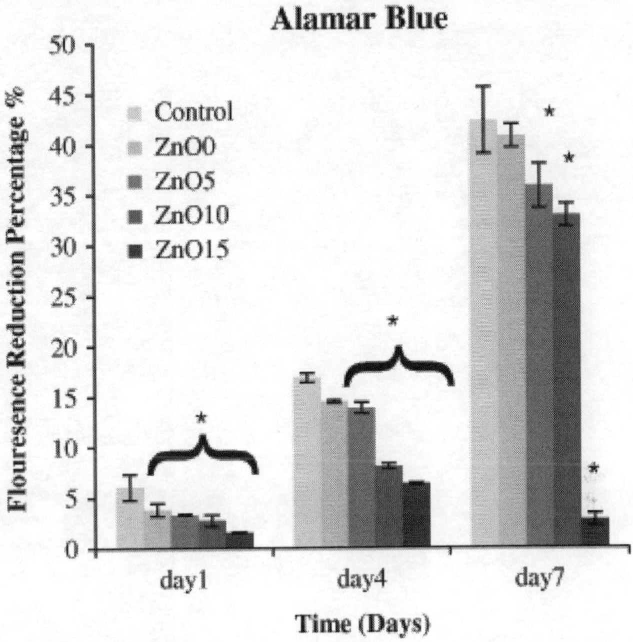

Figure 7. Alamar blue assay showing fluorescence reduction percentage, which is an indicator for metabolic activity. The maximum reduction was in TCP in (1, 4, 7 days), which is insignificantly different from ZnO0, ZnO5, and ZnO10. Zn15 had the lowest rate of fluorescence reduction. Error bars are SD ($n = 3$), while * is significant difference ($p < 0.05$)

SEM images (Fig. 8) showed that MG63 cells have attached to all glass compositions at all time points. After 1 day, the cells seemed to spread around the glass surface and they took an irregular shape. However, the cells on ZnO15 tended to be smaller. On day 4, the cells could be seen to have connected to each other with the appearance of rounded and swollen nucleus, some cells showed longitudinal shape that may be due to cell division. By day 7, cells were evenly distributed over the surface. Moreover, there was a clear increase in cell numbers, which were mostly interconnected with each other forming a semi-complete cellular layer. The only exception was in ZnO15 in which there was a low number of cells which were small in size and separated from each other.

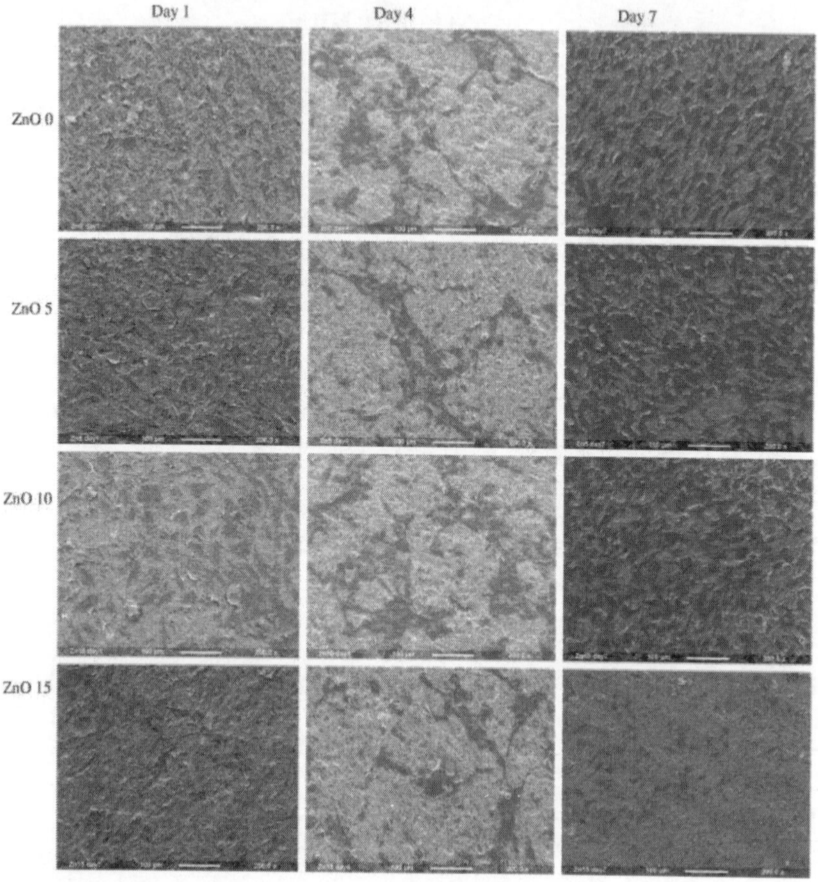

Figure 8. Scanning electron microscope images

4 DISCUSSION

The present study concerned with the development of different phosphate-based glasses for hard tissue (bone) engineering applications, and this was done by substituting the calcium oxide with zinc oxide, since zinc is known to play important roles in the bone metabolism and development [21]. Previous studies have been carried out to assess different zinc phosphate glass properties, these studies assessed (P_2O_5–Na_2O–CaO–ZnO) phosphate glass and showed that these compositions suffered from poor cytocompatibility due to their rapid degradation rate and this rate increased with increasing zinc addition [13, 14]. Other studies were performed with the incorporation of various amount of titanium dioxide to the tertiary based glass to improve its cytocompatibility and enhance its durability. titanium dioxide was added at 3 mol %, 5 mol %, 7 mol % and revealed that as titanium dioxide increased, the degradation rate and ion release rate decreased. Also the 5 mol % titanium dioxide containing glass showed the most favorable cytocompatibility results in comparison to the titanium dioxide free control samples [22] and this was the main rationale for the selection of this titanium dioxide content in the present study.

The thermal properties reduced with incorporation of zinc oxide and this tends to be as a result of the fact that zinc oxide bond enthalpy is about 284 Kj/mol, which is less than that of calcium oxide bond 464 Kj/mol as a result of this less energy is needed to break the zinc oxide bond.

Usually the main factors that control degradation of material are the atom size, bond length and electronegativity of bond. Although both of Zn–O and Ca–O bonds have pure ionic bonds [23,24] as the difference in electronegativity of both Zn–O and Ca–O are more than 2, Ca–O seems to be more electronegative (3.44–1 = 2.44) than that of Zn-O (3.44–1.65 = 1.79) when 3.44, 1.65, and 1 are the electronegativity of O, Zn, and Ca, respectively. As a consequence, Ca–O has more bond polarity and bond strength of Zn–O, which may explain the glass susceptibility to degradation as zinc percentage increase that could affect the whole glass compositional structure and resulted in this pattern of weight loss (ZnO0 < ZnO5 < ZnO10 < ZnO15).

The ion release results were concurrent with the weight loss pattern. It was found that the increase of phosphate anion release increased as the zinc oxide content increased, this may be due to the effect of weak Zn–O–P bond. For the cation release it took the same trend for $Na^{+}$and Zn^{+2} ions, the only exception was for Ca^{+2} ions, which was released at higher levels in the glass with ZnO 0 mol % glass and this result was the opposite to the other ions, but that can be explained by the fact that as the ZnO content decreased, the CaO content increased. however the general information that was obtained from zinc oxide containing glass ion release was that there was a gradual release in all the ions with time which was similar for the compositions ZnO0, 5, 10 mol %, but the release for the ZnO15 mol % was higher and both findings OD mass loss and ion degradation were consistent with that of Salih et al. [13].

Concerning cell studies, live and dead assay was performed by using a template. Previous work was done by subjective image visualization, whereas the main aim in the current study was to objectively determine the real number of cells by using a template. The template was designed to fit the glass discs perfectly (diameter = 15 mm) and was divided into small (1 × 1 mm) squares to facilitate cell counting readings. Only 13 pre chosen squares per each sample were selected for the measurements. Although the total area of these squares constituted only 7 % of the whole disc area, their positions were selected in a way that ensure the full distribution over the whole central area of the disc to prevent repeating readings and give a valid picture of cell distribution over the disc since each of the selected squares was surrounded by non-interpreted squares. Cell seeding was performed with caution to ensure the even cell distribution for all the samples and this was by preparation of the intended cell density in 50 µl of culture media which was found to be the optimum media volume to cover the whole disc surface area. The findings of this study showed that live cell number was insignificantly different between ZnO0 and ZnO5 and the control sample for all time points, it is increased in about the same rate and even if there was slight significant difference in some time points, this does not cancel the positive impact of these two compositions on cells proliferation. Conversely, cell number was clearly low in ZnO15 and ZnO10 and was significantly different from the control group and they have about half the cell number of the control samples , this was furtherly confirmed by the results of the dead cell count which was linked to zinc oxide content (ZnO15 > ZnO10 > ZnO5 > ZnO0 > control) and this showed that the high dissolution glass gave poorer cell data, and for ZnO15 dead cells numbers constitute about 10 % of live cell numbers on the same day for the all time points which indicates cytotoxicity (Fig. 6, Fig. 9). Alamar blue measurements, is the indication of reduction–oxidation reaction (Redox) reaction as the viable metabolic cells tend to form reduced environment, which can reduce the weakly fluorescent resazurin salt to strongly fluorescent resorufin [25]. The data were concurrent with that of live and dead cells as it displayed that Zn15 have very low impacts on MG63 metabolic activity and were with highly significant difference in comparison to the other glass formulations at all time points. Despite there was little difference between ZnO10 and the control sample, it showed high metabolic activity within time. ZnO0 and ZnO5 appeared to have lower levels of effect on metabolic activity compared to the control samples and these results matched that of live cells and had the pattern of (control > ZNO0 > ZNO5 > ZnO10 > ZnO15). Despite the presence of statistical difference ($p < 0.05$) between the control and Zn0 and Zn5, there was evidence of metabolic activity enhancement in the previously mentioned glass compositions which may indicate their benign effect on cellular activity. SEM images emphasized the previous finding and showed differences in cell number for Zn15 and a decline in cellular interconnection. These outcomes may be due to the high rate of degradation in ZnO 15 %, which lead to high Zinc release which could play an unwanted impact on cell proliferation by suppression the expression mRNa as was found in previous studies [26, 27] . Interestingly, these results were very comparable with the previous study which reported that incorporation of zinc content > 1.2 % wt can lead to cytotoxic impacts on osteoblastic MC3T3-E1 cells [28], in the present study Zn composed about 1 % wt of the total glass weight in ZnO15 samples which could be due to the difference in cell type.

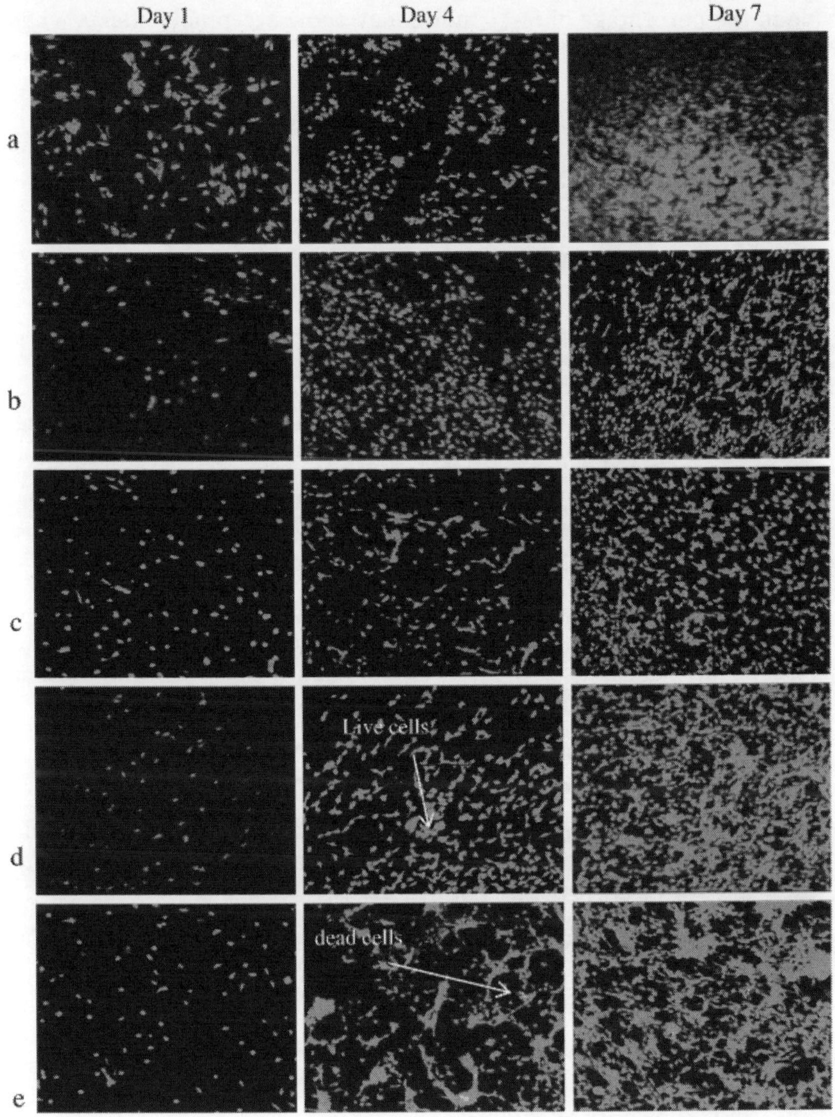

Figure 9. Live and dead images by using confocal. Live looks green while dead are red in color. **a** control, **b** ZnO 0, **c** ZnO5, **d** ZnO 10, **e** ZnO15

5. CONCLUSION

The present study showed that adding zinc to the phosphate based glass can be prepared successfully and displayed acceptable outcomes concerning cellular metabolic activities. However the increase in zinc oxide concentration to 15 mol % may have cytotoxic effects, adding zinc oxide in the percentage of 5–10 mol% may show acceptable effects when compared it with the control sample.

CONFLICT OF INTEREST

The authors declare that they have no competing interests.

REFERENCES

1. Shirtliff VJ, Hench LL. Bioactive materials for tissue engineering, regeneration and repair. J Mater Sci. 2003;38(23):4697–707.
2. Saravanapavan P, Jones JR, Pryce RS, Hench LL. Bioactivity of gel–glass powders in the CaO-SiO_2 system: a comparison with ternary (CaO-P_2P_5-SiO_2) and quaternary glasses (SiO_2-CaO-P_2O_5-Na_2O). J Biomed Mater Res A. 2003;66A(1):110–9.
3. Franks K, Abrahams I, Knowles JC. Development of soluble glasses for biomedical use Part I: in vitro solubility measurement. J Mater Sci Mater Med. 2000;11(10):609–14.
4. Knowles JC. Phosphate based glasses for biomedical applications. J Mater Chem. 2003;13(10):2395–401.
5. Li HJ, Wang CL, Yu HJ, Liang XF, Yang SY. Effects of calcium oxide addition on the structure and thermal properties of iron phosphate glasses. Spectr Lett. 2015;48(3):184–9.
6. Ahmed I, Lewis M, Olsen I, Knowles JC. Phosphate glasses for tissue engineering: part 1. Processing and characterisation of a ternary-based P(2)O(5)-CaO-Na(2)O glass system. Biomaterials. 2004;25(3):491–9.
7. MacDonald RS. The role of zinc in growth and cell proliferation. J Nutr. 2000;130(5):1500s–8s.
8. Herzberg M, Foldes J, Steinberg R, Menczel J. Zinc excretion in osteoporotic women. J Bone Miner Res. 1990;5(3):251–7.
9. Haumont S. Distribution of zinc in bone tissue. J Histochem Cytochem. 1961;9(2):141–5.
10. Masayoshi Y, Oishi H, Suketa Yasunobu. Role of zinc in bone formation and bone resorption. J Trace Elem Exp Med. 1998;11:119–35.

11. Masayoshi Y, Oishi H, Suketa Yasunobu. Zinc stimulation of bone protein synthesis in tissue culture: activation of aminoacyl-tRNA synthetase. Biochem Pharmacol. 1988;37(21):4075–80.

12. Lusvardi G, Malavasi G, Menabue L, Menziani MC, Pedone A, Segre U, et al. Properties of zinc releasing surfaces for clinical applications. J Biomater Appl. 2008;22(6):505–26.

13. Salih V, Patel A, Knowles JC. Zinc-containing phosphate-based glasses for tissue engineering. Biomed Mater. 2007;2(1):11–20.

14. Abou Neel EA, O'Dell LA, Smith ME, Knowles JC. Processing, characterisation, and biocompatibility of zinc modified metaphosphate based glasses for biomedical applications. J Mater Sci Mater Med. 2008;19(4):1669–79.

15. Haimi S, Gorianc G, Moimas L, Lindroos B, Huhtala H, Räty S, et al. Characterization of zinc-releasing three-dimensional bioactive glass scaffolds and their effect on human adipose stem cell proliferation and osteogenic differentiation. Acta Biomater. 2009;5(8):3122–31.

16. Badr-Mohammadi MR, Hesaraki S, Zamanian A. Mechanical properties and in vitro cellular behavior of zinc-containing nano-bioactive glass doped biphasic calcium phosphate bone substitutes. J Mater Sci Mater Med. 2014;25(1):185–97.

17. Navarro M, Ginebra M-P, Clément J, Salvador M, Gloria A, Planell JA. Physicochemical degradation of titania-stabilized soluble phosphate glasses for medical applications. J Am Ceram Soc. 2003;86(8):1345–52.

18. Abou Neel EA, Chrzanowski W, Valappil SP, O'Dell LA, Pickup DM, Smith ME, et al. Doping of a high calcium oxide metaphosphate glass with titanium dioxide. J Non-Cryst Solids. 2009;355(16–17):991–1000.

19. Abou Neel EA, Chrzanowski W, Knowles JC. Effect of increasing titanium dioxide content on bulk and surface properties of phosphate-based glasses. Acta Biomater. 2008;4(3):523–34.

20. Abou Neel EA, Knowles JC. Physical and biocompatibility studies of novel titanium dioxide doped phosphate-based glasses for bone tissue engineering applications. J Mater Sci Mater Med. 2008;19(1):377–86.

21. Seo HJ, Cho YE, Kim T, Shin HI, Kwun IS. Zinc may increase bone formation through stimulating cell proliferation, alkaline phosphatase activity and collagen synthesis in osteoblastic MC3T3-E1 cells. Nutr Res Pract. 2010;4(5):356–61.

22. Lakhkar NJ, Park J-H, Mordan NJ, Salih V, Wall IB, Kim H-W, et al. Titanium phosphate glass microspheres for bone tissue engineering. Acta Biomater. 2012;8(11):4181–90.

23. Ermoshkin AN, Kotomin EA, Evarestov RA. Molecular cluster approach to magnesium and calcium oxide crystals. I. Perfect crystals. Phys Status Solidi B. 1975;72(2):787–94.

24. Ghadah S, Alghamdi AZA. Bonding Formation and Orbitals Nature of ZnO Structure. MiddleEast J Sci Res. 2013;13(9):1144–9.

25. Michael V. Lancaster, Rebecca D. Antibiotic and cytotoxic drug susceptibility assays using resazurin and poising agents. USA patent. 1996;5(501):959

26. Park KH, Park B, Yoon DS, Kwon SH, Shin DM, Lee JW, Lee HG, Shim JH, Park JH, Lee JM. Zinc inhibits osteoclast differentiation bysuppression of Ca^{2+}-Calcineurin-NFATc1 signalingpathway. Cell Commun Signal. 2013;11:74

27. Aina V, Perardi A, Bergandi L, Malavasi G, Menabue L, Morterra C, et al. Cytotoxicity of zinc-containing bioactive glasses in contact with human osteoblasts. Chem Biol Interact. 2007;167(3):207–18.

28. Ito A, Ojima K, Naito H, Ichinose N, Tateishi T. Preparation, solubility, and cytocompatibility of zinc-releasing calcium phosphate ceramics. J Biomed Mater Res. 2000;50(2):178–83.

5

EFFECTS OF SIZE AND SURFACE CHARGE OF POLYMERIC NANOPARTICLES ON IN VITRO AND IN VIVO APPLICATIONS

Sams M. A. Sadat, Sheikh Tasnim Jahan, Azita Haddadi

Division of Pharmacy, College of Pharmacy and Nutrition, University of Saskatchewan, Saskatoon, Canada

ABSTRACT

Biodegradable polymeric materials are the most common carriers for use in drug delivery systems. With this trend, newer drug delivery systems using targeted and controlled release polymeric nanoparticles (NPs) are being developed to manipulate their navigation in complex in vivo environment. However, a clear understanding of the interactions between biological systems and these nanoparticulates is still unexplored. Different studies have been performed to correlate the physicochemical properties of polymeric NPs with the biological responses. Size and surface charge are the two fundamental physicochemical properties that provide a key direction to design an effective NP formulation. In this critical review, our goal is to provide a brief overview on the influences of size and surface charge of different polymeric NPs in vitro and to highlight the challenges involved with in vivo trials.

Keywords: Nanoparticle, Size, Surface Charge, In Vitro, In Vivo

1. INTRODUCTION

Manufacturing effective drug delivery system is a critical challenge in nanomedicine since nanocarriers are expected to reach and accumulate in the site of interest. As a consequence, numerous drug delivery systems have been investigated in vitro and in vivo to deliver a wide range of drugs and molecules. To conquer the challenge, with the aim of avoiding uncontrolled biodistribution, rapid clearance and systemic toxicities in healthy tissues polymeric nanoparticles (NPs) have gained higher interest among all the novel formulations. Research during the past few decades proves their beneficial features in formulation design, characterization, behavior and application [1] . The in vitro and in vivo fate of NPs is particularly depended on uniformity of particle size and zeta potential. Change in these properties has significant biological implications on cellular internalization, pharmacokinetics, and bio-distribution [2] . These characteristics of NPs facilitate the opportunities for therapeutic application, which can be confirmed by in vitro and in vivo studies [3] . The aim of most nano-devices is to prevent the degradation of drug followed by higher bioavailability in cellular level and to regulate its pharmacodynamics profile. Thus, the nanomedicine platforms could serve as a drug delivery system that is able to transport a high dose of therapeutics selectively to the desired site of action. Although very few investigations have been performed, most of the articles related to exploring the effects of size and surface charge of NPs have been discussed in this review. This review provides details on the fate of different polymeric NPs and will discuss how the size and surface charge of polymeric NPs are involved in desired effects for both in vitro and in vivo applications. Moreover, other polymeric NPs using various preparation methods have been also summarized in Table 1, which could be considered for further size and surface charge related experiments.

Abbreviation: BCEC: Brain capillary endothelial cells; BSA: Bovine Serum Albumin; DCs: Dendritic Cells; HASMCs: Human arterial smooth muscle cells; HA-VSMCs: Human aortic vascular smooth muscle cells; HMSCs: Human mesenchymal stem cells; HUVECs: Human umbilical vein endothelial cells; MOEC: Murine ovarian endothelial cells; NAcHis-GC: N-acetyl histidine conjugated glycol chitosan nanoparticles); P (MDS-co-CES): Poly (methyldiethene-aminesebacate)-co-[(cholesterylox-ocarbonylamidoethyl) methylbis (ethylene) ammonium bromide] sebacate; PBMCs: Peripheral blood mononuclear cells; PBS: Phosphate buffer saline; PEG: Poly (ethylene glycol); PEG-PHDCA: Poly (methoxypolyethyleneglycol cyano- acrylate-co-hexadecylcyanoacrylate); PEMA: Poly (ethylene-maleic anhydride); PEO-b-PMA: Poly (ethylene oxide)-b-poly (methacrylic acid); PLA: Poly (lactic acid); PMB: Poly [2-methacryloyloxyethyl phosphorylcholine (MPC)-co-n-butyl methacrylate (BMA); PMBH: Poly [2-methacryloy- loxyethyl phosphorylcholine (MPC)-co-n-butyl methacrylate (BMA)-co-methacryloylhydrazide (MH)]; PVA: Polyvinyl Alcohol; RBECs:

Rat brain endothelial cells; TPGS: Tocopheryl polyethylene glycol succinate; VSMCs: Vascular smooth muscle cells; WGA: Wheat germ agglutinin.

Table 1. Size and surface charge overview of different polymeric NPs.

Polymer type	NPs preparation method	Loaded materials	Stabilizer	Cell line	Size (nm)	Zeta (mV)	References
Chitosan	Chemical reaction	-	-	Caco-2	(418 ± 31)-(531 ± 54)	(13.6 ± 0.8) - (29.4 ± 0.8)	[50]
Chitosan	Ionotropic gelation	Covalent conjugation with fluorescein-5-isothiocyanate	-	A549 cells	195	35.5	[51]
Chitosan (Galactosylated)	Carbodiimide method	DNA	-	HepG2, HeLa	100 - 240	−5	[52]
Chitosan {Arg-Gly-Asp (RGD) peptide-labeled}	Ionic gelation	siRNA	-	SKOV3ip1, HeyA8, A2780, A2780ip2, MOECs	200	−40	[53]
Chitosan, PS, PEG-PLA	Multiple emulsion-solvent evaporation	-	Sodium Cho-late	Caco-2, MTX-E12	(196 ± 20)-(290 ± 7)	(−23.9 ± 1.2) - (1.1 ± 1)	[7]
Chitosan-PEG	Ionic gelation	Caspase inhibitor pebptide	-	Mice	150	16	[54]
Gelatin	Desolvation	Anti CD3 mAb	-	HeLa	250 - 300	−20	[55]
Heparin	Coupling reaction through amide linkage	Paclitaxel, Folic acid	-	KB-3-1, Tu212	60	−16	[56]
Hyaluronic acid (Self-assembled)	Chemical conjugation through Carbodiimide method	-	PBS	SCC7, CV-1	237 - 424	-	[57]
Hyaluronic acid-ceramide	Self-assembly	Docetaxel	Pluronic P85	U87-MG, MCF-7, MCF-7/ADR	110 - 140	−20	[58]
Hydrophobically modified glycol chitosan	Self-assembly	-	-	HeLa	359	22	[59]
Mesoporous silica	-	-	-	3T3-L1, HMSCs	108 - 115	(−4.9) - (19)	[60]
Methyl ether poly(ethylene glycol)-modified poly(beta-amino ester)	Solvent casting	Doxorubicin	-	B16F10	62	-	[61]
NAcHis-GC (self-assembled)	-	Paclitaxel	PBS	HeLa, A549	150 - 250	-	[62]
P(MDS-co-CES)	Self-assembly	DNA, Paclitaxel, Indomethacin, Pyrene	-	HEK293, HepG2, MDA-MB-231	(83 ± 1) - (180 ± 2)	(44 ± 2) - (84 ± 5)	[63]
PCL	Solvent displacement	Tamoxifen	Pluronic® F-68	MCF-7	250 - 300	6 - 25	[64] [65]
PCL	Emulsification diffusion	-	-	-	120 - 125	(−32) - (−60)	[66]

2. POLYMERIC NPS

For an ideal drug delivery system, recognition of the polymer's potentiality has been evaluated since 1960's [4] . Over the past few decades, two main classifications of polymers have been discovered as synthetic and natural, each with various types and sub-types. Synthetic polymers are chemically synthesized based on repeated structural units, whereas natural polymers are obtained from natural sources. Primarily two types of polymeric NPs have been developed for drug delivery purposes i.e. nanocapsules, in which a core of encapsulated drug is surrounded by polymeric membrane or shell;

and nanospheres, where drug is distributed/adsorbed throughout a matrix [5] . The most important feature of polymers is the degree of biodegradability, which is an important criterion to differentiate some slowly biodegradable polymers such as polystyrene (PS), poly (cyanoacrelates) (PCA), polyethylenimine (PEI) and poly (methyl methacrylate) (PMMA) [6] - [11] . On the other hand, some synthetic polymers such as poly (ε-caprolactone) (PCL), poly (lactide) (PL), poly (glycolide) (PGA), poly (D, L-lactide-co-glycolide) (PLGA), and some non-synthetic polymers (e.g. chitosan) are categorized as readily biodegradable materials [12] - [16] . Polymeric NPs are capable to maintain high stability in systemic circulation with enhanced half-life, which can be further optimized by controlling the release of therapeutic agents from the NPs. Moreover, polymeric molecules have various solubility profiles in wide range of solvents. This is advantageous for surface modification or functionalization to achieve different purposes of delivery and targeting. Subsequently, both doses and frequency of administration of therapeutic agents can be reduced due to high payloads into nanocarriers, leading to superior efficacy and minimizing the side effects. Besides, polymeric NPs of desired physicochemical properties are capable of preserving their content from hepatic metabolism, enzymatic degradation and rapid clearance. Specifically, the enormous surface area of polymeric NPs is an attractive feature to control the release kinetics, drug loading capacity and administration route, which can regulate the fate of drug into the body [17] . However, only few of them have been approved by health regulatory agencies for human trial to apply for carrying a wide range of diagnostic and therapeutic agents to the desired site of action [18] .

3. EFFECT OF PARTICLE SIZE AND SURFACE CHARGE BASED ON IN VITRO STUDIES

Different types of NPs have been widely applied as drug delivery vehicles for diagnostic and targeted therapy (active or passive) to achieve maximum cellular uptake and therapeutic bioavailability [19] [20] . Continuous physicochemical changes in the development of polymeric NPs may have substantial implications in the cellular internalization and biological processes [21] . The experiments performed to evaluate the influence of particle size and surface charges of NPs are expected to explain how these physicochemical properties influence the cell uptake through various pathways towards optimum biodistribution.

Cellular internalization or uptake is the most important physicochemical criteria prior to in vivo application. Uptake of small molecules by any cells depends mainly on endocytosis among all other mechanisms (Figure 1). Endocytosis is the bulk active transport process through lipid bilayer wrapping using energy in the form of ATP to form required vesicles. Two main endocytosis mechanisms are reported as phagocytosis and pinocytosis [22] . Phagocytic cells (e.g. macrophages, neutrophils, dendritic cells, etc.) mediated cellular internalization is mostly involved with engulfing the large particles (>1 μm) [23] . Adsorption or receptor dependent internalization is the main mechanism of pinocytosis, which is mainly related to particle uptake by the cells through different pathways such as macro-pinocytosis, clathrin mediated, caveolin dependent or independent pinocytosis [3] . Size and surface charge of polymeric NPs are likely the preliminary physicochemical variables, which govern the endocytosis dependent cellular uptake. Besides, positive charge of the surface of polymeric NPs may endorse more cellular attachment causing higher uptake either by endocytosis or by direct penetration, since cationic surface of polymeric NPs interacts with anionic terminal of phospholipid, proteins and glycans on the cell surface due to the electrostatic interactions [23] .

An interesting experiment by Bhattacharjee et al. demonstrated the effects of size and surface charge of fluorescent, monodisperse tri-block co-polymeric NPs based on cellular uptake through different endocytotic pathways [24] . They synthesized polymeric NPs (PNPs) with two different sizes (45 and 90 nm) and surface charges such as neutral (PNP-OH, −4 mV), positive (PNP-NH_2, +22 mV) and negative (PNP-COOH, −19 mV) to observe the in vitro cellular uptake into NR8383 (rat macrophage) and Caco-2 (human colonic adenocarcinoma) cells. For size dependent cellular uptake, a relative uptake study was carried out, which revealed the higher

intracellular uptake by positively charged polymeric NPs with lower size compared to the other formulations.

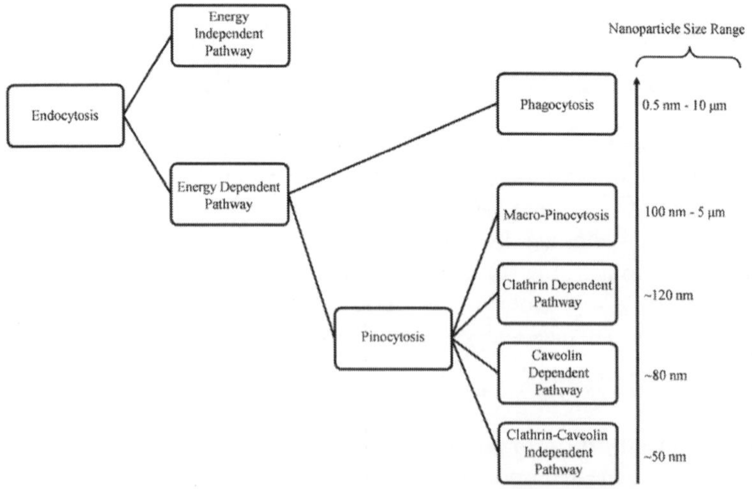

Figure 1. Relative sizes of NPs favorable for ingestion through various endocytotic pathways.

Inhibition of endocytic pathways was adopted to observe the role of endocytosis based cellular internalization of polymeric NPs tracked by two mechanisms such as decreasing temperature to 4°C of experimental unit and exposing cells with 2-deoxyglucose (2-dOG) and sodium azide (NaN_3). Both inhibitory approaches showed considerably lower uptake, which proved the higher uptake by positive charged polymeric NPs compared to that of neutral to more negative charged polymeric NPs. Followed by the same strategy to block the clathrin and caveolin mediated endocytosis, cells were exposed with hypertonic 450 mM sucrose solution and methyl-beta-cyclo- dextran, respectively. Meeting the claimed fenestration sizes of these receptors dependent endocytosis, both inhibitions resulted with reduced uptake with smaller size after treating these cells with polymeric NPs, however the uptake was varied with charge variations. For clathrin dependent endocytosis, uptake by both neutral and negatively charged polymeric NPs was higher (65% and 75%, respectively) than positively charged polymeric NPs (less than 38%), however an opposite result was found for caveolin dependent endocytosis.

On the other hand, Lai et al. investigated that polystyrene (PS) NPs with smaller size (>42 nm) were successfully internalized into HeLa cells following clathrin and caveolin independent endocytic pathways avoiding endosomal or lysosomal accumulation [25] . Recent studies revealed that positively charged NPs uptake was related to energy dependent process such as proteins dynamin and F-actin but negatively charged NPs were not dependent on dynamin proteins around the cell membrane [26] . Moreover, highly positively-charged NPs could cause perforations in the cellular lipid bilayer to enter the cells by-passing endocytic pathways [23] .

Another in vitro study for both fluorescent PS NPs and Coumarin-6 NPs in Caco-2 cells by Win et al. was performed to assess the effects of different polymeric NPs size [27] . Raw Coumarin-6 could not increase the cell uptake, however fluorescent PS NPs of 100 nm to 200 nm size showed the highest percentage of uptake. Smaller particles (50 nm) showed the lowest uptake and particles as large as 1000 nm showed decrease in uptake, which could be attributed to the uptake by other cellular mechanisms.

Optimization of antigen delivery to human dendritic cells (DCs; antigen presenting cells) is a challenge for advanced vaccine delivery systems. To identify the effects of particle size and surface charge on human DCs, in vitro cell uptake study has been investigated by Foged et al. in 2005 [28] . They designed the experiment based on wide size ranges (0.1 μm to 4.5 μm) of fluorescent PS NPs with different surface charges (+12.4 to −66.9 mV) after surface modification. Flow cytometric analysis of DCs after 24 hour incubation showed that lower percentage of DCs had taken up 4.5 μm particles (30%); whereas the highest cellular uptake (60%) was observed for 0.5 μm and 0.1 μm sized particles. To optimize the charge dependent interactions, particles with two sizes (0.1 and 1 μm) were modified by attaching variety poly amino acids/proteins covalently utilizing surface amine and carboxyl groups. Sterically same positive and neutral charges particles were obtained using polypeptides poly-l-lysine (PLL) and poly-d-l-alanine (PA), respectively. After 24 hours incubation, only 10% cellular uptake was observed with negatively charged 1 μm size particles, whereas positive charged particles were accounted for 60% uptake. However, around 90% uptake was observed for lower size (0.1 μm) particles with positive charge.

Prior to in vivo administration, it is essential to consider the compatibility, safety and biodegradability of the particles with the human blood and cells. To investigate the efficiency of particle size and surface charge in in vitro cellular uptake and blood compatibility, recently Dash et al. employed chitosan/polyglutamic acid hollow spheres to treat human umbilical vein endothelial cells (HUVECs) and human umbilical artery smooth muscle cells (HUASMCs) [21] . Enhanced cellular uptake has been observed with 100 nm neutral charged (−4 mV) in both cells such as 76% in HUVECs and 56% in HUASMCs compared to the other larger as well as pegylated particles regardless of surface charge. However, negatively charged particles showed the least cell internalization in both cases. To measure the effects of particles with erythrocytes of human blood, percentage of hemolysis was accounted towards different sizes and charges of particles. All types of particles were partially associated with very insignificant consequence on hemolysis (1% or less) without considering either size or surface charge. But, highly anionic charged particles of smaller size resulted insignificant delayed clotting time and platelet activation profile compared to larger particles and other types of charged particles.

Testing blood compatibility of polymeric NPs with human blood is another way for finding the probable adverse effects, which may happen after in vivo administration. To rationalize the hemo-compatibility test, another research group (Mayer et al.) employed PS NPs with variety of sizes and surface charges in different mediums (such as cell culture medium with different FBS ratio, PBS) [29] . To assess the influence of polymeric NPs' size and surface charge on human blood, the aim of study was to monitor the adverse effects by measuring complement activation, induction of coagulation, thrombocyte activation, membrane integrity, granulocyte activation, and hemolysis using flow cytometric analysis. Complement (C3a and C5a) levels detection is a consideration of the body's immune system activation. Cationic amidine PS particles were involved with high C3a generation (150.8%). Irrespective to size and surface charge, no NPs were involved with prothrombin level induction. CD62P/CD42b labelling was employed to investigate the thrombocyte activation, which was tested for both low (0.5 mg/mL) and high (2 mg/mL) concentrations. But no thromocytic damage was observed, which were confirmed by no lactate-de-hydrogenase (LDH) release for any of the particles. The percentage of CD11b expression (marker for granulocyte activation) for particle's different sizes and surface charges was reported in that study. Increased percentage of hemolysis for all types of particles was reported using high concentration of particle treatment with human blood. However, larger particles were found less hemolytic than smaller particles, and the most important point was that no influence was observed for negatively charged 160 nm size NPs on erythrocytes of human blood by treating with lower concentration. Overall, positively charged larger particles were involved with more hemolysis compared to negatively charged particles and the latter ones larger than 60 nm size appeared to be less hematotoxic than smaller particles. One interesting finding was; particles resuspended in cell culture medium with 10% fetal bovine serum (FBS) showed less negative zeta potential or about to close to neutral charge compared to the particles resuspended in phosphate buffer saline (PBS). The presence of salts and proteins in the dispersion cell medium might be accountable for neutralizing surface charge of polymeric NPs.

Upon exposure of different types of PLGA NPs to different experimental media, Mura et al. also investigated the possible size and zeta potential variations after resuspension of polymeric NPs in different media with time dependent incubation up to 96 hours at 37°C [30] . Three types of medium such as water, cell culture medium plus 10% FBS and PBS have been considered for evaluation in this experiment. Among different media, water and cell culture medium containing 10% FBS were not involved with significant variations in particle size regardless of surface charge, however after incubation of PLGA/chitosan (CS) NOS in PBS the size was increased. Furthermore, upon exposure to serum containing cell culture medium, PLGA/CS, PLGA/polyvinyl alcohol (PVA), and PLGA/pluronic F-68 (PF-68) NPs did not show any noteworthy change in zeta potential values.

They also designed in vitro model to investigate the toxicity of these prepared three types of NPs with Calu-3 cell line derived from human bronchial adenocarcinoma. This cell line could be a representative bronchial epithelial barrier associated with the discharge of airway mucus substances and the moderation of inflammatory reaction [31] [32] . Cell viability responses due to NP treatment with higher concentration after 72 hours incubation demonstrated that only PLGA/PF68 NPs showed progressively decreased cell viability compared to other types of NPs.

From other in vitro studies, it has also been found that NPs with 40 - 50 nm size range are involved with maximum uptake [33] [34] . However, a recent experiment by Schadlich et al. revealed the effect of size for the accumulation of near-infrared (NIR) fluorescent consisting PLA-PEG polymeric NPs in two tumor xenograft models (HT29 colorectal carcinoma and A2780 ovarian carcinoma) utilizing in vivo fluorescence imaging technique [35] . NPs of 111 nm and 141 nm size showed higher biodistribution and accumulation in tumors compared to the larger size (166 nm), which was due to rapid clearance of the larger particles by liver.

4. EFFECT OF PARTICLE SIZE AND SURFACE CHARGE BASED ON IN VIVO STUDIES

To explore the in vivo effects of specifically sized NPs with respect to surface charge, Kulkarni et al. injected the fluorescent modified and unmodified PS NPs into Sprague?Dawley rats after physicochemical characterization [36] . Modification of PS NPs was performed by coating with D-α-tocopheryl polyethylene glycol succinate or Vitamin E TPGS, which was able to switch the zeta potentials of different size NPs to less negative charge.

As previously known, circulating mononuclear phagocytic cells in the bloodstream are the key component of reticuloendothelial system (RES). In addition, RES is also composed of matured cells such as macrophages mainly available in lungs, liver and spleen [37] . Studies have shown that the NPs with the size range of 100 to 200 nm could be the optimum range in order to escape the RES recognition [27] . Due to rapid clearance from systemic circulation, mostly uncoated NPs were distributed to those organs such as liver and spleen, where mononuclear phagocytic system is located. Consequently, 100 and 200 nm size fluorescent PS-TPGS NPs resulted in higher fluorescence concentration in blood plasma regardless of their surface charges. Liver and spleen were the main target organs, where a substantial decrease in NP distribution was observed for all sizes of TPGS modified PS NPs, since hydrophilic coated surface (stealth effect) possesses the ability to prevent the NPs from RES capture.

Moreover, He et al. in 2010 investigated the effects of size and surface charges on the biodistribution of different sizes of rhodamine B (RhB) labeled carboxymethyl chitosan grafted NPs (RhB-CMCNP) and chitosan hydrochloride grafted NPs (RhB-CHNP) having negative and positive surface charges after intravenous administration into H-22 tumor bearing mice [2] . It was clearly demonstrated that biodistribution of 150 nm size NPs having zeta potential of around −15 mV showed higher accumulation at the tumor site and long residence in blood compared to more negative or positive or even larger particles. Due to inflammation and disorder of endothelium along with high demand of nutrient supply, comparatively larger vascular leakage is found in solid tumors, which can provide more access for extravascular targeting macromolecules [38] . Low anionic charge (−15 mV) bearing RhB-CMCNP-PS particles exhibited higher percentage of distribution in tumor, which might be due to enhanced residing time during systemic circulation [39] . On the other hand, more positive charged NPs could leave the interstitium more competently after arriving the tumor's leaky vasculature leading to be up taken by tumor cells or endothelium adjacent to the endothelium [40] . This phenomenon might be the possible reason why high cationic charge (+35 mV) bearing RhB-CHNP particles showed higher percentage of distribution in tumor. Due to enhanced permeability and retention (EPR) effects, smaller particles might be favorable to target the tumor passively due to higher accumulation. However, blood's complement activation system and blood opsonins have been found to fabricate the size of polymeric NPs to a larger extent (>500 nm) resulting rapid blood clearance [39] . RhB-CMCNP and RhB-CHNP with larger particle size resulted in higher hepatic disposition. Such higher hepatic disposition could be explained by the investigation performed by Liu et al. They found that NPs with the size range above 300 nm were inclined to be blocked or captured by RES as well as liver sinusoids [41] . In addition, NPs with the size range from 200 to 500 (nm) was found mainly unaffected by the splenic physical filtration mechanism [42] . This could be attributed to obtain lower percentage of hepatic distribution of RhB-CHNP NPs with the particle size range from 150 to 300 (nm). Besides, no influence was observed in distribution of both RhB-CMCNP and RhB-CHNP particles in kidney owing to their size and surface charge. Due to electrostatic reactivity, positively charge particles had tendency to form aggregates with the cells and proteins present in blood and subsequently the aggregation could be trapped by lung [43] . As this experiment revealed, He et al. demonstrated the similar result where more cationic RhB-CHNP NPs distribution was found in lung.

NPs of 10 - 100 nm size is considered as mainly accepted range to design any NP formulation respective to suitable clearance and biodistribution profile before any in vivo trial [3] . However, the upper range of particle size is dependent on the interactions with body's immune systems and the lower range is determined by the limit of kidney filtration. Opsonization of larger particles by responsible proteins (e.g. plasma complement, immunoglobulins) in blood compartment is common to develop hypersensitivity response comparatively against larger foreign particles [44] [45] . On the other hand, smaller particles (<5.5 nm) have been found with rapid clearance from the body by kidney's glomerular filtration mechanism [46] .

To explore the in vivo effects of different size of NPs, Liu et al. prepared radioisotope labeled liposomes of different sizes (30 - 400 nm) to inject into the mice models to observe the biodistribution in blood, liver, spleen, and tumor [47] . After four hours of post administration, it was found that about 60% of 100 to 200 nm size particles were found in blood, but only 20% of injected particles with size boundary (>250 nm or <50 nm) were detected in blood. In liver, particle size with 100 nm was associated with 20% accumulation, whereas around 25% distribution in liver was detected for larger particles. In spleen, 40% - 50% of the injected dose was detected for larger size (>250 nm) but the percentage of detection was lower for the particle size range below 100 nm. In 2002, Levchenko et al. prepared liposomes of around 200 nm with variety of charged surfaces to evaluate the tissue distribution in mice models [48] . The results from this study showed that the negatively charged liposomes with zeta potential of around

−40 mV were involved with higher clearance rate from the blood in comparison to liposomes with neutral zeta potential.

In addition, Yamamoto et al. investigated the effect of surface charge of poly (ethylene glycol)-poly (D, L-lactide) block copolymer micelles after injecting into male C57/BL6N mice through the tail vein [49] . They prepared the micelles with both neutral (tyrosine) and negative (tyrosineglutamine) functionalities, which did not show any significant variations in blood clearance kinetics. However, the negatively-charged micelles displayed a significant lower distribution in both liver and spleen after four hours of post intravenous injection. Overall effects of NPs size and surface charge could be summarized in Figure 2.

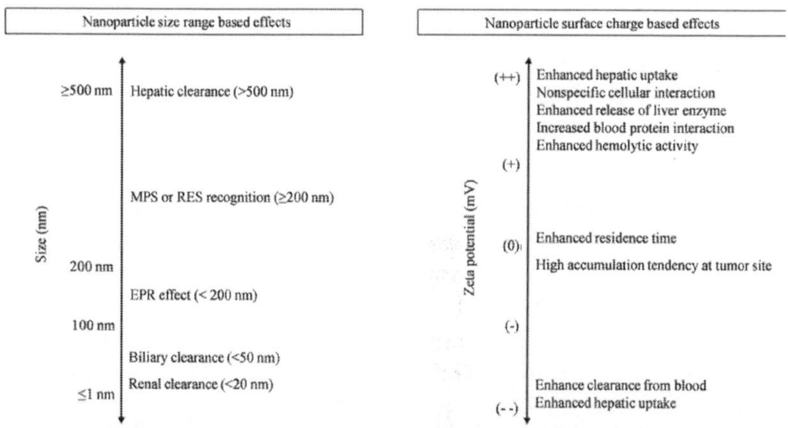

Figure 2. Relative biocompatibility of polymeric NPs based on the effects of size and surface charge. Abbreviation: MPS (Mononuclear Phagocyte System), RES (Reticuloendothelial System), EPR (Enhanced Permeability and Retention).

5. CONCLUSION

In conclusion, polymeric NPs with size range from 10 to 200 nm might not only escape renal filtration and biliary excretion but also accumulate in tumor utilizing EPR effects. Size range above 200 nm may be related to rapid hepatic clearance and RES recognition. Pegylation strategy could be an ideal option to stealth the poly- meric NPs for longer residing time during systemic circulation. After in vivo administration of cationic polymeric NPs, non-specific interaction may occur with non-specific cells or opsonizing protein in blood compartment due to electrostatic bindings, which may involve unexpected cytotoxicity. In order to reduce such non-specific surface reactivity or interaction, relatively less negatively charged anionic (almost neutral) polymeric NPs with desired small size might be more rationale than cationic charged particles for a broad spectrum biological aspect. This review will help researchers to correlate the in vitro and in vivo effects of polymeric NPs based on particle size and charge. Further investigation and correlation of other physicochemical parameters could be performed on polymeric NPs to understand their biological effects.

ACKNOWLEDGEMENTS

This work was funded by a research grant from the Canadian Breast Cancer Foundation (CBCF) and Natural Sciences and Engineering Research (NSERC) Discovery Grant. The authors report no declarations of interest.

CONFLICT OF INTEREST

The authors confirm that this article content has no conflict of interest.

REFERENCES

1. Grottkau, B.E., Cai, X., Wang, J., Yang, X. and Lin, Y. (2013) Polymeric Nanoparticles for a Drug Delivery System. Current Drug Metabolism, 14, 840-846.

2. He, C., Hu, Y., Yin, L., Tang, C. and Yin, C. (2010) Effects of Particle Size and Surface Charge on Cellular Uptake and Biodistribution of Polymeric Nanoparticles. Biomaterials, 31, 3657-3666.

3. Wang, J., Byrne, J.D., Napier, M.E. and DeSimone, J.M. (2011) More Effective Nanomedicines through Particle Design. Small, 7, 1919-1931.

4. Kleber, J.W., Nash, J.F. and Lee, C.C. (1964) Synthetic Polymers as Potential Sustained-Release Coatings. Journal of Pharmaceutical Sciences, 53, 1519-1521.

5. Singh, R. and Lillard Jr., J.W. (2009) Nanoparticle-Based Targeted Drug Delivery. Experimental and Molecular Pathology, 86, 215-223.

6. Lamprecht, A., Schafer, U. and Lehr, C.M. (2001) Size-Dependent Bioadhesion of Micro- and Nanoparticulate Carriers to the Inflamed Colonic Mucosa. Pharmaceutical Research, 18, 788-793.

7. Behrens, I., Pena, A.I., Alonso, M.J. and Kissel, T. (2002) Comparative Uptake Studies of Bioadhesive and Non- Bioadhesive Nanoparticles in Human Intestinal Cell Lines and Rats: The Effect of

Mucus on Particle Adsorption and Transport. Pharmaceutical Research, 19, 1185-1193.

8. Galindo-Rodriguez, S.A., Allemann, E., Fessi, H. and Doelker, E. (2005) Polymeric Nanoparticles for Oral Delivery of Drugs and Vaccines: A Critical Evaluation of in Vivo Studies. Critical Reviews™ in Therapeutic Drug Carrier Systems, 22, 419-464.

9. Graf, A., McDowell, A. and Rades, T. (2009) Poly(alkylcyanoacrylate) Nanoparticles for Enhanced Delivery of Therapeutics—Is There Real Potential? Expert Opinion on Drug Delivery, 6, 371-387.

10. Vinogradov, S.B.E. and Kabanov, A. (1999) Poly(ethylene glycol)-polyethyleneimine NanoGel™ Particles: Novel Drug Delivery Systems for Antisense Oligonucleotides. Colloids and Surfaces B: Biointerfaces, 16, 291-304.

11. Plapied, L., Duhem, N., Rieux, A.D. and Préat, V. (2011) Fate of Polymeric Nanocarriers for Oral Drug Delivery. Current Opinion in Colloid & Interface, 16, 228-237.

12. Ahlin, P., Kristl, J., Kristl, A. and Vrecer, F. (2002) Investigation of Polymeric Nanoparticles as Carriers of Enalaprilat for Oral Administration. International Journal of Pharmaceutics, 239, 113-120.

13. Kumari, A., Yadav, S.K. and Yadav, S.C. (2010) Biodegradable Polymeric Nanoparticles Based Drug Delivery Systems. Colloids and Surfaces B: Biointerfaces, 75, 1-18.

14. Song, X., Zhao, X., Zhou, Y., Li, S. and Ma, Q. (2010) Pharmacokinetics and Disposition of Various Drug Loaded Biodegradable Poly(Lactide-Co-Glycolide) (PLGA) Nanoparticles. Current Drug Metabolism, 11, 859-869.

15. Chaudhury, A. and Das, S. (2011) Recent Advancement of Chitosan-Based Nanoparticles for Oral Controlled Delivery of Insulin and Other Therapeutic Agents. AAPS PharmSciTech, 12, 10-20.

16. Liu, Z., Jiao, Y., Wang, Y., Zhou, C. and Zhang, Z. (2008) Polysaccharides-Based Nanoparticles as Drug Delivery Systems. Advanced Drug Delivery Reviews, 60, 1650-1662.

17. Hu, C.M., Fang, R.H., Luk, B.T. and Zhang, L. (2013) Polymeric Nanotherapeutics: Clinical Development and Advances in Stealth Functionalization Strategies. Nanoscale, 6, 65-76.

18. Kamaly, N., Xiao, Z., Valencia, P.M., Radovic-Moreno, A.F. and Farokhzad, O.C. (2012) Targeted Polymeric Therapeutic Nanoparticles: Design, Development and Clinical Translation. Chemical Society Reviews, 41, 2971-3010.

19. Mundargi, R.C., Babu, V.R., Rangaswamy, V., Patel, P. and Aminabhavi, T.M. (2008) Nano/Micro Technologies for Delivering Macromolecular Therapeutics Using Poly(D,L-Lactide-co-Glycolide) and Its Derivatives. Journal of Controlled Release, 125, 193-209.

20. Dinarvand, R., Sepehri, N., Manoochehri, S., Rouhani, H. and Atyabi, F. (2011) Polylactide-co-Glycolide Nanoparticles for Controlled Delivery of Anticancer Agents. International Journal of Nanomedicine, 6, 877-895.

21. Dash, B.C., Rethore, G., Monaghan, M., Fitzgerald, K., Gallagher, W. and Pandit, A. (2010) The Influence of Size and Charge of Chitosan/Polyglutamic Acid Hollow Spheres on Cellular Internalization, Viability and Blood Compatibility. Biomaterials, 31, 8188-8197.

22. Conner, S.D. and Schmid, S.L. (2003) Regulated Portals of Entry into the Cell. Nature, 422, 37-44.

23. Zhao, F., Zhao, Y., Liu, Y., Chang, X., Chen, C. and Zhao, Y. (2011) Cellular Uptake, Intracellular Trafficking, and Cytotoxicity of Nanomaterials. Small, 7, 1322-1337.

24. Bhattacharjee, S., Ershov, D., Fytianos, K., van der Gucht, J., Alink, G.M., Rietjens, I.M., et al. (2012) Cytotoxicity and Cellular Uptake of Tri-Block Copolymer Nanoparticles with Different Size and Surface Characteristics. Particle and Fibre Toxicology, 9, 11.

25. Lai, S.K., Hida, K., Man, S.T., Chen, C., Machamer, C., Schroer, T.A., et al. (2007) Privileged Delivery of Polymer Nanoparticles to the Perinuclear Region of Live Cells via a Non-Clathrin, Non-Degradative Pathway. Biomaterials, 28, 2876-2884.

26. Dausend, J., Musyanovych, A., Dass, M., Walther, P., Schrezenmeier, H., Landfester, K., et al. (2008) Uptake Mechanism of Oppositely Charged Fluorescent Nanoparticles in HeLa Cells. Macromolecular Bioscience, 8, 1135-1143.

27. Win, K.Y. and Feng, S.S. (2005) Effects of Particle Size and Surface Coating on Cellular Uptake of Polymeric Nanoparticles for Oral Delivery of Anticancer Drugs. Biomaterials, 26, 2713-2722.

28. Foged, C., Brodin, B., Frokjaer, S. and Sundblad, A. (2005) Particle Size and Surface Charge Affect Particle Uptake by Human Dendritic Cells in an in Vitro Model. International Journal of Pharmaceutics, 298, 315-322.

29. Mayer, A., Vadon, M., Rinner, B., Novak, A., Wintersteiger, R. and Frohlich, E. (2009) The Role of Nanoparticle Size in Hemocompatibility. Toxicology, 258, 139-147.

30. Mura, S., Hillaireau, H., Nicolas, J., Le Droumaguet, B., Gueutin, C., Zanna, S., et al. (2011) Influence of Surface Charge on the Potential Toxicity of PLGA Nanoparticles towards Calu-3 Cells. International Journal of Nanomedicine, 6, 2591-2605.

31. Shen, B.Q., Finkbeiner, W.E., Wine, J.J., Mrsny, R.J. and Widdicombe, J.H. (1994) Calu-3: A Human Airway Epithelial Cell Line That Shows cAMP-Dependent Cl-Secretion. American Journal of Physiology, 266, L493-L501.

32. da Paula, A.C., Ramalho, A.S., Farinha, C.M., Cheung, J., Maurisse, R., Gruenert, D.C., et al. (2005) Characterization of Novel Airway Submucosal Gland Cell Models for Cystic Fibrosis Studies. Cellular Physiology and Biochemistry, 15, 251-262.

33. Arap, W., Haedicke, W., Bernasconi, M., Kain, R., Rajotte, D., Krajewski, S., et al. (2002) Targeting the Prostate for Destruction through a Vascular Address. Proceedings of the National Academy of Sciences of the United States of America, 99, 1527-1531.

34. Gratton, S.E., Ropp, P.A., Pohlhaus, P.D., Luft, J.C., Madden, V.J., Napier, M.E., et al. (2008) The Effect of Particle Design on Cellular Internalization Pathways. Proceedings of the National Academy of Sciences of the United States of America, 105, 11613-11618.

35. Schadlich, A., Caysa, H., Mueller, T., Tenambergen, F., Rose, C., Gopferich, A., et al. (2011) Tumor Accumulation of NIR Fluorescent PEG-PLA Nanoparticles: Impact of Particle Size and Human Xenograft Tumor Model. ACS Nano, 5, 8710-8720.

36. Kulkarni, S.A. and Feng, S.S. (2013) Effects of Particle Size and Surface Modification on Cellular Uptake and Biodistribution of Polymeric Nanoparticles for Drug Delivery. Pharmaceutical Research, 30, 2512-2522.

37. Hillery, A.M. and Florence, F.A. (1996) The Effect of Adsorbed Poloxamer 188 and 407 Surfactants on the Intestinal Uptake of 60-nm Polystyrene Particles after Oral Administration in the Rat. International Journal of Pharmaceutics, 132, 123-130.

38. Faraji, A.H. and Wipf, P. (2009) Nanoparticles in Cellular Drug Delivery. Bioorganic & Medicinal Chemistry, 17, 2950-2962.

39. Alexis, F., Pridgen, E., Molnar, L.K. and Farokhzad, O.C. (2008) Factors Affecting the Clearance and Biodistribution of Polymeric Nanoparticles. Molecular Pharmaceutics, 5, 505-515.

40. Nigavekar, S.S., Sung, L.Y., Llanes, M., El-Jawahri, A., Lawrence, T.S., Becker, C.W., et al. (2004) 3H Dendrimer Nanoparticle Organ/Tumor Distribution. Pharmaceutical Research, 21, 476-483.

41. Liu, D., Mori, A. and Huang, L. (1992) Role of Liposome Size and RES Blockade in Controlling Biodistribution and Tumor Uptake of GM1-Containing Liposomes. Biochimica et Biophysica Acta, 1104, 95-101.

42. Akiyama, Y., Mori, T., Katayama, Y. and Niidome, T. (2009) The Effects of PEG Grafting Level and Injection Dose on Gold Nanorod Biodistribution in the Tumor-Bearing Mice. Journal of Controlled Release, 139, 81-84.

43. Ishiwata, H., Suzuki, N., Ando, S., Kikuchi, H. and Kitagawa, T. (2000) Characteristics and Biodistribution of Cationic Liposomes and Their DNA Complexes. Journal of Controlled Release, 69, 139-148.

44. Vonarbourg, A., Passirani, C., Saulnier, P., Simard, P., Leroux, J.C. and Benoit, J.P. (2006) Evaluation of Pegylated Lipid Nanocapsules versus Complement System Activation and Macrophage Uptake. Journal of Biomedical Materials Research Part A, 78, 620-628.

45. Petros, R.A. and DeSimone, J.M. (2010) Strategies in the Design of Nanoparticles for Therapeutic Applications. Nature Reviews Drug Discovery, 9, 615-627.

46. Choi, H.S., Liu, W., Misra, P., Tanaka, E., Zimmer, J.P., Itty Ipe, B., et al. (2007) Renal Clearance of Quantum Dots. Nature Biotechnology, 25, 1165-1170.

47. Li, S.D. and Huang, L. (2008) Pharmacokinetics and Biodistribution of Nanoparticles. Molecular Pharmaceutics, 5, 496-504.

48. Levchenko, T.S., Rammohan, R., Lukyanov, A.N., Whiteman, K.R. and Torchilin, V.P. (2002) Liposome Clearance in Mice: The Effect of a Separate and Combined Presence of Surface Charge and Polymer Coating. International Journal of Pharmaceutics, 240, 95-102.

49. Yamamoto, Y., Nagasaki, Y., Kato, Y., Sugiyama, Y. and Kataoka, K. (2001) Long-Circulating Poly(Ethylene Glycol)-Poly(D,L-Lactide) Block Copolymer Micelles with Modulated Surface Charge. Journal of Controlled Release, 77, 27-38.

50. Ma, Z. and Lim, L.Y. (2003) Uptake of Chitosan and Associated Insulin in Caco-2 Cell Monolayers: A Comparison between Chitosan Molecules and Chitosan Nanoparticles. Pharmaceutical Research, 20, 1812-1819.

51. Huang, M., Ma, Z., Khor, E. and Lim, L.Y. (2002) Uptake of FITC-Chitosan Nanoparticles by A549 Cells. Pharmaceutical Research, 19, 1488-1494.

52. Kim, T.H., Park, I.K., Nah, J.W., Choi, Y.J. and Cho, C.S. (2004) Galactosylated Chitosan/DNA Nanoparticles Prepared Using

Water-Soluble Chitosan as a Gene Carrier. Biomaterials, 25, 3783-3792.

53. Han, H.D., Mangala, L.S., Lee, J.W., Shahzad, M.M., Kim, H.S., Shen, D., et al. (2010) Targeted Gene Silencing Using RGD-Labeled Chitosan Nanoparticles. Clinical Cancer Research: An Official Journal of the American Association for Cancer Research, 16, 3910-3922.

54. Aktas, Y., Yemisci, M., Andrieux, K., Gursoy, R.N., Alonso, M.J., Fernandez-Megia, E., et al. (2005) Development and Brain Delivery of Chitosan-PEG Nanoparticles Functionalized with the Monoclonal Antibody OX26. Bioconjugate Chemistry, 16, 1503-1511.

55. Balthasar, S., Michaelis, K., Dinauer, N., von Briesen, H., Kreuter, J. and Langer, K. (2005) Preparation and Characterisation of Antibody Modified Gelatin Nanoparticles as Drug Carrier System for Uptake in Lymphocytes. Biomaterials, 26, 2723-2732.

56. Wang, X., Li, J., Wang, Y., Cho, K.J., Kim, G., Gjyrezi, A., et al. (2009) HFT-T, a Targeting Nanoparticle, Enhances Specific Delivery of Paclitaxel to Folate Receptor-Positive Tumors. ACS Nano, 3, 3165-3174.

57. Choi, K.Y., Chung, H., Min, K.H., Yoon, H.Y., Kim, K., Park, J.H., et al. (2010) Self-Assembled Hyaluronic Acid Nanoparticles for Active Tumor Targeting. Biomaterials, 31, 106-114.

58. Cho, H.J., Yoon, H.Y., Koo, H., Ko, S.H., Shim, J.S., Lee, J.H., et al. (2011) Self-Assembled Nanoparticles Based on Hyaluronic Acid-Ceramide (HA-CE) and Pluronic(R) for Tumor-Targeted Delivery of Docetaxel. Biomaterials, 32, 7181-7190.

59. Nam, H.Y., Kwon, S.M., Chung, H., Lee, S.Y., Kwon, S.H., Jeon, H., et al. (2009) Cellular Uptake Mechanism and Intracellular Fate of Hydrophobically Modified Glycol Chitosan Nanoparticles. Journal of Controlled Release, 135, 259-267.

60. Chung, T.H., Wu, S.H., Yao, M., Lu, C.W., Lin, Y.S., Hung, Y., et al. (2007) The Effect of Surface Charge on the Uptake and Biological Function of Mesoporous Silica Nanoparticles in 3T3-L1 Cells and Human Mesenchymal Stem Cells. Biomaterials, 28, 2959-2966.

61. Ko, J., Park, K., Kim, Y.S., Kim, M.S., Han, J.K., Kim, K., et al. (2007) Tumoral Acidic Extracellular pH Targeting of pH-Responsive MPEG-Poly(Beta-Amino Ester) Block Copolymer Micelles for Cancer Therapy. Journal of Controlled Release, 123, 109-115.

62. Park, J.S., Han, T.H., Lee, K.Y., Han, S.S., Hwang, J.J., Moon, D.H., et al. (2006) N-Acetyl Histidine-Conjugated Glycol Chitosan Self-Assembled Nanoparticles for Intracytoplasmic Delivery of Drugs:

Endocytosis, Exocytosis and Drug Release. Journal of Controlled Release, 115, 37-45.

63. Wang, Y., Gao, S., Ye, W.H., Yoon, H.S. and Yang, Y.Y. (2006) Co-Delivery of Drugs and DNA from Cationic Core-Shell Nanoparticles Self-Assembled from a Biodegradable Copolymer. Nature Materials, 5, 791-796.

64. Chawla, J.S. and Amiji, M.M. (2002) Biodegradable Poly(Epsilon-Caprolactone) Nanoparticles for Tumor-Targeted Delivery of taMoxifen. International Journal of Pharmaceutics, 249, 127-138.

65. Chawla, J.S. and Amiji, M.M. (2003) Cellular Uptake and Concentrations of Tamoxifen upon Administration in Poly(Epsilon-Caprolactone) Nanoparticles. AAPS PharmSci, 5, 28-34.

66. Cade, D., Ramus, E., Rinaudo, M., Auzely-Velty, R., Delair, T. and Hamaide, T. (2004) Tailoring of Bioresorbable Polymers for Elaboration of Sugar-Functionalized Nanoparticles. Biomacromolecules, 5, 922-927.

67. Wang, J., Tian, S., Petros, R.A., Napier, M.E. and Desimone, J.M. (2010) The Complex Role of Multivalency in Nanoparticles Targeting the Transferrin Receptor for Cancer Therapies. Journal of the American Chemical Society, 132, 11306-11313.

68. Lee, H., Fonge, H., Hoang, B., Reilly, R.M. and Allen, C. (2010) The Effects of Particle Size and Molecular Targeting on the Intratumoral and Subcellular Distribution of Polymeric Nanoparticles. Molecular Pharmaceutics, 7, 1195-1208.

69. Xin, H., Jiang, X., Gu, J., Sha, X., Chen, L., Law, K., et al. (2011) Angiopep-Conjugated Poly(Ethylene Glycol)-co-Poly(Epsilon-Caprolactone) Nanoparticles as Dual-Targeting Drug Delivery System for Brain Glioma. Biomaterials, 32, 4293-4305.

70. Kim, H.R., Gil, S., Andrieux, K., Nicolas, V., Appel, M., Chacun, H., et al. (2007) Low-Density Lipoprotein Receptor-Mediated Endocytosis of PEGylated Nanoparticles in Rat Brain Endothelial Cells. Cellular and Molecular Life Sciences: CMLS, 64, 356-364.

71. Harush-Frenkel, O., Debotton, N., Benita, S. and Altschuler, Y. (2007) Targeting of Nanoparticles to the Clathrin-Mediated Endocytic Pathway. Biochemical and Biophysical Research Communications, 353, 26-32.

72. Lu, W., Tan, Y.Z., Hu, K.L. and Jiang, X.G. (2005) Cationic Albumin Conjugated Pegylated Nanoparticle with Its Transcytosis Ability and Little Toxicity Against Blood-Brain Barrier. International Journal of Pharmaceutics, 295, 247-260.

73. Harush-Frenkel, O., Rozentur, E., Benita, S. and Altschuler, Y. (2008) Surface Charge of Nanoparticles Determines Their Endocytic and Transcytotic Pathway in Polarized MDCK Cells. Biomacromolecules, 9, 435-443.

74. Aryal, S., Hu, C.M. and Zhang, L. (2010) Polymer-Cisplatin Conjugate Nanoparticles for Acid-Responsive Drug Delivery. ACS Nano, 4, 251-258.
Danhier, F., Lecouturier, N., Vroman, B., Jerome, C., Marchand-Brynaert, J., Feron, O., et al. (2009) Paclitaxel-Loaded PEGylated PLGA-Based Nanoparticles: In Vitro and In Vivo Evaluation. Journal of Controlled Release, 133, 11-17.

75. Mao, S., Germershaus, O., Fischer, D., Linn, T., Schnepf, R. and Kissel, T. (2005) Uptake and Transport of PEG- Graft-Trimethyl-Chitosan Copolymer-Insulin Nanocomplexes by Epithelial Cells. Pharmaceutical Research, 22, 2058- 2068.

76. Nukolova, N.V., Oberoi, H.S., Cohen, S.M., Kabanov, A.V. and Bronich, T.K. (2011) Folate-Decorated Nanogels for Targeted Therapy of Ovarian Cancer. Biomaterials, 32, 5417-5426.

77. Cho, C.S., Cho, K.Y., Park, I.K., Kim, S.H., Sasagawa, T., Uchiyama, M., et al. (2001) Receptor-Mediated Delivery of All Trans-Retinoic Acid to Hepatocyte Using Poly(L-Lactic Acid) Nanoparticles Coated with Galactose-Carrying Polystyrene. Journal of Controlled Release, 77, 7-15.

78. Banquy, X., Leclair, G., Rabanel, J.M., Argaw, A., Bouchard, J.F., Hildgen, P., et al. (2008) Selectins Ligand Decorated Drug Carriers for Activated Endothelial Cell Targeting. Bioconjugate Chemistry, 19, 2030-2039.

79. Nasongkla, N., Bey, E., Ren, J., Ai, H., Khemtong, C., Guthi, J.S., et al. (2006) Multifunctional Polymeric Micelles as Cancer-Targeted, MRI-Ultrasensitive Drug Delivery Systems. Nano Letters, 6, 2427-2430.

80. Prabha, S. and Labhasetwar, V. (2004) Critical Determinants in PLGA/PLA Nanoparticle-Mediated Gene Expression. Pharmaceutical Research, 21, 354-364.

81. Musumeci, T., Ventura, C.A., Giannone, I., Ruozi, B., Montenegro, L., Pignatello, R., et al. (2006) PLA/PLGA Nanoparticles for Sustained Release of Docetaxel. International Journal of Pharmaceutics, 325, 172-179.

82. Prabha, S., Zhou, W.Z., Panyam, J. and Labhasetwar, V. (2002) Size-Dependency of Nanoparticle-Mediated Gene Transfection: Studies with Fractionated Nanoparticles. International Journal of Pharmaceutics, 244, 105-115.

83. Sahoo, S.K., Panyam, J., Prabha, S. and Labhasetwar, V. (2002) Residual Polyvinyl Alcohol Associated with Poly (D,L-Lactide-co-Glycolide) Nanoparticles Affects Their Physical Properties and Cellular Uptake. Journal of Controlled Release, 82, 105-114.

84. Sahoo, S.K. and Labhasetwar, V. (2005) Enhanced Antiproliferative Activity of Transferrin-Conjugated Paclitaxel-Loaded Nanoparticles Is Mediated via Sustained Intracellular Drug Retention. Molecular Pharmaceutics, 2, 373-383.

85. Obermajer, N., Kocbek, P., Repnik, U., Kuznik, A., Cegnar, M., Kristl, J., et al. (2007) Immunonanoparticles—An Effective Tool to Impair Harmful Proteolysis in Invasive Breast Tumor Cells. The FEBS Journal, 274, 4416-4427.

86. Davda, J. and Labhasetwar, V. (2002) Characterization of Nanoparticle Uptake by Endothelial Cells. International Journal of Pharmaceutics, 233, 51-59.

87. Mo, Y. and Lim, L.Y. (2005) Preparation and in Vitro Anticancer Activity of Wheat Germ Agglutinin (WGA)-Conjugated PLGA Nanoparticles Loaded with Paclitaxel and Isopropyl Myristate. Journal of Controlled Release, 107, 30-42.

88. Mo, Y. and Lim, L.Y. (2005) Paclitaxel-Loaded PLGA Nanoparticles: Potentiation of Anticancer Activity by Surface Conjugation with Wheat Germ Agglutinin. Journal of Controlled Release, 108, 244-262.

89. Mo, Y. and Lim, L.Y. (2004) Mechanistic Study of the Uptake of Wheat Germ Agglutinin-Conjugated PLGA Nanoparticles by A549 Cells. Journal of Pharmaceutical Sciences, 93, 20-28.

90. McCarron, P.A., Marouf, W.M., Quinn, D.J., Fay, F., Burden, R.E., Olwill, S.A., et al. (2008) Antibody Targeting of Camptothecin-Loaded PLGA Nanoparticles to Tumor Cells. Bioconjugate Chemistry, 19, 1561-1569.

91. Tosi, G., Fano, R.A., Bondioli, L., Badiali, L., Benassi, R., Rivasi, F., et al. (2011) Investigation on Mechanisms of Glycopeptide Nanoparticles for Drug Delivery across the Blood-Brain Barrier. Nanomedicine, 6, 423-436.

92. Fonseca, C., Simoes, S. and Gaspar, R. (2002) Paclitaxel-Loaded PLGA Nanoparticles: Preparation, Physicochemical Characterization and in Vitro Anti-Tumoral Activity. Journal of Controlled Release, 83, 273-286.

93. Xie, J. and Wang, C.H. (2005) Self-Assembled Biodegradable Nanoparticles Developed by Direct Dialysis for the Delivery of Paclitaxel. Pharmaceutical Research, 22, 2079-2090.

94. Esmaeili, F., Dinarvand, R., Ghahremani, M.H., Ostad, S.N., Esmaily, H. and Atyabi, F. (2010) Cellular Cytotoxicity and In-Vivo Biodistribution of Docetaxel Poly(Lactide-co-Glycolide) Nanoparticles. Anti-Cancer Drugs, 21, 43-52.

95. Betancourt, T., Brown, B. and Brannon-Peppas, L. (2007) Doxorubicin-Loaded PLGA Nanoparticles by Nanoprecipitation: Preparation, Characterization and in Vitro Evaluation. Nanomedicine, 2, 219-232.

96. Manchanda, R., Fernandez-Fernandez, A., Nagesetti, A. and McGoron, A.J. (2010) Preparation and Characterization of a Polymeric (PLGA) Nanoparticulate Drug Delivery System with Simultaneous Incorporation of Chemotherapeutic and Thermo-Optical Agents. Colloids and Surfaces B: Biointerfaces, 75, 260-267.

97. Yallapu, M.M., Gupta, B.K., Jaggi, M. and Chauhan, S.C. (2010) Fabrication of Curcumin Encapsulated PLGA Nanoparticles for Improved Therapeutic Effects in Metastatic Cancer Cells. Journal of Colloid and Interface Science, 351, 19-29.

98. Yan, F., Zhang, C., Zheng, Y., Mei, L., Tang, L., Song, C., et al. (2010) The Effect of Poloxamer 188 on Nanoparticle Morphology, Size, Cancer Cell Uptake, and Cytotoxicity. Nanomedicine, 6, 170-178.

99. Song, X.R., Cai, Z., Zheng, Y., He, G., Cui, F.Y., Gong, D.Q., et al. (2009) Reversion of Multidrug Resistance by Co-Encapsulation of Vincristine and Verapamil in PLGA Nanoparticles. European Journal of Pharmaceutical Sciences: Official Journal of the European Federation for Pharmaceutical Sciences, 37, 300-305.

100. Zeisser-Labouebe, M., Lange, N., Gurny, R. and Delie, F. (2006) Hypericin-Loaded Nanoparticles for the Photodynamic Treatment of Ovarian Cancer. International Journal of Pharmaceutics, 326, 174-181.

101. Shah, N., Chaudhari, K., Dantuluri, P., Murthy, R.S. and Das, S. (2009) Paclitaxel-Loaded PLGA Nanoparticles Surface Modified with Transferrin and Pluronic((R))P85, an in Vitro Cell Line and in Vivo Biodistribution Studies on Rat Model. Journal of Drug Targeting, 17, 533-542.

102. Win, K.Y. and Feng, S.S. (2006) In Vitro and in Vivo Studies on Vitamin E TPGS-Emulsified Poly(D,L-Lactic-co-Glycolic Acid) Nanoparticles for Paclitaxel Formulation. Biomaterials, 27, 2285-2291.

103. Dong, Y. and Feng, S.S. (2007) Poly(D,L-Lactide-co-Glycolide) (PLGA) Nanoparticles Prepared by High Pressure Homogenization for Paclitaxel Chemotherapy. International Journal of Pharmaceutics, 342, 208-214.

104. Bondioli, L., Costantino, L., Ballestrazzi, A., Lucchesi, D., Boraschi, D., Pellati, F., et al. (2010) PLGA Nanoparticles Surface Decorated with the Sialic Acid, N-Acetylneuraminic Acid. Biomaterials, 31, 3395-3403.

105. Zheng, Y., Yu, B., Weecharangsan, W., Piao, L., Darby, M., Mao, Y., et al. (2010) Transferrin-Conjugated Lipid-Coated PLGA Nanoparticles for Targeted Delivery of Aromatase Inhibitor 7alpha-APTADD to Breast Cancer Cells. International Journal of Pharmaceutics, 390, 234-241.

106. Kou, G., Gao, J., Wang, H., Chen, H., Li, B., Zhang, D., et al. (2007) Preparation and Characterization of Paclitaxel-Loaded PLGA Nanoparticles Coated with Cationic SM5-1 Single-Chain Antibody. Journal of Biochemistry and Molecular Biology, 40, 731-739.

107. Kocbek, P., Obermajer, N., Cegnar, M., Kos, J. and Kristl, J. (2007) Targeting Cancer Cells Using PLGA Nanoparticles Surface Modified with Monoclonal Antibody. Journal of Controlled Release, 120, 18-26.

108. Acharya, S., Dilnawaz, F. and Sahoo, S.K. (2009) Targeted Epidermal Growth Factor Receptor Nanoparticle Bioconjugates for Breast Cancer Therapy. Biomaterials, 30, 5737-5750.

109. Sun, B., Ranganathan, B. and Feng, S.S. (2008) Multifunctional Poly(D,L-Lactide-co-Glycolide)/Montmorillonite (PLGA/MMT) Nanoparticles Decorated by Trastuzumab for Targeted Chemotherapy of Breast Cancer. Biomaterials, 29, 475-486.

110. Gao, J., Kou, G., Wang, H., Chen, H., Li, B., Lu, Y., et al. (2009) PE38KDEL-Loaded Anti-HER2 Nanoparticles Inhibit Breast Tumor Progression with Reduced Toxicity and Immunogenicity. Breast Cancer Research and Treatment, 115, 29-41.

111. Bicho, A., Peca, I.N., Roque, A.C. and Cardoso, M.M. (2010) Anti-CD8 Conjugated Nanoparticles to Target Mammalian Cells Expressing CD8. International Journal of Pharmaceutics, 399, 80-86.

112. Cruz, L.J., Tacken, P.J., Fokkink, R., Joosten, B., Stuart, M.C., Albericio, F., et al. (2010) Targeted PLGA Nano- but Not Microparticles Specifically Deliver Antigen to Human Dendritic Cells via DC-SIGN in Vitro. Journal of Controlled Release, 144, 118-126.

113. Alshamsan, A., Haddadi, A., Hamdy, S., Samuel, J., El-Kadi, A.O., Uludag, H., et al. (2010) STAT3 Silencing in Dendritic Cells by siRNA Polyplexes Encapsulated in PLGA Nanoparticles for the Modulation of Anticancer Immune Response. Molecular Pharmaceutics, 7, 1643-1654.

114. Ghotbi, Z., Haddadi, A., Hamdy, S., Hung, R.W., Samuel, J. and Lavasanifar, A. (2011) Active Targeting of Dendritic Cells with Mannan-Decorated PLGA Nanoparticles. Journal of Drug Targeting, 19, 281-292.

115. Haddadi, A., Elamanchili, P., Lavasanifar, A., Das, S., Shapiro, J. and Samuel, J. (2008) Delivery of Rapamycin by PLGA Nanoparticles Enhances Its Suppressive Activity on Dendritic Cells. Journal of Biomedical Materials Research Part A, 84, 885-898.

116. Panyam, J., Williams, D., Dash, A., Leslie-Pelecky, D. and Labhasetwar, V. (2004) Solid-State Solubility Influences Encapsulation and Release of Hydrophobic Drugs from PLGA/PLA Nanoparticles. Journal of Pharmaceutical Sciences, 93, 1804-1814.

117. Panyam, J. and Labhasetwar, V. (2003) Dynamics of Endocytosis and Exocytosis of Poly(D,L-Lactide-co-Glycolide) Nanoparticles in Vascular Smooth Muscle Cells. Pharmaceutical Research, 20, 212-20.

118. Panyam, J., Zhou, W.Z., Prabha, S., Sahoo, S.K. and Labhasetwar, V. (2002) Rapid Endo-Lysosomal Escape of Poly- (DL-Lactide-co-Glycolide) Nanoparticles: Implications for Drug and Gene Delivery. FASEB Journal: Official Publication of the Federation of American Societies for Experimental Biology, 16, 1217-1226.

119. Vasir, J.K. and Labhasetwar, V. (2007)Biodegradable nanoparticles for cytosolic delivery of therapeutics. Advanced drug delivery reviews, 5, 718-728.

120. Zhang, N., Chittasupho, C., Duangrat, C., Siahaan, T.J. and Berkland, C. (2008) PLGA Nanoparticle-Peptide Conjugate Effectively Targets Intercellular Cell-Adhesion Molecule-1. Bioconjugate Chemistry, 19, 145-152.

121. Townsend, S.A., Evrony, G.D., Gu, F.X., Schulz, M.P., Brown Jr., R.H. and Langer, R. (2007) Tetanus Toxin C Fragment-Conjugated Nanoparticles for Targeted Drug Delivery to Neurons. Biomaterials, 28, 5176-51784.

122. Cheng, J., Teply, B.A., Sherifi, I., Sung, J., Luther, G., Gu, F.X., et al. (2007) Formulation of Functionalized PLGA-PEG Nanoparticles for in Vivo Targeted Drug Delivery. Biomaterials, 28, 869-876.

123. Dhar, S., Gu, F.X., Langer, R., Farokhzad, O.C. and Lippard, S.J. (2008) Targeted Delivery of Cisplatin to Prostate Cancer Cells by Aptamer Functionalized Pt(IV) Prodrug-PLGA-PEG Nanoparticles. Proceedings of the National Academy of Sciences of the United States of America, 105, 17356-17361.

124. Li, J., Feng, L., Fan, L., Zha, Y., Guo, L., Zhang, Q., et al. (2011) Targeting the Brain with PEG-PLGA Nanoparticles Modified with Phage-Displayed Peptides. Biomaterials, 32, 4943-4950.

125. Gryparis, E.C., Hatziapostolou, M., Papadimitriou, E. and Avgoustakis, K. (2007) Anticancer Activity of Cisplatin-Loaded PLGA-mPEG Nanoparticles on LNCaP Prostate Cancer Cells. European Journal of Pharmaceutics and Biopharmaceutics: Official Journal of Arbeitsgemeinschaft fur Pharmazeutische Verfahrenstechnik eV, 67, 1-8.

126. Wang, Z., Chui, W.K. and Ho, P.C. (2009) Design of a Multifunctional PLGA Nanoparticulate Drug Delivery System: Evaluation of Its Physicochemical Properties and Anticancer Activity to Malignant Cancer Cells. Pharmaceutical Research, 26, 1162-1171.

127. Danhier, F., Vroman, B., Lecouturier, N., Crokart, N., Pourcelle, V., Freichels, H., et al. (2009) Targeting of Tumor Endothelium by RGD-Grafted PLGA-Nanoparticles Loaded with Paclitaxel. Journal of Controlled Release, 140, 166-173.

128. Patil, Y., Sadhukha, T., Ma, L. and Panyam, J. (2009) Nanoparticle-Mediated Simultaneous and Targeted Delivery of Paclitaxel and Tariquidar Overcomes Tumor Drug Resistance. Journal of Controlled Release, 136, 21-29.

129. Farokhzad, O.C., Cheng, J., Teply, B.A., Sherifi, I., Jon, S., Kantoff, P.W., et al. (2006) Targeted Nanoparticle-Ap-tamer Bioconjugates for Cancer Chemotherapy in Vivo. Proceedings of the National Academy of Sciences of the United States of America, 103, 6315-6320.

130. Esmaeili, F., Ghahremani, M.H., Ostad, S.N., Atyabi, F., Seyedabadi, M., Malekshahi, M.R., et al. (2008) Folate-Receptor-Targeted Delivery of Docetaxel Nanoparticles Prepared by PLGA-PEG-Folate Conjugate. Journal of Drug Targeting, 16, 415-423.

131. Gu, F., Zhang, L., Teply, B.A., Mann, N., Wang, A., Radovic-Moreno, A.F., et al. (2008) Precise Engineering of Targeted Nanoparticles by Using Self-Assembled Biointegrated Block Copolymers. Proceedings of the National Academy of Sciences of the United States of America, 105, 2586-2591.

132. Bivas-Benita, M., Romeijn, S., Junginger, H.E. and Borchard, G. (2004) PLGA-PEI Nanoparticles for Gene Delivery to Pulmonary Epithelium. European Journal of Pharmaceutics and Biopharmaceutics: Official Journal of Arbeitsgemeinschaft fur Pharmazeutische Verfahrenstechnik eV, 58, 1-6.

133. Patil, Y.B., Swaminathan, S.K., Sadhukha, T., Ma, L. and Panyam, J. (2010) The Use of Nanoparticle-Mediated Targeted Gene Silencing

and Drug Delivery to Overcome Tumor Drug Resistance. Biomaterials, 31, 358-65.

134. Chen, H., Zheng, Y., Tian, G., Tian, Y., Zeng, X., Liu, G., et al. (2011) Oral Delivery of DMAB-Modified Docetaxel-Loaded PLGA-TPGS Nanoparticles for Cancer Chemotherapy. Nanoscale Research Letters, 6, 4.

135. Zhang, Z.P., Lee, S.H. and Feng, S.-S. (2007) Folate-Decorated Poly(Lactide-co-Glycolide)-Vitamin E TPGS Nano-particles for Targeted Drug Delivery. Biomaterials, 28, 1889-1899.

136. Ma, Y., Zheng, Y., Liu, K., Tian, G., Tian, Y., Xu, L., et al. (2010) Nanoparticles of Poly(Lactide-Co-Glycolide)-d-a-Tocopheryl Polyethylene Glycol 1000 Succinate Random Copolymer for Cancer Treatment. Nanoscale Research Letters, 5, 1161-1169.

137. Iwasaki, Y., Maie, H. and Akiyoshi, K. (2007) Cell-Specific Delivery of Polymeric Nanoparticles to Carbohydrate-Tagging Cells. Biomacromolecules, 8, 3162-3168.

138. Shenoy, D., Little, S., Langer, R. and Amiji, M. (2005) Poly(Ethylene Oxide)-Modified Poly(Beta-Amino Ester) Nanoparticles as a pH-Sensitive System for Tumor-Targeted Delivery of Hydrophobic Drugs: Part 2. In Vivo Distribution and Tumor Localization Studies. Pharmaceutical Research, 22, 2107-2114.

139. Green, J.J., Chiu, E., Leshchiner, E.S., Shi, J., Langer, R. and Anderson, D.G. (2007) Electrostatic Ligand Coatings of Nanoparticles Enable Ligand-Specific Gene Delivery to Human Primary Cells. Nano Letters, 7, 874-879.

140. Ding, H., Inoue, S., Ljubimov, A.V., Patil, R., Portilla-Arias, J., Hu, J., et al. (2010) Inhibition of Brain Tumor Growth by Intravenous Poly(β-L-Malic Acid) Nanobioconjugate with pH-Dependent Drug Release. Proceedings of the National Academy of Sciences of the United States of America, 107, 18143-18148.

141. Inoue, S., Ding, H., Portilla-Arias, J., Hu, J., Konda, B., Fujita, M., et al. (2011) Polymalic Provides Efficient Systemic Breast Cancer Treatment by Inhibiting Both HER2/Neu Receptor Synthesis and Activity. Cancer Research, 71, 1454-1464.

142. Liang, H.F., Chen, S.C., Chen, M.C., Lee, P.W., Chen, C.T. and Sung, H.W. (2006) Paclitaxel-Loaded Poly(Gamma-Glutamic Acid)-Poly(Lactide) Nanoparticles as a Targeted Drug Delivery System against Cultured HepG2 Cells. Bioconjugate Chemistry, 17, 291-299.

143. Lai, S.K., Hida, K., Chen, C. and Hanes, J. (2008) Characterization of the Intracellular Dynamics of a Non-Degradative Pathway Accessed by Polymer Nanoparticles. Journal of Controlled Release, 125, 107-111.

144. Bellocq, N.C., Pun, S.H., Jensen, G.S. and Davis, M.E. (2003) Transferrin-Containing, Cyclodextrin Polymer-Based Particles for Tumor-Targeted Gene Delivery. Bioconjugate Chemistry, 14, 1122-1132.

6

STUDY OF NANOSKIN ECM-BACTERIAL CELLULOSE WOUND HEALING/UNITED ARAB EMIRATES

Saqer Al Mualla[1], Raed Farahat[1], Pierre Basmaji[2], Gabriel Molina de Olyveira[3]*, Ligia Maria Manzine Costa[4], José Domingos da Costa Oliveira[2], Gino Bruno Francozo[2]

[1]Al Qassimi Hospital/Kuwait Hospital, Sharjah, United Arab Emirates

[2]Innovatec's-Biotechnology Research and Development, São Carlos, Brazil

[3]Department of Physical Chemistry, Universidade Estadual Paulista (UNESP), Araraquara, Brazil

[4]Materials Engineering Department, São Carlos School of Engineering, University of São Paulo, São Carlos, Brazil

ABSTRACT

Natural extracellular matrices (ECMs) perform the tasks necessary for tissue formation, maintenance, regulation and function, providing a powerful means of controlling the biological performance of regenerative materials. In addition, biomedical materials have claimed attention because of the increased interest in tissue engineering materials for wound care and regenerative medicine. Moreover, the nanostructure and morphological similarities with collagen make BC attractive for cell immobilization, cell support and Natural Extracellular Matrix (ECM) Scaffolds. In this work, we present the extracellular matrix (ECM) using the bacterial cellulose (Nanoskin®) which regulates cell behavior by influencing cell proliferation, survival, shape, migration and differentiation. Bacterial cellulose fermentation process is modified before the bacteria are inoculated for mimicking ECM to cells support and built new local material for wound healing. Chemical groups influences and thermal behavior in bacterial cellulose were analyzed using transmission infrared spectroscopy (FTIR) and thermogravimetric analysis (TGA), respectively. Besides, In vivo analysis was evaluated with clinical study at Sharjah Kuwait Hospital.

Keywords: Bacterial Cellulose (Nanoskin), Natural Nanocomposites, Regenerative Medicine, Stem Cells

1. INTRODUCTION

Tissue engineering is a recent field that creates functioning artificial tissues and organs. Major considerations in tissue engineering include both the type of cell and the substrate (scaffold) to be used. Many strategies use an artificial scaffold that functions as the ECM to facilitate both organization and differentiation of implanted cells into a functional 3D tissue [1] .

Wounds are injuries that result in an ulcer or break of the skin. Healing is a complex process in response to an injury to restores the integrity of wound. Wound healing involves cell-cell and cell matrix interactions to allow the process [2] [3] . Besides, wounds are followed by inflammatory processes with various mediators, such as eicosanoids, prostaglandins or cytokines [4] . NF-B is one central protein that regulates several inflammatory cytokines with IB, its inhibitory subunit. However, there are many factors which help NF-B activity break IB linkage. Because of this, to avoid inflammatory action, inhibition of NF-B activation is crucial to treat wounds inflammation [4] .

The extracellular matrix (ECM) contains an abundant variety of signals that are received by cell surface receptors and contribute to cell adhesion and cell fate, via regulation of cellular activities such as proliferation, migration and differentiation. As such, regenerative medicine studies often rely on mimicking the natural ECM to promote the formation of new tissue by host cells, and characterization of natural ECM components is vital for the development of new biomimetic approaches [5] [6] .

The failure of re-ephitelialization in a poorly wound healing is initially due a deficit in the epithelial cell itself, typically in slow to heal wounds like diabetic ulcers, however, lack of normal ECM is now recognized to be the problem as fibronectin and collagen are required for the cells to attach to the surface and migrate [7] - [11] .

In this scope, Natural ECM is the ideal biological scaffold since it contains all the components of the tissue. Constructive remodeling can be performed using such natural ECM scaffolds and vegetal/animal stem cells since, the cells can be delivered to the site of infraction and then cells help wound healing process. The development of niche mimicking biomaterials and hybrid biomaterial can further advance directed differentiation without specific induction [12] [13] .

Bacterial cellulose (BC) is a natural cellulose produced by bacterial synthesis by biochemical steps and self- assembling of the secreted cellulose fibrils on the medium [14] [15] . Shaping of BC materials in the culture medium can be controlled by the type of cultivation that changes chain size, origin of strains which produces different proportions of crystalline phase of BC and the kind of bioreactor. BC hydrogel or BC in dry state is then obtained by methods, such as freeze-drying [16] [17] . The structural features of microbial cellulose, its properties and compatibility as a biomaterial for regenerative medicine can be changed by modifying its culture medium [18] or surface modification by physical [19] [20] ; chemical methods [21] and genetic modifications [22] to obtain a biomaterial with less rejection when in contact to the body [23] [24] .

In this work, novel studies of natural nanocomposites with Bacterial cellulose (Nanoskin®) for functional materials are reported. In order to produce scaffolds with drug delivery ability, porous structure and better cell adhesion, fermentation changes in gel bacterial cellulose with chondroitin 4-sulfate, hyaluronic acid and vegetal stem cells were performed with its in vivo cell behavior.

2. MATERIALS AND METHODS

2.1. Materials

The Bacterial cellulose (Nanoskin®) raw material was provided from Innovatec's (São Carlos SP, Brazil). Chon- droitin 4-sulfate and hyaluronic acid sodium salt from Streptococcus equi (bacterial glycosaminoglycan polysaccharide) were purchased from Sigma Aldrich. Vegetal stem cells were obtained from Brazillian environment, Waltheria Douradinha.

2.2. Methods

2.2.1. Synthesis of Bacterial Cellulose and Bacterial Cellulose/Chondroitin Sulfate/Hyaluronic Acid

Bacterial cellulose (BC) is produced by Gram-negative bacteria Gluconacetobacter xylinus, which can be obtained consisting of an ultra fine network of cellulose nanofibers [19] . BC hydrogel or BC in dry state is then obtained by freeze-drying. The acetic fermentation process is achieved by using glucose as a carbohydrate source. Results of this process are vinegar and a nanobiocellulose biomass. The modified process is based on the addition of chondroitin 4-sulfate (1% w/w) and hyaluronic acid (1% w/w) to the culture medium (green tea) before the bacteria are inoculated.

2.2.2. Bionanocomposite Preparation

In the present study, a novel biomaterial has been explored and different bacterial cellulose nanocomposites have been prepared; BC/chondroitin 4-sulfate/hyaluronic acid. Samples were washed and it's medium was changed with cells culture medium as illustrated in Figure 1.

2.3. Vegetal Stem Cells

The material of the plant of interest is collected and induced damage to causing the formation of scar tissue called callus. This tissue consists of totipotent cells, undifferentiated (stem cells) are collected and grown on agar plates to complete differentiation and generation of a homogeneous culture (2 - 10 days).

Cultures of these stem cells are grown in bioreactors and the batch is collected after all the sugar was metabolized. The cells are washed and homogenized to release secondary metabolites. Soluble metabolites in oil and water are collected and, if you need the isomalt-based spraying is performed.

2.4. Characterization

Transmission infrared spectroscopy (FTIR, Perkin Elmer Spectrum 1000)- Influences of hyaluronic acid (HA) and chondroitin 4-sulfate (CS) in bacterial cellulose were analyzed in the range between 250 and 4000 cm^{-1} and with 2 cm^{-1} resolution with samples.

Thermogravimetric analysis (TGA) was carried out for biocomposites using a NETZSCH TG 209F1 in oxygen environment, with a heating rate of 10C/min. The temperature range scanned was from 25 celsius degree to

650 celsius degree. The weight of all specimens was maintained around 10 mg.

In vivo analysis-Evaluation-Clinical study at Sharjah Kuwait Hospital under supervision of Dr. Saqer Al Mualla and Dr. Raed Farahat. Evaluation model-Crushed injury with amputation of the index, Middle and ring of the right hand during work with skin loss since 5 days and all wounds shows necrotic tissue with black parts, bad offensive odor and small amount of pus.

3. RESULTS AND DISCUSSION

3.1. FTIR-Interaction between Bacterial Cellulose with Hyaluronic Acid and Chondroitin 4-Sulfate

Influences of chondroitin 4-sulfate (CS) and hyaluronic acid in bacterial cellulose (BC) were analyzed by ATR- FTIR in the range of 4000 - 2400 cm^{-1} (Figure 2) with resolution of 2 cm^{-1}. The main features of the bacterial cellulose in mid-infrared spectroscopy are at 3349 cm^{-1} (sharp band; O-H stretching vibration associated to intramolecular hydrogen bonding), 3240 cm^{-1} (medium band; OH Stretching characteristic of hydroxyl

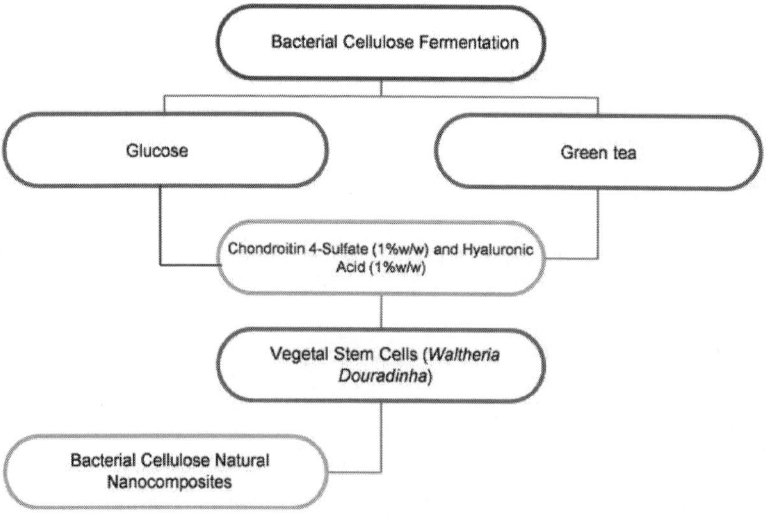

Figure 1. Materials and methods draft.

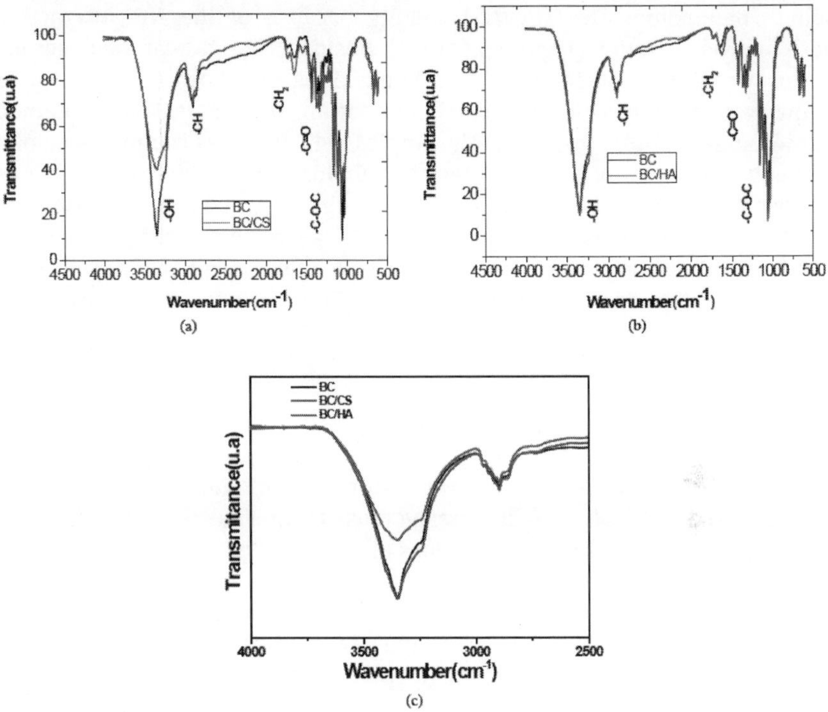

Figure 2. FTIR spectra of bacterial cellulose nanocomposites.

bonding in cellulose structure due to Iα crystalline phase), 2896 cm^{-1} (C-H stretching vibration of alkane and asymmetric CH_2 stretching), 2853 cm^{-1} (CH_2 symmetric stretching), 1732 cm^{-1}(stretching vibration of saturated aliphatic aldehyde carbonyl of BC), 1640 cm^{-1} (bending mode of absorbed water), 1428 cm^{-1} (CH_2 or O-H in plane bending), 1373 cm^{-1} (C-H bending), 1340 cm^{-1}(OH in-plane bending), 1237 cm^{-1} (O-H bending), 1203 cm^{-1} (C-O-H in plane bending at C-6), 1163 cm^{-1}(strong band; antisymmetric bridge C-O-C strentching), 1110 cm^{-1} (strong band; antisymmetric in-phase ring stretching), 1065 cm^{-1} (strong band; skeletal vibrations involving C-O stretching at C-3), 1035 cm^{-1} (strong band; skeletal vibrations involving C-O stretching at C-6),1015 cm^{-1} (medium band; skeletal vibrations involving C-O stretching), 897 cm^{-1} (b-glucosidic linkages between the glucose units) and 748 cm^{-1} (OH stretching characteristic of hydroxyl bonding in cellulose structure due to Iα crystalline phase) [23] - [26] .

It can be observed from Figures 2(a)-(c), the transmittance intensity is different of bacterial cellulose and bacterial cellulose nanocomposites, which means the exposed groups are interacting with bacterial cellulose components. Similar OH stretching (at 2900 cm^{-1}) can be observed in bacterial cellulose/hyaluronic nanocomposites (BC/HA) and chondroitin 4-

sulfate nanocomposites (BC/CS), mainly because of the NH_2 interaction with hydroxyl groups (Figures 2(a)-(c)). Besides, changes can be observed in the symmetrical stretching of CH_2 bonds of bacterial cellulose structures at the absorption peak of 1640 cm^{-1}. Another absorption peak was obtained in the range of 1490 cm^{-1} on both samples, which shows the presence of a carbonyl group in the bacterial cellulose together with bonds corresponding to those of glycoside, including C-O-C at 1162 cm^{-1}(as in the case of natural cellulose) [23] - [26] . These results clearly show one possible interaction between bacterial cellulose and chondroitin 4-sulfate/hyaluronic acid, mainly by hydrogen interactions between hydroxyl and carbonyl groups.

3.2. TGA

In order to analyze thermal behavior for bionanocomposites are characterized typical weight loss versus temperature plots. The TG spectrum (Figure 3) shows a weak loss of weight due to the evaporation of water (at temp. 85 celsius) and also quick drop in weight at a temperature of approx. 300 celsius is mainly attributed to thermal depolymerization of cellulose and the cleavage of glycosidic linkages of cellulose [27] [28] , complete degradation of cellulose take place between 275 and 400 celsius [29] [30] .

All system has similar thermal behavior in bacterial cellulose showed significant alterations. A carbonaceous residue was similar in BC membranes, around 0% at 600 celsius, however sample with hyaluronic acid has little differences in thermal behavior than tested others mainly because there is higher hydrogen bond between bacterial cellulose groups (hydroxyl) and hyaluronic acid (acetyl) which changes bacterial cellulose fibers formation and thermal properties (Figure 4).

3.3. In Vivo Analysis

Patient enter in Hospital on 12/10/2014 under supervision of Dr. Saqer Al Mualla Consultant Head of Plastic Surgeons Dept' and Dr. Raed Farahat Plastic Surgeon Specialist, the case planed for skin graft after getting good clean granulation tissue. It was performed immediate intervention with antibiotics and clinical protocols for this disease. In 12/15/2014, started treatment with Bacterial cellulose membranes (Nanoskin®), one membrane of Nanoskin applied to the wound after removal of all air bubbles under dry gauze used as secondary dressing and light pressure bandage over (Figure 5).

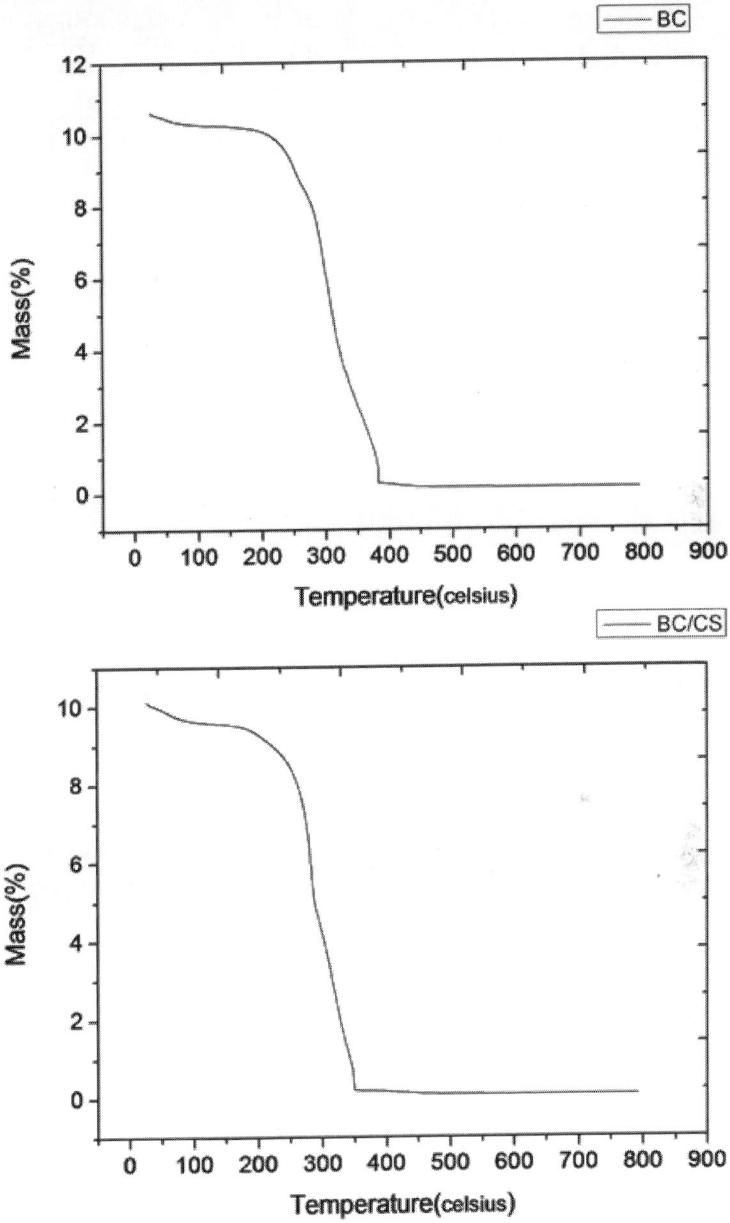

Figure 3. TGA thermogram of bacterial cellulose/chon- droitin 4-sulfate nanocomposites samples.

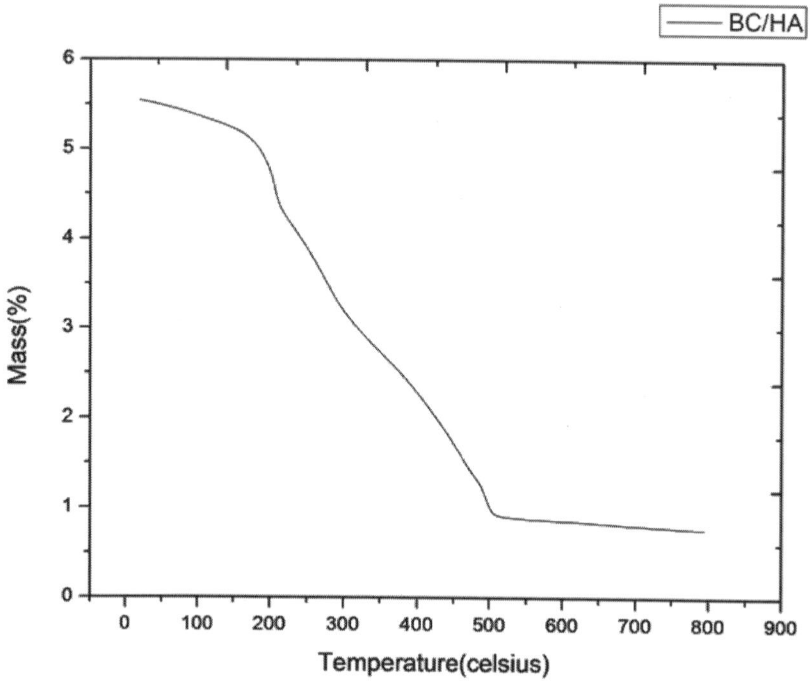

Figure 4. TGA thermogram of bacterial cellulose/hyaluronic acid nanocomposites samples.

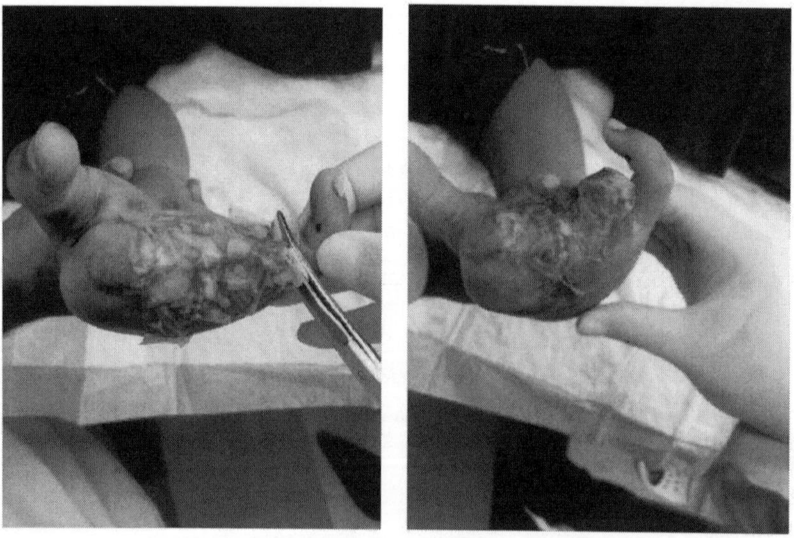

Figure 5. Bacterial cellulose (Nanoskin®) uses like biological wound dressing.

After using the Bacterial cellulose material in alternate days, it can be observed a recovery of the edge and bottom of the wound, however, after 1st week the wound was better and with necrotic, so, debridement was done and applied Nanoskin (Figure 6).

After, in 2nd week, wound getting better with minimal exudate and better wound bed. The edge looks health and normal health margin, healthy granulation tissue and no odor and decrease the size of wound bed (Figure 7).

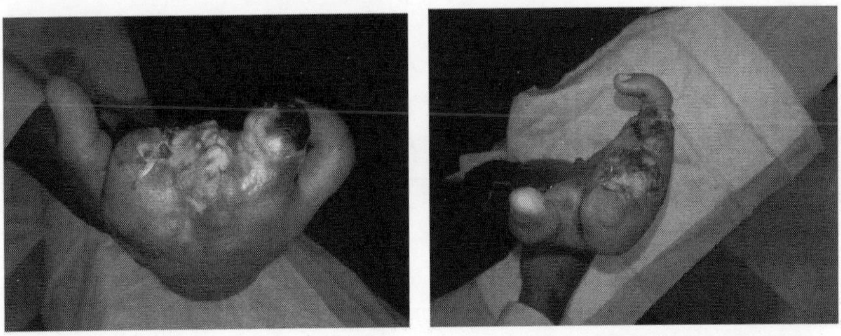

Figure 6. Wound healing evolution in 1weeks with bacterial cellulose (Nanoskin®).

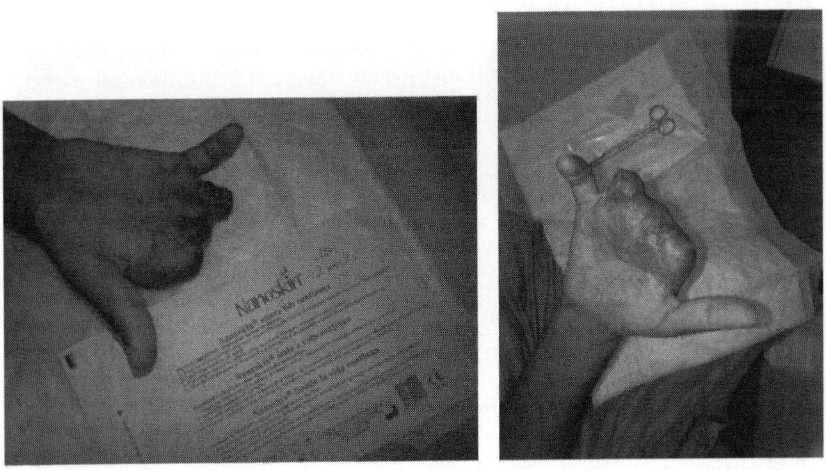

Figure 7. Wound healing evolution in 2 weeks with bacterial cellulose (Nanoskin®).

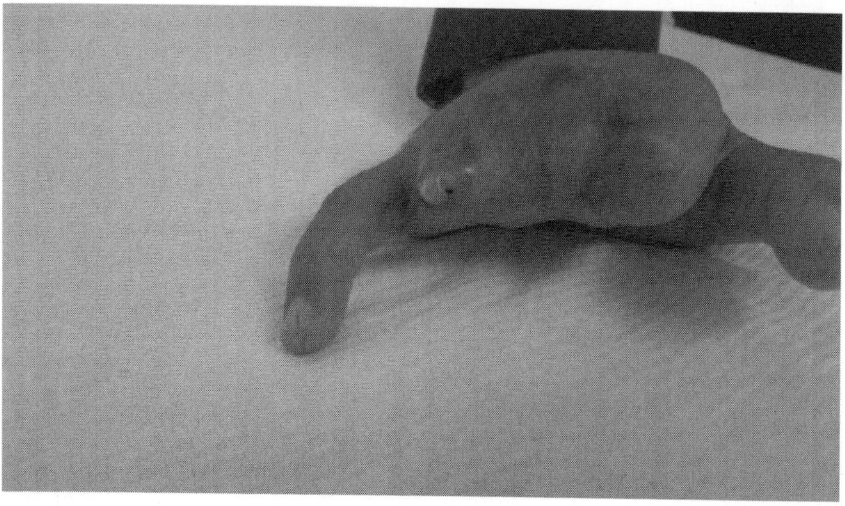

Figure 8. Completed wound healing in 19 days with bacterial cellulose (Nanoskin®).

The simple application of dressing, only required the association of saline, gauze and bandage, decrease patient stay and operating room use, resulting in a better cost-benefit. All wounds gets better with Nanoskin® dressing, dramatic changes was noted after application of Nanoskin®dressing, patient rarely complain of pain. Healing happened by normal health tissue with less fibrous tissue, normal skin and no pigmentation, besides fast epithelization was observed. The wound healed in 19 days, then the wound coming better even Ring of the hand healed completely after 5 weeks (Figure 8).

4. CONCLUSIONS

Bacterial cellulose (Nanoskin®) was successfully modified by changing the fermentation medium as shown by FTIR and TGA, which produced suitable scaffolds for use in surface morphology applications with promising cell viability/attachment. Nanoskin Natural extracellular matrix (ECMs) perform the tasks necessary for tissue formation, maintenance, regulation and function, providing a powerful means of controlling the biological performance of regenerative materials. However, understanding how cells interact with these to assemble their own ECM and how the scaffolds can be used to control delivery of signals in a temporal and spatial manner to guide or maintain cell differentiations need future investigation. But, undoubtedly, natural-origin polymers or nature- inspired materials appear as the natural and desired choice for medical applications.

In conclusion, Bacterial cellulose membrane (Nanoskin®) applies to the protective surface and sutures, in exudate lesions, promoting the healing process with no rejection body, besides decreasing recovery time and the treatment cost.

CITE THIS PAPER

Saqer Al Mualla, Raed Farahat, Pierre Basmaji, Gabriel Molina de Olyveira, Ligia Maria Manzine Costa, José Domingos da Costa Oliveira, Gino Bruno Francozo (2016) Study of Nanoskin ECM-Bacterial Cellulose Wound Healing/United Arab Emirates. *Journal of Biomaterials and Nanobiotechnology*,**07**,109-117.

REFERENCES

1. Thakur, R., Jain, N., Pathak, R. and Sandhu, S.S. (2011) Practices in Wound Healing Studies of Plants. Evidence-Based Complementary and Alternative Medicine, 2011, Article ID: 438056, 1-17.
2. Basmaji, P., de Olyveira, G.M., Manzine Costa, L.M., Francozo, G.B. and da Costa Oliveira, J.D. (2015) Nanoderm Extracellular Matrix for Reconstructive Surgery Applications. Peertechz Journal of Biomedical Engineering, 1, 021-024.
3. Basmaji, P., Ambrizzi, D.R., Olyveira,G.M., Costa,L.M.M., Francozo, G.B., Oliveira, J.D.C.(2016) Natural Extracellular Matrix Modified for Tissue Engineering Applications. Advanced Science, Engineering and Medicine, 8, 62-67.
4. Basmaji, P., Olyveira, G.M., Costa, L.M.M., Souza, J.A.M., Lucchesi, G.L., Amariz, G.A.S., Affonso, M.E.R.M., Amaral, N.C.P., Oliveira, J.D.C. and Francozo, G.B. (2015) Natural ECM-Biological Wound Dressing. Aperito Journal of Nanoscience Technology, 2, 105-110. [Paper reference:2]
5. Olyveira, G.M., Santos, M.L., Daltro, P.B., Basmaji, P., Daltro, G.C. and Guastaldi, A.C. (2014) Bacterial Cellulose/ Chondroitin Sulfate for Dental Materials Scaffolds. Journal of Biomaterials and Tissue Engineering, 4, 150-154.
6. Olyveira, GM., Santos, M.L., Costa, L.M.M., Daltro, P.B., Basmaji, P., Daltro, G.C. and Guastaldi, A.C. (2014) Bacterial Cellulose Nanobiocomposites for Dental Materials Scaffolds. Journal of Biomaterials and Tissue Engineering, 4, 536-542.

7. Olyveira, G.M., Santos, M.L., Costa, L.M.M., Daltro, P.B., Basmaji, P., Daltro, G.C. and Guastaldi, A.C. (2014) Bacterial Biocomposites for Guided Tissue Regeneration. Science of Advanced Materials, 6, 2673-2678.

8. Olyveira, G.M., Santos, M.L., Costa, L.M.M., Daltro, P.B., Basmaji, P., Daltro, G.C. and Guastaldi, A.C. (2015) Physically Modified Bacterial Cellulose Biocomposites for Guided Tissue Regeneration. Science of Advanced Materials, 7, 1657-1664.

9. Olyveira, G.M., Santos, M.L., Costa, L.M.M., Riccardi, C.S., Daltro, P.B., Basmaji, P., Daltro, G.C. and Guastaldi, A.C. (2015) Physically Modified Bacterial Cellulose Biocomposites for Dental Materials Scaffolds. Materials Focus, 4, 111-117.

10. Olyveira, G.M., Acasigua, G.A.X., Costa, L.M.M., Scher, C.R., Filho, L.X., Pranke, P.H.L. and Basmaji, P. (2013) Human Dental Pulp Stem Cell Behavior Using Natural Nanotolith/Bacterial Cellulose Scaffolds for Regenerative Medicine. Journal of Biomedical Nanotechnology, 9, 1370-1377.

11. Acasigua, G.A.X., Olyveira, G.M., Costa, L.M.M., Braghirolli, D.A., Guastaldi, A.C., Pranke, P. and Basmaji, P. (2014) Novel Natural Bacterial Cellulose Nanocomposites as Potential Biomaterial for Stem Cell Therapy. Current Stem Cell Research & Therapy, 9, 117-123.

12. El-Hoseny, S.M., Basmaji, P., Olyveira, G.M., Costa, L.M.M., Alwahedi, A.M., Oliveira, J.D.C. and Francozo, G.B. (2015) Natural ECM-Bacterial Cellulose Wound Healing—Dubai Study. Journal of Biomaterials and Nanobiotechnology, 6, 237-246.

13. Faria, E., Blanes, L., Hochman, B., Filho, M.M. and Ferreira, L. (2011) Health-Related Quality of Life, Self-Esteem, and Functional Status of Patients with Leg Ulcers. Wounds, 23, 4-10.

14. Siddiqui, A.R. and Bernstein, J.M. (2010) Chronic Wound Infection: Facts and Controversies. Clinics in Dermatology, 28, 519-526.

15. Han, A., Zenilman, J.M., Melendez, J.H., Shirtliff, M.E., Agostinho, A., James, G., Stewart, P.S., Mongodin, E.F., Rao, D., Rickard, A.H. and Lazarus, G.S. (2011) The Importance of a Multi-Faceted Approach to Characterizing the Microbial Flora of Chronic Wounds. Wound Repair and Regeneration, 19, 532-541.

16. Costa, L.M.M., Olyveira, G.M., Basmaji, P., Valido, D.P., Góis, P.B.P., Júnior, R.L.A.C. and Filho, L.X. (2012) Novel Otoliths/Bacterial Cellulose Nanocomposites as a Potential Natural Product for Direct Dental Pulp Capping. Journal of Biomaterials and Tissue Engineering, 2, 48-53.

17. de Olyveira, G.M., Luiz dos Santos, M., dos Santos Riccardi, C., Costa, L.M.M., Daltro, P.B., Basmaji, P., de Cerqueira Daltro, G. and Guastaldi, A.C. (2015) Bacterial Cellulose Biocomposites for Periodontology Treatment. Advanced Science, Engineering and Medicine, 7, 409-414.

18. Costa, L.M.M., Olyveira, G.M., Basmaji, P. and Filho, L.X. (2011) Bacterial Cellulose towards Functional Green Composites Materials. Journal of Bionanoscience, 5, 167-172.

19. Costa, L.M.M., Olyveira, G.M., Basmaji, P. and Filho, L.X. (2012) Bacterial Cellulose towards Functional Medical Materials. Journal of Biomaterials and Tissue Engineering, 2, 185-196.

20. Molina de Olyveira, G., Maria Manzine Costa, L. and Basmaji, P. (2013) Physically Modified Bacterial Cellulose as Alternative Routes for Transdermal Drug Delivery. Journal of Biomaterials and Tissue Engineering, 3, 227-232.

21. Olyveira, G.M., Riccardi, C.S., Santos, M.L., Costa, L.M.M., Daltro, P.B., Basmaji, P., Daltro, G.C. and Guastaldi, A.C. (2014) Bacterial Cellulose Nanobiocomposites for Periodontal Disease. Journal of Bionanoscience, 8, 319-324.

22. Gois, P.B.P., Olyveira, G.M., Costa, L.M.M., Chianca, C.F., Fraga, I.I.S., Basmaji, P., Cordoba, C.V. and Filho, L.X. (2013) Influence of Symbioses Culture between Microorganisms/Yeast Strain on Cellulose Synthesis. International Review of Biophysical Chemistry, 3, 48-54.

23. Filho, L.X., OLyveira, G.M., Costa, L.M.M. and Basmaji, P. (2013) Novel Electrospun Nanotholits/PHB Scaffolds for Bone Tissue Regeneration. Journal of Nanoscience and Nanotechnology, 13, 4715-4719.

24. Olyveira, G.M., Costa, L.M.M., Basmaji, P. and Filho, L.X. (2011) Bacterial Nanocellulose for Medicine Regenerative. Journal of Nanotechnology in Engineering and Medicine, 2, 34001-34009.

25. Salmen, L., Akerholm, M. and Hinterstoisser, B. (2005) Two-Dimensional Fourier Transform Infrared Spectroscopy Applied to Cellulose and Paper. In: Dumitriu, S., Ed., Polysaccharides: Structural Diversity and Functional Versatility, 2nd Edition, Marcel Dekker, New York, 159-187.

26. Kondo, T. (1998) Hydrogen Bonds in Cellulose and Cellulose Derivatives. In: Dumitriu, S., Ed., Polysaccharides: Structural Diversity and Functional Versatility, Marcel Dekker, New York, 131-172.

27. Manfredi, L.B., Rodriguez, E.S., Wladyka-Przybylak, M. and Vazquez, A. (2006) Thermal Degradation and Fire Resistance of Unsaturated Polyester Modified Acrylic Resins and Their Composites with Natural Fibers. Polymer Degradation and Stability, 91, 255-261.

28. Ouajai, S. and Shanks, R.A. (2006) Solvent and Enzyme Induced Recrystallization of Mechanically Degraded Hemp Cellulose. Cellulose, 13, 31-44.

29. Alvarez, V.A., Ruseckaite, R.A. and Vazquez, A. (2006) Degradation of Sisal Fibre/Mater Bi-Y Biocomposites Buried in Soil. Polymer Degradation and Stability, 91, 3156-3162.

30. Deepa, B., Abraham, E., Cherian, B.M., Bismarc, A., Blaker, J.J. and Pothan, L.A. (2011) Structure, Morphology and Thermal Characteristics of Banana Nano Fibers Obtained by Steam Explosion. Bioresource Technology, 102, 1988-1997.

7

APPLICATION POTENTIAL OF BONE MARROW MESENCHYMAL STEM CELL (BMSCS) BASED TISSUE-ENGINEERING FOR SPINAL CORD DEFECT REPAIR IN RAT FETUSES WITH SPINA BIFIDA APERTA

Xiaoshuai Li[1] , Zhengwei Yuan[1], Xiaowei Wei[1] , Hui Li[1] , Guifeng Zhao[1] , Jiaoning Miao[1] , Di Wu[1] , Bo Liu[1] , Songying Cao[1] , Dong An[1] , Wei Ma[1] , Henan Zhang[1] , Weilin Wang[2] , Qiushi Wang[3] , Hui Gu[1]

[1]Key Laboratory of Health Ministry for Congenital Malformation, Shengjing Hospital, China Medical University, No.36, Sanhao Street, Heping District, Shenyang 110004, China

[2]Department of Pediatric Surgery, Shengjing Hospital, China Medical University, Shenyang, China

[3]Department of Blood Transfusion, Shengjing Hospital, China Medical University, Shenyang, China

ABSTRACT

Spina bifida aperta are complex congenital malformations resulting from failure of fusion in the spinal neural tube during embryogenesis. Despite surgical repair of the defect, most patients who survive with spina bifida aperta have a multiple system handicap due to neuron deficiency of the defective spinal cord. Tissue engineering has emerged as a novel treatment for replacement of lost tissue. This study evaluated the prenatal surgical approach of transplanting a chitosan–gelatin scaffold seeded with bone marrow mesenchymal stem cells (BMSCs) in the healing the defective spinal cord of rat fetuses with retinoic acid induced spina bifida aperta. Scaffold characterisation revealed the porous structure, organic and amorphous content. This biomaterial promoted the adhesion, spreading and in vitro viability of the BMSCs. After transplantation of the scaffold combined with BMSCs, the defective region of spinal cord in rat fetuses with spina bifida aperta at E20 decreased obviously under stereomicroscopy, and the skin defect almost closed in many fetuses. The transplanted BMSCs in chitosan–gelatin scaffold survived, grew and expressed markers of neural stem cells and neurons in the defective spinal cord. In addition, the biomaterial presented high biocompatibility and slow biodegradation in vivo. In conclusion, prenatal transplantation of the scaffold combined with BMSCs could treat spinal cord defect in fetuses with spina bifida aperta by the regeneration of neurons and repairmen of defective region.

1 INTRODUCTION

Neural tube defects (NTDs) are common congenital malformation. Spina bifida and anencephaly are the most common and severe forms of NTD affecting about 1 in 2000 live births worldwide [1]. In China, approximately 80,000–100,000 new-borns each year are diagnosed to have NTDs [2]. So far, Neural tube defects treatment has no breakthrough progress, in addition, Hydrocephalus, lower limb dysfunction and urinary incontinence is still serious postoperative complications [3]. So it is important to give appropriate treatment at early stage of gestational period to minimize continuous damages of the exposed spinal cord.

Fetal cellular therapy is a treatment option for a variety of birth defects, and it has been employed in the treatment of congenital haematologic disorders and immunodeficiency disease [4–6], and also in experimental foetal lamb model of myelomeningocele (MMC) [7]. We have previously shown that the deficiency of sensory and motor neuron is a primary anomaly in spinal cord of spina bifida aperta, which suggests that neuron replacement therapy based on neuron regeneration or cell replacement of the defective spinal cord is a promising approach to achieve a better functional outcome in spina bifida [8, 9]. Bone marrow-derived mesenchymal stem cells (BMSCs) have the capacity to self-renew, easy isolation, implanted less adverse reaction and multipotent differentiation potential such as skeletal muscle, lung, vascular, neuron, astrocyte, bone, intestinal and liver cells in response to different factors [10, 11]. A large number of studies have shown that BMSCs can trans-differentiated into neurons in vitro and in vivo [12–16]. In recent studies, BMSCs have also shown healing capability by preventing fibrosis and improving angiogenesis, which could have a role in tissue repair and tissue regeneration [17]. Therefore, we successfully established a new approach to inject BMSCs into defective spinal cord for the potential treatment of spina bifida aperta. Our data indicated that BMSCs survived, migrated, and differentiated into neurons in the spinal cord [18]. In addition, injected BMSCs expressed and induced the surrounding spinal tissue to express neurotrophic factors [19]. Our results suggest that prenatal BMSCs transplantation can be used to treat spinal neuron deficiency in NTDs and might serve as a potential treatment option for other congenital anomalies. However, due to the limited injection space, the amount of surviving BMSCs was not enough to reconstruct the defective spinal cord. In this context, tissue engineering has emerged as an effective alternative therapy to repair the spinal cord defects. Tissue engineering is an interdisciplinary field involving cell biology and materials science that works toward the development of biological substitutes to restore, maintain or improve tissue function [20, 21]. The basic principle of tissue engineering involves seeding cells onto a biodegradable scaffold to generate new tissue [22–24]. The choice of the biomaterial is critical for the success of such an approach in spinal cord repair. The ideal scaffold should possess suitable mechanical properties, biocompatibility, biodegradability, three-dimensional structure, and promote cell growth. A biocompatible material does not cause toxic or injurious effects on biological systems [25]. Because of its acceptable biodegradability [26], high biocompatibility [27], chemical similarity to the structure of extracellular matrix (ECM), anti-microbial activity, [28] and capacity to produce porous scaffolds, [29] chitosan has been widely studied as a biomaterial for tissue-engineering applications [30–32]. Its association with gelatin, a source of collagen type I, enhances the mechanical and cell adhesion in culture, due to the affinity of the cells for the adhesion proteins [33–35]. Chitosan has also other useful biological properties such as hemostasis and acesodyne activity, wound healing property, permeability to oxygen, reducing scars, antifungal and bacteriostasis character which make it important as a dermal substitute, and has similar structure to glycosaminoglycans (GAGs) to help wound healing [36,37].The chitosan/gelatin/PAG scaffold as a suitable material having proper conductivity, mechanical properties and biocompatibility that may be appropriate for different biomedical applications such as scaffold material in tissue engineering for neural repair [38]. Tissue engineering materials could potentially lead to new therapeutic strategies for nervous tissue injuries as well as provide novel investigative tools for biological research [39]. In a word, tissue engineering are changing the way we think about how we might treat patients born with serious congenital malformations [40].

Therefore, in present study, we proposed a strategy for tissue engineering, by the implantation of a chitosan–gelatin scaffold seeded with labeled eGFP-BMSCs in fetal rats with spina bifida by the combined techniques of microsurgery and fetal surgery.

2 MATERIALS AND METHODS

2.1 Experiment animals

Outbred Wister rats of 10–12 weeks of age (250–300 gm) were purchased from the animal centre of China Medical University. The appearance of vaginal plugs in the female rat the morning after mating was timed as the embryonic day 0 (E0). Spina bifida aperta were induced with a single intragastric retinoic acid (140 mg/kg body weight; Sigma) administration on E10 as previously described [9, 18, 19, 41]. All the animal experiments were performed with the approval obtained from the ethics committee of China Medical University.

2.2 Isolation, culture expansion and transfection of BMSCs

BMSCs were isolated from 4-week-old Wister rat following previously published protocol [14,19]. Rat BMSCs were cultured in DMEM/F12 (Gibco) supplemented with 10 % fetal bovine serum (FBS; Hyclone) and 100 IU/ml penicillin to 100 μg/ml streptomycin (Gibco) on 25 cm^2tissue culture flasks (BD Biosciences). Primarily isolated BMSCs were defined as P0, at confluency, cells were passaged (1 in 2 dilution) with fresh medium. Cultured BMSCs expressed CD90, CD44, CD73 and CD29 but not CD34 and CD45 as revealed by flow cytometry using specific antibodies (551401, 550974, 551123, 562154, 560932, 554878; BD Biosciences) following previously published methods. After primary antibodies incubation, sections were washed three times with PBS followed by incubation with Alexa Fluor 488-conjugated goat anti-rabbit IgG antibody (Invitrogen) and TRITC-conjugatedgoat anti-mouse IgG (AP124R; Millipore) in 10 % FBS-PBST for 1 h at room temperature. After washing, sections were stained with DAPI (C1002; Beyotime Institute of Biotechnology), then mounted with anti-fade mounting medium (P0126; Beyotime Institute of Biotechnology). The expression of CD90 was used to estimate the purity of BMSCs [36, 42]. Twenty-four hours before transplantation, BMSCs were transfected with eGFP expression adeno-5 vector (SinoGenoMax.Co., Ltd, Beijing, China) (100 pfu/1 cell), for the visualization of BMSCs after transplantation into rat fetus.

2.3 Preparation and characterization of the chitosan-gelatin scaffolds and cell-scaffold construct

2.3.1 Preparation of chitosan–gelatin scaffolds

Chitosan–gelatin scaffolds were obtained via a freeze-drying technique and genipin (Sigma) cross-linking as previously described [43]. In this method, the solvent was removed by sublimation, and the corresponding empty spaces became the pores of the scaffold. The scaffolds were produced from natural polymers, chitosan (Sigma) with a degree of de-acetylation of 85 % and type A porcine skin gelatin (Vetec,RJ, Brazil). Chitosan 0.7 % (w/v) solution and gelatin 0.7 % (w/v)solution were dissolved in 0.1 M acetic acid (Vetec, RJ, Brazil) separately. These two distinct solutions rested for 24 h at room temperature, and they were then mixed at a 3:1 (chitosan/gelatin) ratio. Subsequently, the chitosan–gelatin blend was cross-linked with 25 % glutaraldehyde solution at a 0.1 % concentration. The final solution was under agitation for 1 h. One millilitre of the homogenous mixture of chitosan, gelatin and glutaraldehyde was poured into each well of cell culture plate (24-well), used as a template to mould cylindrical standard discs of chitosan–gelatin sponges. After overnight air drying, the plates were incubated at 37 °C for 24 h. The plates were then frozen at −20 °C for an additional 24 h, and the scaffolds were lyophilised. Next, the lyophilized scaffolds were resuspended in a 25 % glutaraldehyde solution at a 0.1 % concentration (1 mL/well), and the plates were shaken for 30 min. The procedure (heat, freezing and lyophilisation) was then repeated. Finally, 1 mL/well of 100 % alcohol was added, and the samples were left to completely dry. Eight-millimetre diameter discs, 2.5-mm thick, with a spongy texture, were cut into 4 parts. The standard-size scaffolds, slices of 4 mm in radius (Fig. 1a), were used in vitro and in vivo procedures. The samples were placed at the bottom of a 24-well cell culture plates, sealed and sterilized at Nuclear Development Technology Centre/Nuclear Energy National Commission (CDTN/CNEN) by γirradiation, ^{60}Co, 20 Grays.

2.3.2 Characterization of swelling degree and surface density

The scaffold was dried to constant weight, the length and width was measured by micrometer and the thickness was measured by spiral micrometer and finally calculate the volume as V_{dry}, the weight was measured by photoelectric analytical balance as W_{dry}, and then be placing in the PBS buffer solution for 48 h to fully swelling, using filter paper to absorb surface moisture and rapidly weighing as Wwet, the formula of swelling rate and apparent density calculation is as follows:

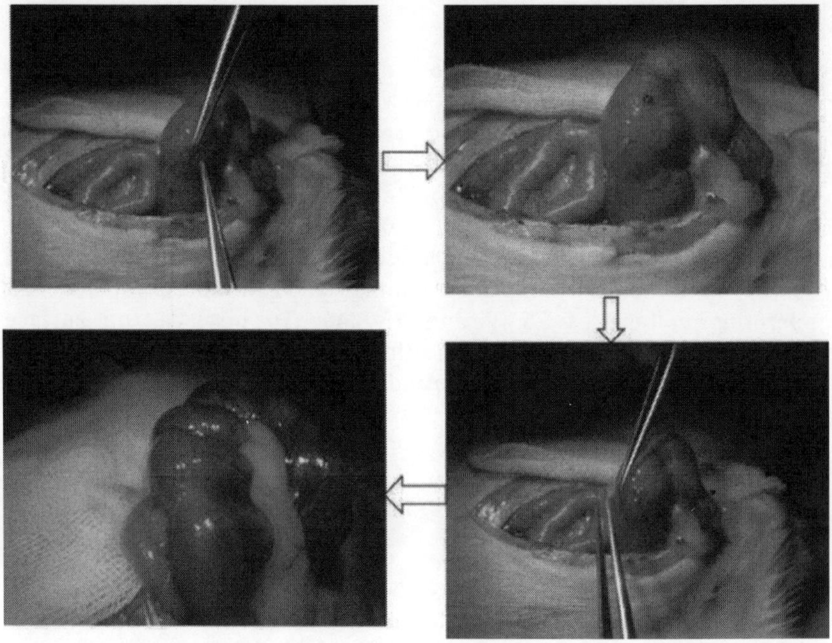

Figure 1. Operation process of fetal surgery and implantation. The *black arrow* indicate the cell-scaffold construct

$$SW_{swellingrate} = W_{wet} - W_{dry} / W_{dry} \times 100\ \% \tag{1}$$

$$\rho_{apparent-density} = W_{dry} / V_{dry} \times 100\ \% \tag{2}$$

$$V_{dry} = D_{dry\_thickness} \times L_{dry\_length} \times W_{dry\_width} \tag{3}$$

2.3.3 Measurement of porosity

Four scaffolds were randomly selected, soaking in anhydrous alcohol, until they were filled out and using filter paper to absorb the alcohol on the surface and rapid weighing, measuring three times to get the average value is denoted as Q. The density of anhydrous ethanol denoted as P_{eth}, the calculation formula of porosity is as follows:

$$P_d = (Q - W_{dry}) / \rho_{eth} \cdot V_{dry} \times 100\ \% \tag{4}$$

2.3.4 Measurement of aperture

The morphological characterization of scaffolds was performed by scanning electronic microscopic (SEM) analysis. The samples of chitosan–gelatin scaffolds and cell-scaffold constructs were fixed in 2.5 % glutaraldehyde (0.1 mol/L phosphate buffer—PBS, pH 7.4)for 48 h and post fixed with 1 % osmium tetroxide for 2 h. After dehydration, the critical-point dried in liquid CO_2 (CPD-020 Balzers) was performed. The samples were mounted on metallic holders, sputter-coated with gold and observed in a Zeiss DSM 950 scanning electron microscope, at an accelerating voltage of 15 kV and 750 mA. To observe the surface morphology of chitosan–gelatin scaffolds on the console and image acquisition, selecting 20 apertures randomly and calculates the size, and calculate the average value.

2.3.5 Seeding and culture of BMSCs in chitosan–gelatin scaffold

Before seeding, cells were trypsinized, centrifuged, resuspended in small aliquot of fresh medium, GFP positive cell numbers were counted. By the method of microinjection, the cells are more evenly distributed inside the scaffold. And then they were plated in 24-well culture plates (Nunc) in DMEM/F12 (Gibco) supplemented with 10 % fetal bovine serum. Before transplantation, they were incubated for 48 h at 37 °C with 5 % CO_2.

2.3.6 Proliferation assay

To evaluate the viability of cells in the scaffold. The scaffold seeded with BMSCs was cultured in DMEM/F12 (Gibco) supplemented with 10 % fetal bovine serum (FBS; Hyclone) and 100 IU/ml penicillin to 100 μg/ml streptomycin (Gibco), and medium changes everyday. After 1 and 2 week, the scaffold was stained by fluorescein diacetate (FDA) (BioBas)-propidium iodide (Pl) (Sigma). The scaffold was removed from the culture, and it was washed with PBS 10 min twice. The scaffold was stained with FDA (1ug/ml), it was incubated for 15 min at 37 °C. After washing with PBS 10 min twice, the scaffold was stained with PI (1 mg/ml) for 1 h at room temperature. After washing with PBS 10 min twice, image was taken with fluorescence microscope connected to a CCD camera (Nikon).

2.4 Foetal surgery and implantation

We developed a new approach to transplant the cell-scaffold construct into fetuses with spina bifida aperta using combined techniques of fetal surgery and microsurgery. Retinoic acid treated pregnant rats of E16 were anaesthetized with pentobarbitone sodium (40 mg/kg body weight).

An incision was made in the abdominal wall, and the uterine horn was exteriorized. To relieve uterine spasm, the uterus was covered with wet gauze immersed with warm physiologic saline and atropine (0.1 mg/kg body weight) was given intraperitoneally. Under the microscope the position of defective spinal dord of fetus was identified through the wall of uterus. Then 7–0-nylon purse-string suture and a small incision were made on the wall of uterus. The amniotic sac was opened and the defective region of spinal cord was exposed. The cell-scaffold construct was implanted into the defective region of spinal cord under the microscope. After implantation, the fetuses were returned to the uterus, and the wound of the uterus was closed (Fig. 1). In average, 2–3 fetuses could be implanted without compromising the survival of the fetuses. The pregnant rats recovered from the anesthesia within 1 h and were returned to their home cage.

2.5 Sample collection

The pregnant rats on E20 were reanesthetized with an overdose of pento-barbitone sodium, and the fetuses that had been transplanted were harvested by caesarean section. Fetuses were perfused transcardially with 15 ml physiologic saline followed by 25 ml 4 % paraformaldehyde. The lumbosacral spinal column containing muscle, spinal cord and subcuta-neous tissue was dissected and post-fixed in the same fixative for 24 h at 4 °C, followed by cryopreservation in 20 % sucrose for 24 h. The chitosan–gelatin scaffolds seeded with labeled eGFP-BMSCs in the spinal column were observed under stereomicroscope (Leica). The spinal column were then sectioned into 30 μm serial transverse sections, and all GFP positive BMSCs in the spinal column were observed and counted under fluorescence microscope (Nikon, Tokyo, Japan). All the sections were also stained with DAPI, and only those GFP positive cells that contained a nucleolus were counted to avoid the same GFP cell being counted more than once. Image was taken with C1 confocal microscope (Nikon). Sections with GFP positive BMSCs observed were marked and kept at −80 °C in the dark for further immuno-fluorescence staining.

2.6 Immunofluorescence

To evaluate the differentiation potentials of transplanted BMSCs in the spinal column, sections were analysed by immunofluorescence using antibodies against molecular markers of neural stem cell and neuron. Primary antibodies used were mouse anti-nestin (1:100) (MAB353;Millipore) and mouse anti-Tubulin (1:200) (MAB3402; Millipore). After primary antibodies incubation, sections were washed three times with PBS followed by incubation with Alexa Fluor 488-

conjugated goat anti-rabbit IgG antibody (Invitrogen) and TRITC-conjugated goat anti-mouse IgG (AP124R; Millipore) in 10 % FBS-PBST for 1 h at room temperature. After washing, sections were stained with DAPI (C1002; Beyotime Institute of Biotechnology), then mounted with anti-fade mounting medium (P0126; Beyotime Institute of Biotechnology).

3 RESULTS

3.1 BMSCs culture characters and eGFP transfection

BMSCs presented typical spindle-shape morphology with adherence to plastic and organisation in monolayer (Fig. 2). The phenotypic characterisation of the BMSCs indicated high expression of non-hematopoietic markers, CD90 (95 %) and CD44 (94 %). Additionally, no CD34 and CD45 expression was present in 97 and 98 % of the cells (Fig. 2c). Thus, the isolated cells met the criteria used to define BMSCs.

3.2 Characterization of the physicochemical properties of chitosan gelatin scaffolds

Through the measurement, calculating the porosity of chitosan–gelatin scaffold was above 90 %, and the degree of swelling was 25 %, the apparent density was 2.0 mg/mm^3.The pore size of scaffold was between 100 and 400 μm. The macroscopic aspect and the three-dimensional architecture of the scaffold were similar to a sponge. Similar to trabecular bone, it consisted of a homogeneous porous-like arrangement (Fig. 3c). The SEM analysis revealed interconnected pores of different sizes and flat, relatively smooth walls (Fig. 3a, b).

3.3 Characterization of the cell-scaffold construct

After 1 weeks' culture, the scaffold was stained by FDA-Pl. Compared to a number of viable cells emitting green light, the loss of activity of the cells emitting red light are relatively few (Fig. 4a, b). A lot of dyed green fluorescent cells were clinging on the aperture, and showed a long typical fusiform as the morphology of stem cell peculiar. The PI was staining the nuclei of death cell, they were relatively few from Fig. 4, suggesting that morphology of BMSCs were normal after co-culturing, growing in good condition. BMSCs fused into cell clusters, covering internal and surface of the entire scaffold, so it could be that the chitosan–gelatin scaffold has non-toxic side effects on BMSCs.

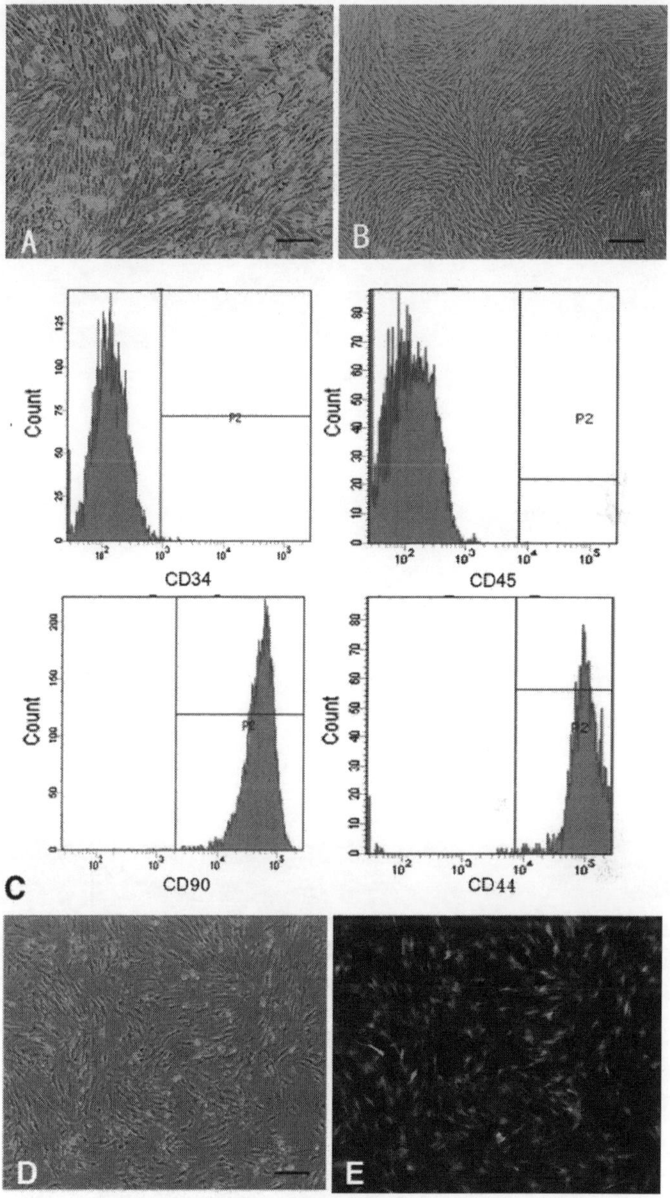

Figure 2. a The LM micrograph of BMSCs after adherent 24 h; **b** the LM micrograph of BMSCs to P3 generation showing the characteristic of plastic adherence and spindle shape. **c** The result of flow cytometry histograms demonstrating high expression of non-hematopoietic markers of CD90 and CD44 and low expression for CD34 and CD45. **d,e** The figure **d**, **e** was in the same perspective, after GFP adenovirus transfection, the efficiency rate can reach above 90 %. The scale of all the pictures are 100 μm

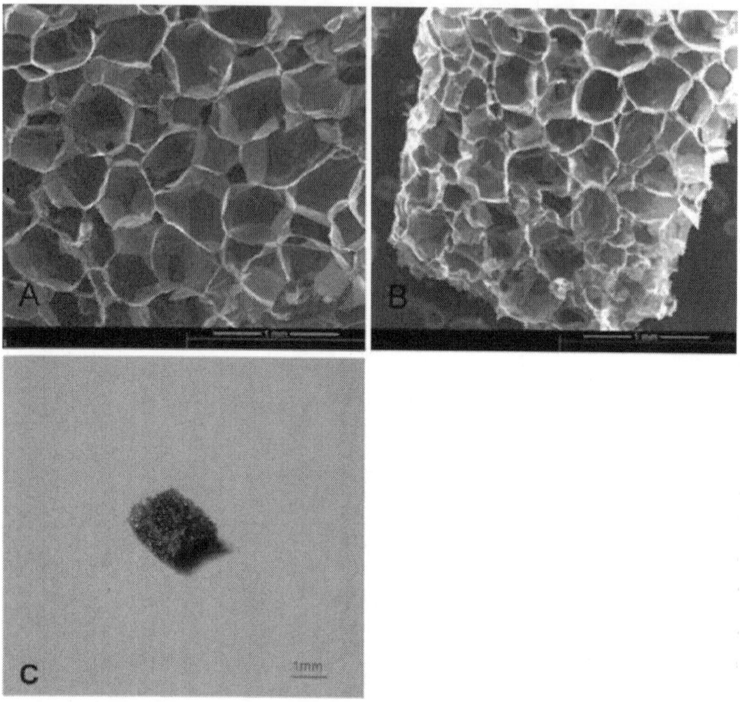

Figure 3. Scaffold characterisation. **a**, **b** SEM photomicrographs of the chitosan–gelatin scaffold. **c** Macroscopic aspects similar to a sponge

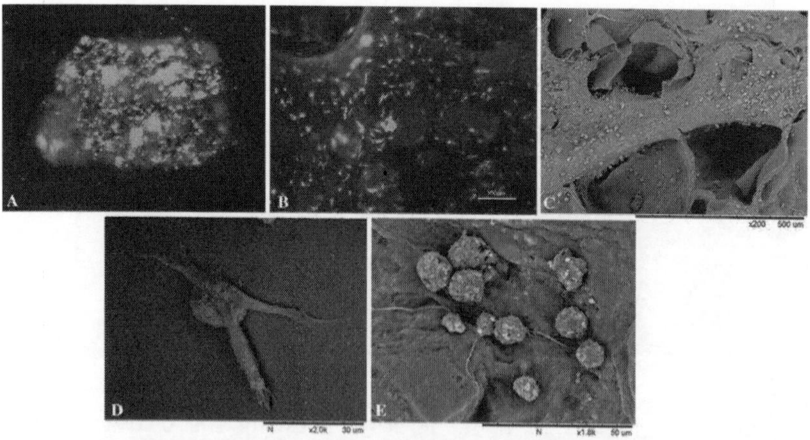

Figure 4. Characterization of the cell-scaffold construct. **a** The chitosan-gelatin scaffold seeded with GFP postive BMSCs. Before transplantation, the cell-scaffold construct was incubated for 48 h at 37 °C with 5 % CO_2. **b** FDA- PI staining of the cell-scaffold construct. **c**–**e** Scanning electron microscopy of the cell-scaffold construct

SEM showed lots of cells attached to the scaffold widely, tight junction between cells and forming neurospheres similar to neural stem cells (Fig. 4c). Some cells were dividing, some cells secreted extracellular matrix, some cells extended many pseudopodia (Fig. 4b, c), so it could be that BMSCs were growing in a good condition.

3.4 The repair of the defective spinal cord after cell-scaffold transplantation

The experiment was operated through intrauterine transplantation of cell-scaffold constructs on congenital spina bifida aperta fetal rats at E16. A total of 52 pregnant rats underwent surgery, one of them died in E20, another 51 pregnant rats which underwent surgery though cesarean section to obtain the fetal rats. A total of 134 fetal rats received cell-scaffold construct transplantation, of which 69 survived (survival rate was 51.5 %). The results found that the defective region of spinal cord in rat fetuses with spina bifida aperta at E20 decreased obviously after transplantation of the scaffold combined with BMSCs, and the skin defect almost closed compared with the fetus without transplantation under stereomicroscopy (Fig. 5a, c). Under fluorescent stereomicroscope most of GFP labelled BMSCs were stayed in the scaffold, and many GFP labelled BMSCs migrate outside of the scaffold and spread to the entire defective region of spina bifida aperta (Figs. 5b, 6). The results of frozen transverse sections showed that the defective tissue were filled with not only the transplanted chitosan–gelatin scaffold and GFP labeled BMSCs, but also many kinds of surrounding niche cells (Fig. 5d).

3.5 Differentiation of BMSCs in transplanted cell-scaffold construct in spinal column of spina bifida aperta

To determine whether BMSCs in the scaffold differentiated into different cell types after transplantation in the defective region of fetal rats with spinal bifida aperta, the expression of specific cell markers were analyzed. During the 4 days′ time of embryonic development, transplanted BMSCs expressed neural stem cell specific surface markers (Nestin) and early neuron specific surface markers (Tubulin). These results indicated that the transplanted BMSCs in the rat fetuses could differentiate into neural stem cell and neurons, suggesting that prenatal transplantation of the scaffold combined with MSCs could have the potential to treat spinal cord defect in fetuses with spina bifida aperta by the regeneration of neurons.

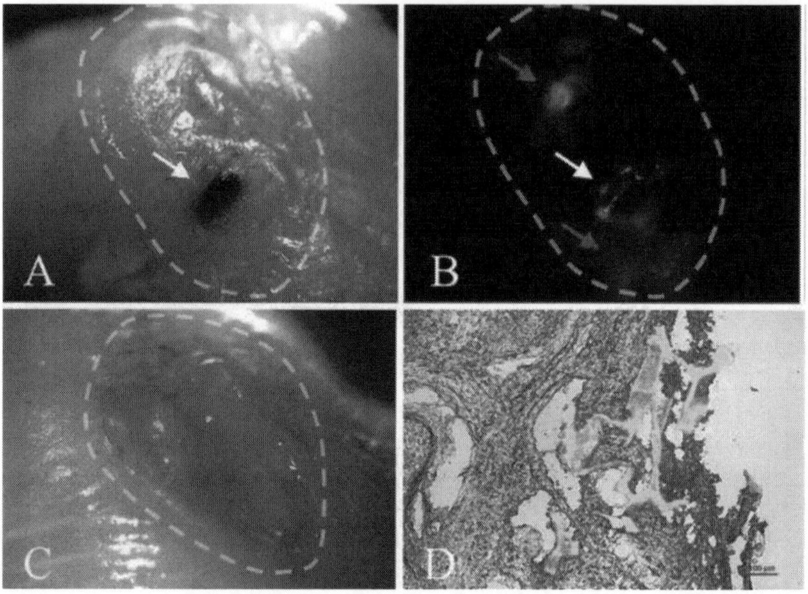

Figure 5. The repair of the defective spinal cord after cell-scaffold transplantation. **a** Stereomicroscopic imaging of fetus with spina bifida aperta showed the skin defect almost closed after cell-scaffold transplantation. The *black* object indicated by *arrow* was cell-scaffold construct. **b** Fluorescent stereomicroscopic imaging (same vision with figure **a**) showed the most of GFP labelled BMSCs were stayed in the scaffold (*white arrow*), and many GFP labelled BMSCs migrate outside to the scaffold and spread to the defective region of spina bifida aperta (*red arrow*). **c** Stereomicroscopic imaging of fetus with spina bifida aperta without cell-scaffold transplantation showed obvious defective spinal cord(view multiple was ×2). **d** Frozen transverse section of defective spinal cord with cell-scaffold (same fetus with Fig. 6a, the thickness of 20 μm), the *red* was the scaffold, the *green* was GFP labelled BMSCs (the scale was 100 μm)

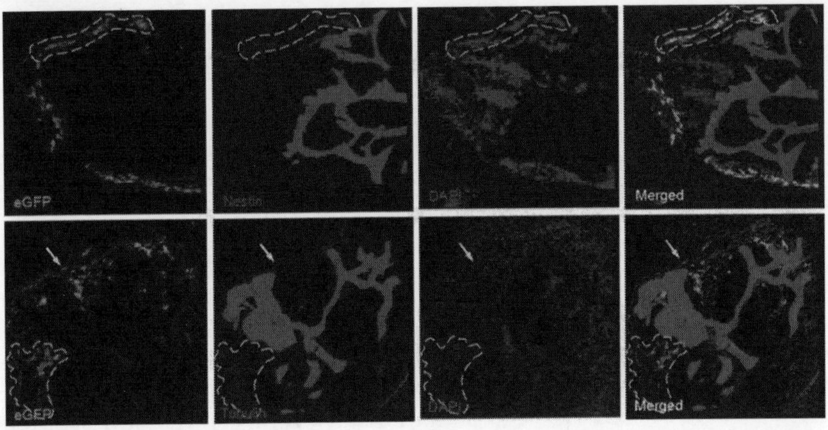

Figure 6. The differentiation of BMSCs in transplanted cell-scaffold construct in spinal column. The linear region is double positive cells, namely the GFP-BMSCs after transplantation in vivo differentiated into neural stem cells (Nestin) and neurons (Tubulin). (the vision are all in ×20)

4 DISCUSSION

Biological material is an important field of tissue engineering, it serves as a carrier of seed cells, providing a three-dimensional space for nutrition metabolism, proliferation and differentiation and the secretion of extracellular matrix of seed cells. As tissue engineering material, there must be lots of characteristics, besides stable quality, non-toxic and no adverse reaction, there are also compatibility and proper pore size and porosity [33]. The pore size of scaffold affects the organization reconstruction, reports in the literature have indicated aperture diameter of 20 μm in favor of the growth of some fibroblast cells, aperture of 20–125 μm in favor of the skin regeneration in adult mammals [11], the pore size of chitosan–gelatin scaffolds was 300 μm, can fully meet the demand of the tissue regeneration. High porosity of scaffold are conducive to nutrients' entering and discharge of metabolic waste, the uniform pore size, particular size of the scaffold is suitable for the growth of different cells. The pore size has a direct influence on the cell, pore size and connectivity is the key for the design of material structure and cell adhesion tissue [29, 35]. In addition, chemical composition of chitosan–gelatin scaffolds consisted uniquely of carbon and oxygen, which indicates that degradation product of scaffold is harmless to the organization [36]. FDA as a fluorochrome, can make active cell cytoplasm staining, the experiment choose FDA to evaluate the activity of cells in the scaffold, while PI can make the nuclear of cell death staining, the results found that the vision under the fluorescence microscope is full of green fluorescent spindle cells, and only visible minority death cells dying red. The cells with normal morphology in attached to the scaffold can also be found under electron microscope, around the cells with a large amount of extracellular matrix, and mitotic cells can be observed. This series showed that BMSCs in the scaffold grew in good condition, the chitosan–gelatin scaffold had good biocompatibility.

Congenital malformations affect approximately 5 % of all live births every year [44, 45]. Several of these are associated with in utero foetal demise, significant neonatal morbidity and mortality. Despite early diagnosis of the condition in most of cases, treatment options for the affected fetuses are still limiting and not satisfactory. Currently, MMC repair operation in utero is being offered to selected mothers in several centres, and over 330 such interventions have been performed since 2000 [46–51]. It was reported that prenatal repair of the spinal cord defect greatly reduced the mortality of the birth defect, suggesting that attenuation of the progress of the damage of defective spinal cord at the early stage of pregnancy is beneficial to the fetus survival. However, if the defect involves too large an area, MMC repair operation cannot achieve an ideal result. The use of a tissue engineering scaffold seeded with stem cells can largely overcome this issue. In the present study, we first combined application of tissue engineering scaffolds and microsurgery technology in utero transplantation, we transplanted cell-scaffold construct to the spinal cord of congenital spina bifida aperta fetal rat to study whether this method can play a role in defect tissue regeneration of the spinal bifida fetal rat. The results found that the defective region of spinal cord in rat fetuses with spina bifida aperta at E20 decreased obviously, and the skin defect almost closed compared with the fetus without transplantation. Under fluorescent stereomicroscope we found that most of GFP labelled BMSCs were stayed in the scaffold, and many of them migrate outside of the scaffold and spread to the entire defective region of spina bifida aperta. This characteristic would be valuable for the repair of serious defects, especially for the spina bifida aperta which have the defects of not only spinal cord but also skin, muscle and vertebra. Chitosan–gelatin scaffold characterization revealed the porous structure, which indicates that these matrices are highly interconnected to allow nutrient exchange and waste removal [52]. The scaffold has some biological characteristics, such as bleeding, pain, wound healing, and has the permeability to oxygen, can reduce scar formation, help wound healing [53–55]. In a certain degree, the large range of spinal cord repair of spina bifida fetal rats are related to the properties of the scaffold. Mesenchymal stem cells have the potential to differentiate into different cell types, being induced by the surrounding microenvironment to differentiate into many kinds of cell type after being transplanted in vivo. Our study used immunofluorescence in frozen transverse sections of spinal vertebra found that transplanted BMSCs in chitosan–gelatin scaffolds in vivo can be respectively differentiation into neuronal stem cells and neurons, and the defective tissue were filled with not only the transplanted chitosan–gelatin scaffold and GFP labelled BMSCs, but also many kinds of surrounding niche cells. Mesenchymal stem cells can differentiate into neuronal cells on a tissue engineered nerve scaffold [56]. Our previous studys' results show that MSCs survived, migrated and was able to differentiate into neural lineage cells [18, 19]. This study further confirms the characteristics of bone marrow mesenchymal stem cells. Neural damages in fetus with spina bifida aperta were occurred at early stage of development, so it is necessary to use stem cell replacement therapy as early as possible. In the present study, we performed intrauterine transplantation of cell-scaffold constructs at E16, in the future study it would performed at earlier stage of development (E14 or E15). It can make stem cells have more time to migration and differentiation after transplantation, and the spinal cord defect of spina bifida can be better reconstructed. In future studies, the long-term study about the function of rat fetuses with spina bifida aperta should be performed.

With the development of stem cell technology and tissue engineering, major components of regenerative medicine have made significant progress in many fields, providing a new approach and tremendous potential for spina bifida aperta repair and reconstruction. Although the application of mesenchymal stem cells further expands the source of seed cells, the legal and ethical issues remain unsolved. Therefore, the applications of tissue engineering and stem cell technology are beginning to enter clinical practice, and some clinical trials should see a wide prospect. However, the studies should take time to provide realistic follow-up data before the method can be applied to human disorders. We maintain the view that stem cell technology and tissue engineering -based strategies for the treatment of spina bifida aperta are in the early stages and the move from basic experiments to clinical applications will remain a long and difficult conversion.

ACKNOWLEDGMENTS

This study was supported by the National Natural Foundation of China. (Grant Numbers: 81370717,81171072). National Basic Research Program of China (973 Program, No. 2013CB945402). Program for Liaoning Innovative Research Team in University (LT2013016).

REFERENCES

1. Danzer E, Ernst LM, Rintoul NE, et al. In utero meconium passage in fetuses and newborns with myelomeningocele. J Neurosurg Pediatr. 2009;3:141–6.
2. Zhu L, Ling H. National neural tube defects prevention program in China. Food Nutr Bull. 2008;29:S196–204.
3. Copp AJ, Greene NDE. Genetics and development of neural tube defects. J Pathol. 2010;220(2):217–30.
4. Roybal JL, Santore MT, Flake AW. Stem cell and genetic therapies for the fetus. Semin Fetal Neonatal Med. 2010;15:6.
5. Merianos D, Heaton T, Flake AW. In utero hematopoietic stem cell transplantation: progress toward clinical application. Biol Blood Marrow Transplant. 2008;14:729–40.
6. Muench MO. In utero transplantation: baby steps towards an effective therapy. Bone Marrow Transplant. 2005;35:537–47.
7. Fauza DO, Jennings RW, Teng YD, et al. Neural stem cell delivery to the spinal cord in an ovine model of fetal surgery for spina bifida. Surgery. 2008;144:367–73.
8. Guan K, Li H, Fan Y, et al. Defective development of sensory neurons innervating the levator ani muscle in foetal rats with anorectal malformation. Birth Defects Res A Clin Mol Teratol. 2009;85:583–7.
9. Yuan ZW, Lui VC, Tam PK. Deficient motor innervation of the sphincter mechanism in foetal rats with anorectal malformation: a quantitative study by fluorogold retrograde tracing. J Pediatr Surg. 2003;38:1383–8.
10. Karp JM, Teo GSL. Mesenchymal stem cell homing: the devil is in the details. Cell Stem Cell. 2009;4:206–16.
11. Prockop DJ, Kota DJ, Bazhanov N, et al. Evolving paradigms for repair of tissues by adult stem/progenitor cells (MSCs). J Cell Mol Med. 2010;14:2190–9.
12. Torrente Y, Polli E. Mesenchymal stem cell transplantation for neurodegenerative diseases. Cell Transplant. 2008;17:1103–13.
13. Brazelton TR, Rossi FM, Keshet GI, et al. From marrow to brain: expression of neu-ronal phenotypes in adult mice. Science. 2000;290:1775–9.

14. Chen X, Wang XD, Chen G, et al. Study of in vivo differentiation of rat bone marrow stromal cells into schwann cell-like cells. Microsurgery. 2006;26:111–5.

15. Hofstetter CP, Schwarz EJ, Hess D, et al. Marrow stromal cells form guiding strands in the injured spinal cord and promote recovery. Proc Natl Acad Sci USA. 2002;99:2199–204.

16. Pisati F, Bossolasco P, Meregalli M, et al. Induction of neurotrophin expression via human adult mesenchymal stem cells: implication for cell therapy in neu-rodegenerative diseases. Cell Transplant. 2007;16:41–55.

17. Norambuena GA, Khoury M. Mesenchymal stem cells in osteoarticular pediatric diseases: an update. Nature. 2012;71(4 Pt 2):452–8.

18. Li H, Gao F, Ma L, et al. Therapeutic potential of in utero mesenchymal stem cell (MSCs) transplantation in rat foetuses with spina bifida aperta. J Cell Mol Med. 2012;16(7):1606–17.

19. Li H, Miao J, Zhao G, Wu D, Liu B, Wei X, Cao S, Gu H, Zhang Y, Wang L, Fan Y, Yuan Z. Different expression patterns of growth factors in rat fetuses with spina bifida aperta after in utero mesenchymal stromal cell transplantation. Cytotherapy. 2014;16(3):319–30.

20. Xiao Y, Qian H, Young WG, Bartold PM. Tissue engineering for bone regeneration using differentia ted alveolar bone cells in collagen scaffolds. Tissue Eng. 2003;9(11):67–1177.

21. Shin M, Yoshimoto H, Vacanti JP. In vivo bone tissue engineering using mesenchymal stem cells on a novel electrospun nanofibrous scaffold. Tissue Eng. 2004;10(33–4):1.

22. Zhu L, Liu W, Cui L, Cao Y. Tissue engineered bone repair of goatfemur defects with osteogenically induced bone marrow stromal cells. Tissue Eng. 2006;12:423–33.

23. Burdick JA, Vunjak-Novakovic G. Engineered microenvironments for controlled stem cell differentiation. Tissue Eng Part A. 2009;15:205–19.

24. Puppi D, Chiellini F, Piras AM, Chiellini E. Polymeric materials for bone and cartilage repair. Prog Polym Sci. 2010;35(4):03–440.

25. Keong LC, Halim AS. In vitro models in biocompatibility assessment for biomedical-grade chitosan derivatives in wound management. Int J Mol Sci. 2009;10:1300–13.

26. Varum KM, Myhr MM, Hjerde RJ, Smidsrod O. In vitro degradation rates of partially N-acetylated chitosans in human serum. Carbohydr Res. 1997;299:99–101.

27. Lahiji A, Sohrabi A, Hungerford DS, Frondoza GG. Chitosan supports the expression of extracellular matrix proteins in human osteoblasts and chondrocytes. J Biomed Mater Res. 2000;51:586–95.

28. Khor E, Lim LY. Implantable applications of chitin and chiotosan. Biomaterials. 2003;24:2339–49.

29. Huang Y, Onyeri S, Siewe M, Moshfeghian A, Madihally SV. In vitro characterization of chitosan-gelatin scaffolds for tissue engineering. Biomaterials. 2005;26:7616–27.

30. Wang XH, Li DP, Wang WJ, Feng QL, Cui FZ, Xu YX, et al. Cross-linked collagen/chitosan matrix for artificial livers. Biomaterials. 2003;24:3213–20.

31. Ma L, Gao C, Mao Z, Zhou J, Shen J, Hu X, et al. Collagen/chitosan porous scaffolds with improved biostability for skin tissue engineering. Biomaterials. 2003;24:4833–41.

32. Xia W, Liu W, Cui L, Liu Y, Zhong W, Liu D, et al. Tissue engineering of cartilage with the use of chitosan-gelatin complex scaffolds. J Biomed Mater Res B. 2004;71:373–80.

33. Le Trong I. McDe vitt TC, Nelson KE, Stayton PS, Stenkamp RE. Structural characterization and comparison of RGD cell-adhesion recognition sites engineered into streptavidin. Acta Crystallogr. 2003;59(82):8–834.

34. Huang Y, Onyeri S, Siewe M, Moshfeghian A, Madihally SV. In vitro characterization of chitosan-gelatin scaffolds for tissue engineering. Biomaterials. 2005;2(6):7616–27.

35. Lawrence BJ, Madihally SV. Cell colonization in degradable 3D porous matrices. Cell AdhES Migr. 2008;2:9–16.

36. Miranda SC. Three-dimensional culture of rat BMMSCs in a porouschitosan-gelatin scaffold: a promising association for bonetissue engineering in oral reconstruction. Arch Oral Biol. 2011;56(1):1–15.

37. Jayakumar R, Prabaharan M, Kumar PTS, Nair SV, Tamura H. Biomaterials based on chitin and chitosan in wound dressing applications. Biotechnol Adv. 2011;29:322–37.

38. Baniasadi H. Fabrication and characterization of conductive chitosan/gelatin-based scaffolds for nerve tissue engineering. Int J Biol Macromol. 2015;74:360–6.

39. Potas JR. Interleukin-10 conjugated electrospun polycaprolactone (PCL) nanofibre scaffolds for promoting alternatively activated

(M2) macrophages around the peripheral nerve in vivo. J Immunol Methods. 2015;420:38–49.

40. George J. Christ, Engineering muscle tissue for the fetus: getting ready for a strong life. Front Pharmacol. 2015;10(6):53.

41. Danzer E, Schwarz U, Wehrli S, et al. Retinoic acid induced myelomeningocele in fetal rats: characterization by histopathological analysis and magneticresonance imaging. Exp Neurol. 2005;194:467–75.

42. Cipriani P, Guiducci S, Miniati I, et al. Impairment of endothelial cell differentiation from bone marrow-derived mes-enchymal stem cells: new insight into the pathogenesis of systemic sclerosis. Arthritis Rheum. 2007;56:1994–2004.

43. Yin Y, Ye F, Cui J, Zhang F, Li X, Yao K. Preparation and characterization of macroporous chitosan-gelatin/beta-tricalcium phosphate composite scaffolds for bone tissue engineering. J Biomed Mater Res A. 2003;67:844–55.

44. Pathak S, Lees C. Ultrasound structural fetal anomaly screening: an update. Arch Dis Child Fetal Neonatal Ed. 2009;94:F384–90.

45. Romosan G, Henriksson E, Rylander A, et al. Diagnostic performance of routineultrasound screening for fetal abnormalities in an unselected Swedish population in2000–2005. Ultrasound Obstet Gynecol. 2009;34:526–33.

46. Kohl T, Hering R, Heep A, et al. Percutaneous fetoscopic patch coverage ofspina bifida aperta in the human–early clinical experience and potential. Fetal DiagnTher. 2006;21:185–93.

47. Sutton LN. Fetal surgery for neural tube defects. Best Pract Res Clin Obstet Gynaecol. 2008;22:175–88.

48. Farmer DL, von Koch CS, et al. In utero repair of myelomeningocele: experimental pathophysiology, initial clinical experience, and outcomes. Arch Surg. 2003;138:872–8.

49. Johnson MP, Sutton LN, Rintoul N, et al. Fetal myelomeningocele repair: short-termclinical outcomes. Am J Obstet Gynecol. 2003;189(482–7):38.

50. Zambelli H, Barini R, Iscaife A, et al. Successful developmental outcome inintrauterine myelomeningocele repair. Childs Nerv Syst. 2007;23(123–6):39.

51. Tulipan N, Sutton LN, Bruner JP, et al. The effect of intrauterine myelomeningocele repair on the incidence of shunt-dependent hydrocephalus. Pediatr Neurosurg. 2003;38:27–33.

52. Bhat S. Cell proliferation on three-dimensional chitosaneagaroseegelatin cryogel scaffolds for tissue engineering applications. Biosci Bioeng. 2012;.

53. Mohamad Pezeshki-Modaress. Cell loaded gelatin/chitosan scaffolds fabricated by salt-leaching/lyophilization (SLL) for skin tissue engineering: in vitro and in vivo study. J Biomed Mater Res A. 2014;102(11):3908–17.

54. Tseng H-J. Characterization of chitosan–gelatin scaffolds for dermal tissue engineering. J Tissue Eng Regen Med. 2013;7(1):20–31.

55. Mohamed KR. In vitro study of nano-hydroxyapatite/chitosan–gelatin composites for bio-applications. J Adv Res. 2014;5(2):201–8.

56. Feng Y. Differentiation of mesenchymal stem cells into neuronal cells on fetal bovine acellular dermal matrix as a tissue engineered nerve scaffold. Neural Regen Res. 2014;9(22):1968–78.

8

OSTEOGENESIS OF MC3T3 PREOSTEOBLASTS ON 3D BIOACTIVE PEPTIDE MODIFIED NANO-MACROPOROUS BIOACTIVE GLASS SCAFFOLDS

Colin Przybylowski[1,2], Mohamed Ammar[3], Courtney LeBlon[4], Sabrina S. Jedlicka[2,3,5]

[1]Integrated Degree in Engineering, Arts & Sciences (IDEAS) Program, Lehigh University, Bethlehem, PA, USA

[2]Bioengineering Program, Lehigh University, Bethlehem, PA, USA

[3]Department of Materials Science & Engineering, Lehigh University, Bethlehem, PA, USA

[4]Department of Mechanical Engineering & Mechanics, Lehigh University, Bethlehem, PA, USA

[5]Center for Advanced Materials & Nanotechnology, Lehigh University, Bethlehem, PA, USA

ABSTRACT

Biointerface design that targets osteogenesis is a growing area of research with significant implications in biomedicine. Materials known to either support or stimulate osteogenesis are composed of a biomimetic ceramic material, such as bioactive glass. Bioactive glass is osteoproductive, and the potential for osteoproductivity can be enhanced by the addition of proteins or other additives designed to alter functionality. In addition, soluble growth factors are often added to osteogenic culture on bioactive glasses, further intensifying the effects of the material. In this paper, synthetic peptide combinations, covalently bound to a three-dimensional bioactive glass network, are used to mimic the effects of the whole fibronectin and bone morphogenetic proteins (BMP) 2 and 9. Peptide-silanes possessing critical binding sequences from each of these proteins are synthesized and used to decorate the surface of three-dimensional (3D) nano-macroporous bioactive glass. MC3T3 preosteoblast cells are then assessed for differentiation on the materials in the absence of soluble differentiation cues. MC3T3 preosteoblasts undergo enhanced differentiation on the peptide-silane samples over the standard nano-macroporous bioactive glass, and the differentiation capacity of the cells exposes only to peptide-silane surfaces approaches that of cells grown in chemical differentiation induction media.

Keywords: Bioactive Glass, Bioactive Peptide, MC3T3, Osteogenesis

1. INTRODUCTION

Biomaterials designed for bone repair benefit from an interface that is designed to increase osteogenesis. Bioactive glasses have been shown to be osteoproductive, stimulating osteogenesis via surface and solution-mediated mechanisms [1] - [3] . Therefore, bioactive glasses have received considerable attention from researchers aiming to enhance the treatment of bony defects in orthopedics and dentistry [1] [4] - [12] . Perhaps the most useful characteristic of bioactive glasses is the rapid surface reaction rate [13] - [18] . This rapid reaction rate leads to direct bone attachment. Bioactive glasses are quite flexible with regards to manufacturing, and also are amenable to surface modification, opening up a wide array of potential formats and compositions that offer versatility and unique properties [5] - [8] [19] - [23] .

Osteogenic cells exhibit enhanced growth and biological activity in the presence of 45S5 bioactive glass substrates [3] [5] [9] [14] [24] - [33] . Studies performed in vivo have shown that bioactive glasses lead to enhanced proliferation and differentiation of progenitor cells [1] [25] [28] [29] [32] . This enhancement of osteogenic behavior has been linked to an enhanced production of biological growth factors and other proteins implicated in cell proliferation and tissue organization [3] [13] [24] [34] .

One common research goal for bioactive glass as a clinical treatment for bone defects is to increase the tissue engineering potential. Several potential improvements have been suggested, including surface modification with bioactive proteins [23] [35] [36] and incorporation of growth factors [4] [5] [36] into the materials for slow release into the extracellular surroundings. While these approaches are promising, controlling the conformation of proteins that contact the highly reactive surface is challenging. Therefore, many researchers have chosen to use highly specific synthetic peptides that mimic protein motifs, such as the RGD group from Fibronectin Type III repeats [37] - [39] . This study examines the potential of mixed peptide motifs on three-dimensional, nano-macro- porous bioactive glass scaffolds. Specifically, peptides derived from growth factors that are implicated in osteogenesis were chosen. The peptides were incorporated through an established peptide-silane technique, via covalent modification of the N-terminus with an alkoxysilane moiety. The peptides chosen were the RGD group to enhance cell adhesion [37] [38] [40] - [43] , and sequence motifs from BMP-2 and BMP-9, for their potency to induce osteogenic differentiation [5] [44] - [48] . These peptides were covalently linked to a bioactive glass network using silane chemistry [49] , and MC3T3 cells were used to examine the influence of bioactive peptides on the induction of osteogenesis on 3D bioactive glass monoliths.

2. MATERIALS AND METHODS

2.1. Cell Culture

MC3T3-E1 cells (ATCC CRL-2593) (MC3T3) were maintained in Modifed Eagle's Medium (alpha modification) (α-MEM) + 7.5% bovine calf serum (BCS) + 2.5% fetal bovine serum (FBS) in a 5% CO_2, 37°C humidified atmosphere. Media was replaced three times a week and cells were split 1:50 when reaching 80% confluency.

2.2. Peptide-Silane Synthesis

Three peptide-silanes were synthesized (Table 1), including AYAVTGRGDSPAS (RGD), ACGGKVGKACC- VPTKLSPISVLYK (CGG), and ACKIPKASSVPTELSAISTLYL (CKI). Peptides were synthesized using solid-phase synthesis in an automated peptide synthesizer (ResPep, Intavis, Germany) using 9-fluorenylmethy- loxycarbonyl (Fmoc) chemistry. Peptides were then conjugated with 3'-aminopropyl trimethoxysilane (APTMS) using carbonyldiimadazole (CDI) as a linker. Resulting peptide-silanes were cleaved using a cleavage cocktail consisting of 89% trifluoracetic acid, 5% phenol, 5% dd H_2O, and 1% triisopropylsilane. Peptides were then

Table 1. Peptide-silanes used in this study.

Name	Sequence	Origin
RGD	AYAVTGRGDSPASA [37]-[39] [43]	Fibronectin Type III$_{10}$
CGG	ACGGKVGKACCVPTKLSPISVLYKA [5]	BMP-9
CKI	ACKIPKASSVPTELSAISTLYLA [48]	BMP-2

ether precipitated and purified. Peptide conjugation is routinely confirmed using matrix assisted laser desorption ionization - mass spectrometry (MALDI-MS), with α-cyano-4-hydroxycinnamic acid as the matrix [50] .

2.3. Material Preparation

Bimodal (nano-macroporous) 3D bioactive glass scaffolds (70% SiO_2 - 30% CaO) were synthesized as previously described [17] [51] -[53] . This process is summarized in Figure 1. Briefly, water soluble polymer polyethylene oxide (PEO, Mw = 100,000, 2.1 g), was dissolved in 0.05 N acetic acid (CH_3COOH, 30 mL). The mixture was stirred for 20 minutes to dissolve the polymer. Urea (1.35 g) and tetramethoxysilane (TMOS) (AcrosInorganics, 13.5 mL) were added and vigorously stirred for 20 minutes. Calcium nitrate tetrahydrate (9.27 g) was then added and the mixture stirred for 10 more minutes. To accelerate gelation, aqueous hydrofluoric acid (HF, 6%) was added. The sol was then cast into molds (which were sized to fit a 24 well tissue culture plate) and left to gel at 40°C for 24 hours. The resulting gel was dried at 60°C for 1 day (at a relative humidity of 90%), followed by 180°C for 2 days. Following the drying, the gels were heat treated sequentially at 600°C for one hour and 700°C for two hours, with heating and cooling (to room temperature) rates of 100°C/h. The resulting scaffolds are shown in Figure 1.

Following synthesis, the scaffolds were prepared for cell culture. The scaffolds were autoclaved (121°C, 15 minutes) and then submerged in Simulated Body Fluid (SBF) overnight (SBF: 138 μM NaCl, 4 μM $NaHCO_3$, 3 μM KCl, 1.4 μM K_2HPO_4, 1.5 μM $MgCl_2$, 2.64 μM $CaCl_2$, 507 nM Na_2SO_4, and 50.56 μM Tris, adjusted to pH 7.4 with 20 mL of 1.0 M HCl). The peptide-silanes were dissolved in 50 μL of dimethylsulfoxide. The peptide- silanes were then suspended in SBF, and pipetted on top of the scaffolds at 0.01% mole peptide: 1 mole SiO_2. The experimental samples included bioactive glass samples with no peptides; with RGD (0.01 mole% total peptide-silane: 1 mole Si); with RGD and CGG (0.02 mole% total peptide-silane: 1 mole Si); and with RGD, CGG and CKI (0.03 mole% total peptide-silane: 1 mole Si). Cells grown on scaffolds in differentiation media represented a positive control. All scaffolds were submerged in media overnight before cell seeding.

2.4. Peptide Tagging with Fluorescent Dye

To ensure that peptides were diffusing into the scaffolds, the RGD peptide was labeled using an amine reactive fluorescent peptide-labeling reagent (Cy3 mono-reactive NHS ester). The labeling reagent was used as specified by the manufacturer (GE Healthcare, Princeton, NJ). Labeled peptides were separated from free dye by dialysis.

2.5. X-Ray Photoelectron Spectroscopy (XPS)

Jedlicka et al. have previously described the analysis of sol-gel silica surfaces modified with peptide-silanes [49] . XPS data were obtained by a Kratos Ultra DLD spectrometer using monochromatic AlKR radiation of 1486.58 eV. The survey and high-resolution spectra were collected at a fixed analyzer pass energy of 160 and 20 eV, respectively. The spectra were collected at a 0° and 60° angle with respect to the surface normal. The atomic concentrations of the chemical elements in the near-surface region were estimated after the subtraction of a Shirley- type background, taking into account the corresponding Scofield atomic sensitivity factors and inelastic mean free pass (IMFP) of photoelectrons as a standard procedure of the CasaXPS software.

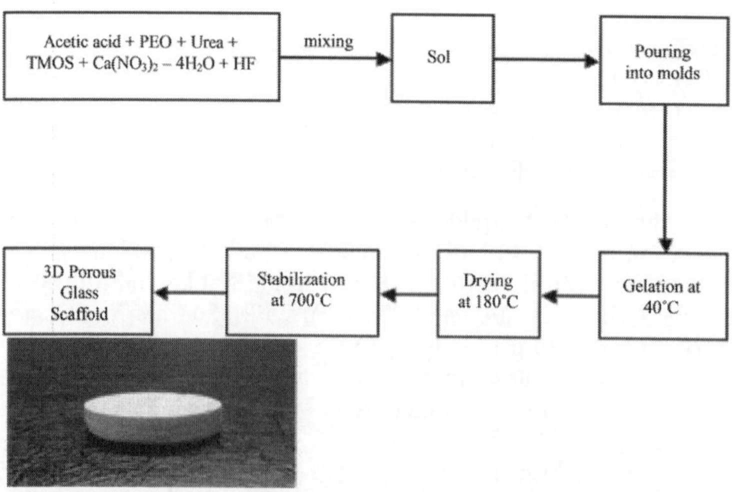

Figure 1. Three-dimensional nano-macroporous glass scaffold preparation. The scaffolds were made to fit in a 24 well tissue culture plate.

2.6. Cell Culture on Materials

MC3T3-E1 cells (ATCC CRL 2593), at passage numbers 7 - 9, were used to examine the osteogenic potential of the 3D peptide-modified scaffolds. MC3T3 cells were routinely maintained on tissue culture treated polystyrene (TCPS) dishes at 37°C in a humidified atmosphere containing 5% CO2. Media was changed twice weekly. To seed cells on materials (Table 2), cells were removed from TCPS and seeded at a density of 1000 cells/mm^3 onto sterilized scaffolds, via addition of a concentrated cell suspension in a drop-wise manner to ensure direct cell loading onto scaffolds. Samples were incubated at 37°C for 30 min to allow for cell attachment before new media was added to flood the samples. Cells were analyzed at variable time points (2, 3, 4, 5 weeks).

Table 2 summarizes the scaffolds/treatments examined in this study. The positive control (denoted Diff) was maintained on an unmodified 3D bioactive glass scaffold in differentiation media (α-MEM + 7.5% BCS + 2.5% FBS + 0.01 mol/L β-glycerol phosphate + 50 μg/mL of L-ascorbic acid) throughout the experimental timeline. The experimental samples (RGD, CGG and CKI) were maintained in standard maintenance medium (α-MEM + 7.5% BCS + 2.5% FBS), to assess the potential of the peptides to induce differentiation on the 3D scaffolds in the absence of soluble differentiation cues. The negative control cells were grown in proliferation media on the 3D bioactive glass scaffolds (BIO). Triplicate samples were prepared for each successive experiment. Given the 3D nature of the scaffolds, daily cell monitoring was not possible; however, media color and clarity were checked on a regular basis to monitor for contamination.

2.7. Cell Viability Assay

Cell viability was assessed using a non-destructive Alamar Blue Assay, per manufacturer's instructions (Invitrogen). This assay was chosen over the MTT assay, due to a slightly higher sensitivity, and the non-destructive nature of the dyes [54] . Samples were removed from cell media and incubated for 14.5 hours in 10% Alamar Blue solution. This timing was empirically determined for the MC3T3 cells. Solution absorbance was read at 620 nm and 590 nm, and the percentage of reduced Alamar Blue was determined. Cell counts were determined from this reduction through an experimentally derived standard curve.

2.8. Quantitative Real-Time Polymerase Chain Reaction (qRT-PCR)

RNA was extracted from the samples using the RNeasy Plus Micro Kit (Qiagen), per manufacturer's instructions, with slight modifications. Cell lysate was obtained by crushing the scaffolds while they were submerged in lysis buffer. This was followed by centrifugation. Initial RNA concentrations were very low, so the process was modified by then adding RNase-free water to the scaffold debris and mixing (to dissolve the RNA that had bound to the silica particles from the scaffold), centrifuging and adding this supernatant to the cell lysate. The last step of the protocol was modified to elute a total of 20 μl of RNA by twice adding 12 μl of RNase-free water to the spin column membrane and centrifuging.

qRT-PCR was performed using the Rotor-Gene™ SYBR Green RT-PCR Kit, with a Rotor-Gene qRT-PCR cycler. The genes of interest were glyceraldehyde 3 phosphate dehydrogenase (GAPDH) (control), Osteocalcin (BGLAP), Transcription factor RunX2 (RunX), Alkaline Phosphatase (AP) and SMAD family member 4

Table 2. Materials/Specimens examined.

Experimental Samples	Positive Control	Negative Control
Bioactive Glass + RGD (RGD) Bioactive Glass + RGD + CGG (CGG) Bioactive Glass + RGD + CGG + CKI (CKI)	Bioactive Glass Only + Differentiation Media (Diff)	Bioactive Glass Only + Regular Media (BIO)

(SMAD4) (Table 3). The primers used in this study are listed in Table 3. Analysis of the cycle data was assessed using the ΔΔ Ct method, with a threshold value of 20% above the background. Efficiency values were empirically derived from 3 plates of differentiated cells (grown in the same media as the positive control) for the purposes of this analysis. Fold changes were determined through Ct comparison to the negative control. All data was subject to normalization to a housekeeping gene (GAPDH) corresponding to each experimental condition and specimen. All data was then averaged and subject to ANOVA statistical analysis.

2.9. Scanning Electron Microscopy (SEM)

To examine cell attachment and scaffold morphology, scanning electron microscopy (SEM) in environmental mode was used. Scaffolds were washed three times in 0.1 M phosphate buffer and fixed in 12% gluteraldehyde overnight. Following fixation, specimens were washed again with phosphate buffer and dehydrated with an ethanol gradient: [25% (10 minutes), 50% (10 minutes), 75% (15 minutes), 95% (15 minutes) and 100% (3 × 10 minutes)]. Samples were then covered in hexamethyldisilazane (HMDS) for 10 minutes before drying in a dessicator overnight. Specimens were imaged in a Philips XL30 ESEM, at 0.2 torr, with an accelerating voltage of 5 kV.

3. Results and Discussion

In this study, we used mixed peptide silanes from fibronectin, BMP-2 and BMP-9 grafted onto three-dimensional nano-macroporous bioactive glass to examine the potential of these short peptides to influence osteogenic differentiation. SEM was used to assess that the surface morphology after grafting of the peptides was not significantly altered. SEM imaging (Figure 2) revealed that the surface morphology of the scaffolds was not significantly altered with the addition of peptide-silanes to the surface.

Specimens were also imaged after SBF submersion, prior to cell seeding. A difference in the hydroxyapatite layer after SBF incubation was observed in the SEM studies in our experiment between the control samples and the peptide grafted samples (Figure 3). This can be attributed to the presence of negatively charged groups within some of the amino acids in the peptides chains, such as glutamic acid in the CKI peptide sequence and aspartic acid in the RGD peptide sequence [55] . These amino acids carry carboxyl groups with negative charges that act like a nucleation sites for the hydroxyapatite crystals through the binding of the dissolved calcium ions to them [56] .

Diffusion of the peptides through the 3D scaffolds was tracked using fluorescent microscopy. To visualize the peptide-silanes, the RGD peptide-silane was conjugated using a Cy3 monoreactive NHS ester dye for this experiment only (Figure 4). The conjugated peptide-silane was added at 0.01 mole%: 1 mole Si to the scaffold structure in the same manner used in cell culture studies. Using the z-stacking feature, several levels of the 3D scaffold were inspected for fluorescence signature, confirming that the depth of penetration exceeded the microscope objectives working depth (800 μm). While the additional fluorescent dye may have led to altered interactions between the peptide-silane and the bioactive glass surface, the peptide-silanes appear to diffuse freely through the bioactive glass network.

To confirm our fluorescent peptide-silane diffusion study, XPS was performed on the bioactive glass surfaces both before and after peptide additions to the materials. These data are summarized inTable 4. The unmodified

Table 3. qRT-PCR primers.

Gene	Description	Primer	Tm
GAPDH	Glyceraldehyde 3-phosphate dehydrogenase (reference gene)	Forward: 5' TCGGTGTGAACGGATTTGG 3' Reverse: 5' TCTCCACTTTGCCACTGCA 3'	55.4C 57.1C
BGLAP	Osteocalcin	Forward: 5' CTGACAAAGCCTTCATGTCCAA 3' Reverse: 5' GCCGGAGTCTGTTCACTACCTT 3'	55.6C 58.7C
Runx	Transcription Factor RunX2	Forward: 5' CTTCAAGGTTGTAGCCCTCGGA 3' Reverse: 5' ATGACGGTAACCACAGTCCCA 3'	58.8C 58.2C
AP	Alkaline Phosphatase	Forward: 5' AAGGATATCGACGTGATCATGG 3' Reverse: 5' GGCCTTCTCATCCAGTTCGTA 3'	54.3C 56.5C
SMAD4	SMAD family member 4	Forward: 5' CACAAGTCAGCCGGCCAGTATT 3' Reverse: 5' TTCCAGTCCAGGTGGTAGT 3'	59.8C 60.1C

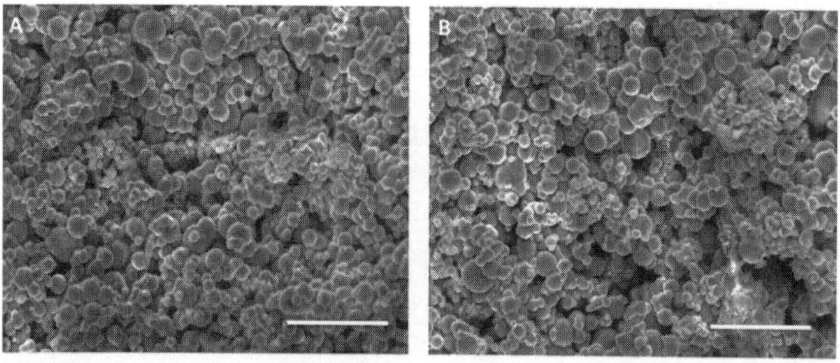

Figure 2. SEM image of nano-macroporous bioactive glass (A) before and (B) after covalent peptide modification with RGD, CGG and CKI peptide-silanes added at equal ratios for a total concentration of 0.03 mole%: mole Si (CKI material). Scale bar = 50 µm.

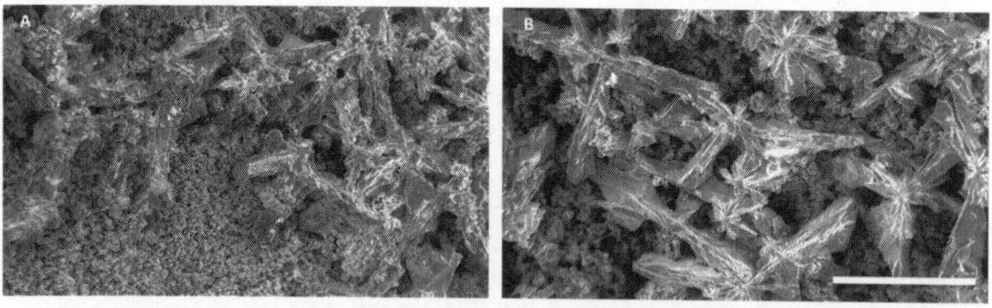

Figure 3. SEM image hydroxyapatite layer formation after SBF incubation on nano-macroporous bioactive glass (A) withouth and (B) with covalent peptide modifcation with RGD, CGG, and CKI added at equal ratios for a total concentration of 0.03 mole%: mole % Si (CKI material). Scale bar = 200 µm.

Figure 4. Fluorescent microscope image of fluorescent peptide modified nano-macroporous bioactive glass (A) at the surface, (B) 800 μm into the surface and (C) z-stack image of 10 images between the surface and the upper working depth of the microscope objective of a scaffold modified with 0.01 mole % fluorescent RGD peptide-silane: 1 mole Si. Scale bar = 50 μm.

bioactive glass, when analyzed with XPS, appears to possess a small degree of carbon contamination; therefore, the carbon percentages are not used to determine peptide-silane conjugation. The Nitrogen 1 s and Sulfur 2p are not seen on standard bioactive glass, however, as the peptide-silanes are added to the scaffolds, these elements begin to appear in the spectra. The RGD peptide-silane used in this study does not possess a cysteine residue, and no sulfur was detected on these specimens. However, as the CGG and CKI peptides are added, both of which possess a cysteine residue, S 2p is detected in the spectra, indicating that the peptide-silanes are present

Table 4. XPS Analysis. BIO is unmodified three-dimensional nano-macroporous bioactive glass. RGD is BIO modified with 0.01 mole % RGD peptide-silane: 1 mole of Si. CGG is RGD modified with an additional 0.01 mole % CGG peptide-silane: 1 mole of Si (total of 0.02 mole %: 1 mole of Si). CKI is CGG modified with an additional 0.01 mole % CKI peptide-silane: 1 mole of Si (total of 0.03 mole %: 1 mole of Si).

Sample	Location	Atomic %					
		Ca 2p	O 1s	Si 2p	C 1s	N 1s	S 2p
BIO	Surface	3.92	61.75	29.12	5.21	0	N/D
BIO	Internal	4.98	65.2	23.6	6.22	0	N/D
RGD	Surface	1.96	63.35	28.06	6.19	0.44	N/D
RGD	Internal	1.8	67.42	26.37	4.31	0.1	N/D
CGG	Surface	0.92	63.75	26.33	4.61	0.53	3.86
CGG	Internal	1.56	63.35	27.76	3.19	0.45	3.69
CKI	Surface	1.05	42.79	20.82	26.97	5.03	3.34
CKI	Internal	2.35	52.58	21.11	17.4	3.19	3.37

on these specimens. In addition, the Nitrogen 1 s content also increases as a function of added peptide-silanes to the bioactive glass monoliths. Finally, the surface of the scaffolds experiences a 4 - 6 fold increase in C 1 s content when all three peptide-silanes are added collectively to the scaffolds. This could indicate some degree of peptide intermolecular association at the surface of the scaffolds, potentially preventing full diffusion of the three peptide-silanes throughout the inner structure of the monolith.

To augment the surface analysis and support the RGD fluorescence analysis, an internal XPS analysis on the peptide-silane modified specimens was performed. These are denoted as "Internal" in Table 4. To obtain these specimens, the original specimens were removed from the XPS and the top surface (~1 mm) was scraped off with a sterile scalpel blade. The dust from the scraping was gently removed with air, and the specimens further analyzed. Of specific interest are the N 1 s and S 2p atomic percentages. Even after removal of ~1 mm of the top surface of the bioactive glass, there remains nitrogen and sulfur in the XPS spectra, indicating diffusion and binding of the peptides within the bioactive glass inner structure. This is key in the modification, as the cells do infiltrate the scaffolds, and ideally, the peptides of interest would be present for the cells to bind to. This XPS data, combined with the fluorescent microscopy study, confirms that the peptide-silanes are distributed throughout the bulk of the scaffold, with the highest concentration at the surface, but with appreciable peptide-silane presentation throughout.

Following confirmation of successful conjugation, MC3T3 cells were seeded onto the materials and assessed for cell adhesion, cell proliferation and osteogenic differentiation. To examine cell morphology on the scaffolds and confirm cell adhesion on specimens, SEM analysis was performed. The SEM analysis confirmed that the cells adhered to all samples, and persisted at least three weeks into the experiment, when the images shown in Figure 5 were collected. The cells appear homogeneously distributed on the scaffolds, with cell extensions between the walls of the scaffolds. There are no significant morphological differences between the different specimens examined.

Proliferation analysis was performed using Alamar Blue (Figure 6). Alamar Blue is a non-destructive means to monitor cell growth using the reduction of Alamar Blue as an indicator of cell metabolism. All specimens allowed for significant cell adhesion, as demonstrated by the cell counts at Day 2. Within one week, the cell number had increased on all specimens, although not appreciably on the RGD and CKI materials. By Day 14, the cell number had declined, potentially indicating a degree of cell die-off as the cells filled the scaffold pores and began to undergo differentiation. Three weeks into the experiment, the cell numbers had increased again, which could be attributed to the overall size of the MC3T3 cells decreasing as cells undergo the early stages of differentiation. Finally, by Day 35, the cell number decreased again, possibly indicating a slowed degree of metabolism as calcification occurs. This apparent decrease in cell numbers was not accompanied by a reduction in the total mRNA isolated for qRT-PCR analysis, possibly indicating that the standard curve used in generating our cell numbers may not be applicable if the MC3T3 cells are reaching a late stage of differentiation.

Figure 5. Environmental SEM images of MC3T3 cells adhered to the scaffolds three weeks after seeding. The specimen types are labeled on the images. Scale bar = 50 μm.

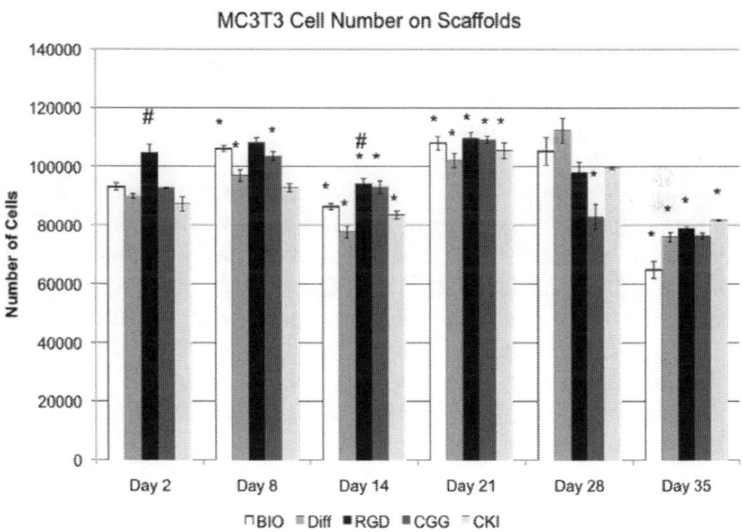

Figure 6. Cell counts on scaffolds using an Alamar Blue assay. Cells were treated with Alamar Blue for 14.5 hours starting at 48 hours after initial cell seeding. Cell counts were generated from an experimentally derived standard curve with undifferentiated cells. * = Significant Cell number change from previous timepoint examined, $p < 0.05$. # = Significant cell number difference over Diff. at same time point, $p < 0.05$.

qRT-PCR was performed to assess the differentiation of the MC3T3 cells on the scaffolds on interest, in the absence of differentiation media. mRNA was extracted from the specimens at Days 21 and 35 (Weeks 3 and 5) after seeding. Genes analyzed included BGLAP (osteocalcin), AP (alkaline phosphatase), SMAD4, RunX, and GAPDH was used as a housekeeping gene. Osteocalcin is a late stage marker of osteogenesis implicated in bone mineralization [57] -[60] . The expression of this protein increases during osteogenesis, reaching a maximum concentration at four weeks during in vitro differentiation experiments. Alkaline phosphatase is an enzyme that plays a role in bone mineralization and is considered an early stage marker of osteogenesis [57] . SMAD4 and RunX were chosen due to their known involvement in osteogenesis of MC3T3 pre-osteoblast cells. Specifically, RunX is essential in regulating osteocalcin and osteopontin expression during osteogenesis and is considered to be the central control gene in osteogenesis [1] [30] [46] [57] [59] [61] [62] . Smad4 is a transcription factor involved in transforming growth factor-β (TGF-β) signaling during bone growth and remodeling. It is also linked to osteocalcin expression via the activity of Smad2 [44] [63] [64] . qRT-PCR data was analyzed using the $\Delta\Delta$ Ct method, using GAPDH as a control gene, and the mRNA from the cells grown on nano-macroporous bioactive glass in the absence of any differentiation media (BIO) as the reference specimens. The data is presented as the fold change over the BIO specimens in Figure 7.

BGLAP (osteocalcin) expression is significantly upregulated in the cells growing on unmodified bioactive glass treated with differentiation media by Week 3. This upregulation persists through Week 5. The RGD, CGG, and CKI specimens have a moderate, statistically significant increase in osteocalcin expression at Week 3 (compared to the BIO specimen) as well; however, the upregulation is not as apparent as in the cells treated with differentiation media (Diff). As bioactive glass is known to be osteoinductive by itself [2] [10] [12] [24] , this increase in the late stage marker on the peptide-modified substrates without the addition of any additional osteoinductive compounds is considered significant. Future experiments, combining the peptide-silane modified surfaces and differentiation media will be performed to assess how these scaffolds would perform in a natural environment.

Alkaline phosphatase was not significantly upregulated over the BIO samples in any of the treatments at Week 3. AP mRNA has been shown to be upregulated in osteoblasts grown on bioactive glass, therefore, this result is not surprising [65] [66] . The CKI and CGG specimens both exhibit a jump in AP mRNA levels, potentially indicating the synthesis of fresh enzyme to augment bone growth. Typically AP mRNA expression will peak about 3 weeks into differentiation; however, there is an increase in AP activity by Week 5 of osteogenesis.

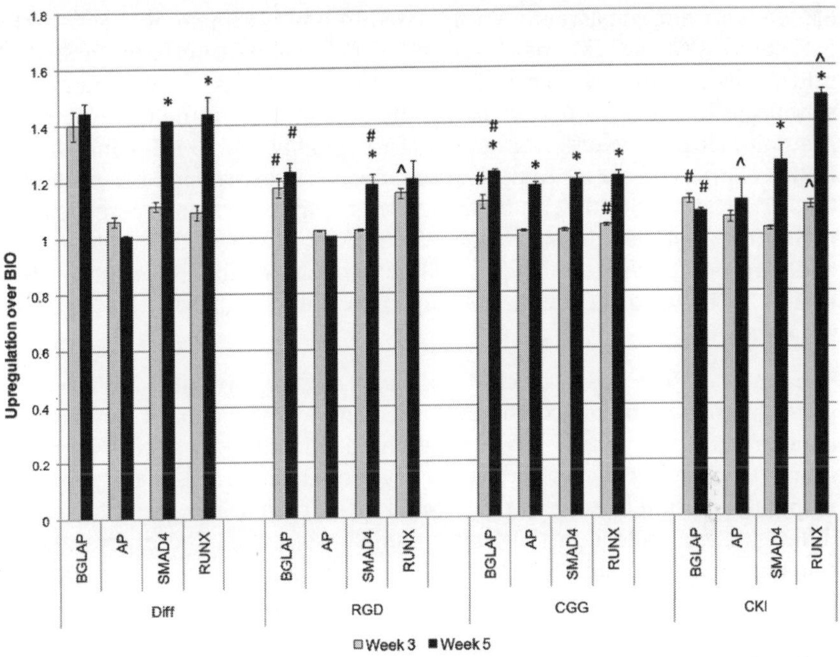

Figure 7. qRT-PCR analysis. Data was analyzed using the ΔΔCt method using GAPDH as a control gene and the BIO specimens (cells grown on bioactive glass in standard proliferation medium) as the reference sample. Data is presented as upregulation over the reference specimen, corrected for GAPDH expression. * = Upregulation over 3 weeks, $p < 0.01$, # = Downregulation over Differentiation medium on Bioactive Glass, $p < 0.01$, ^ = Upregulation over Differentiation medium on Bioactive Glass, $p < 0.05$.

Our data indicates that on bioactive glass samples that are treated with differentiation media, expression of AP peaks at Week 3; however, the CGG and CKI specimens appear to also undergo significant increases in AP expression, albeit a little later in the timeline. This could be indicative of either fresh enzyme production or a slight delay in differentiation over specimens that are treated with both soluble differentiation cues and an osteoproductive scaffold.

RunX is significantly upregulated in the Diff and RGD specimens at Week 3. RunX is involved with the control of osteocalcin expression; therefore, it is expected that these genes are consistently regulated. However, the CGG and CKI do not undergo any appreciable increase in RunX expression until Week 5, when all specimens demonstrate at least a 20% increase of RunX over the BIO specimens. In the CGG specimen, the expression levels of RunX rival those of the cells treated with differentiation media.

SMAD4 was not consistently upregulated in any specimen at Week 3. The Diff, RGD, CGG, and CKI specimens all exhibit a significant jump in SMAD4 mRNA expression by Week 5. The Diff specimens exhibit the most significant increase, as expected, but the cells grown on the CGG specimen are quite close in expression levels. Most notably, the SMAD4 increase in mRNA expression seems to peak after osteocalcin expression has peaked in these specimens. SMAD4, an important transcription factor in osteogenesis, is linked to osteocalcin expression via SMAD2, however, the exact timing and nature of this relationship is still not well understood, specifically as it relates to in vitro differentiation and BMP interactions in this signaling network.

The interactions of transcription factors and proteins involved in osteogenesis and bone growth are quite complex. The data presented here demonstrates that when cells are presented with solid-state cues (peptide-silanes) that mimic the active sites of key growth factors, the cells appear to respond in a similar way as they would if the cells are exposed to soluble differentiation factors in the form of small molecules. The differentiation timeline and expression profiles, while not perfectly in sync, are consistent across the different types of specimens treated, with the CGG specimen, containing three mixed peptide-silane cues, showing the most promise at potentially replacing media spiked with differentiation cues. This finding has broad implications across a wide variety of possible tissue engineering applications, from incorporating solid-state peptides into hydrogel "injectable" bone therapies, to coatings on permanent implants to facilitate attachment and site compatibility. This approach could surpass the potency of soluble growth factors in vitro, due to the extended and constant profile of the peptide-silane presentation. Future work should include a direct comparison of cells treated with soluble BMP-2 and BMP-9 with the cells grown on the solid-state networks.

4. CONCLUSION

This study examined the influence of bioactive peptides on the differentiation of MC3T3 cells grown on nano- macroporous bioactive glass. The approach described in this study represents a future enabling strategy in protein-free stem cell-based regenerative medicine. When compared to cells on non-modified bioactive glass that were treated with differentiation media, the cells on the peptide-silane treated samples underwent differentiation that was in line with cells treated with differentiation media, indicating that these solid-state cues were sufficient to enhance differentiation on bioactive glass in the absence of soluble signals. This is crucial in developing biomimetic bone scaffolds that do not contain the whole proteins or other difficult to control and possibly expensive molecular modifications. Ideally, these scaffolds would be placed in a location that provides the cells with necessary cues to regrow, and the peptides would enhance the re-growth of osteoblasts into the damaged area. To develop this study further, future research will include an analysis of the peptide-silane modified scaffolds in the presence of differentiation media and/or in the presence of the whole BMP-2, BMP-9 and fibronectin to determine if the multi-level approach of solid-state cues and soluble small molecules will enhance differentiation of the adherent osteoblasts.

ACKNOWLEDGEMENTS

Colin Przybylowski, Mohamed Ammar, Courtney LeBlon, Sabrina S. Jedlicka,, (NSF GRANT # DMR-0844014) for summer undergraduate research funding for C. Przybylowski (2010). We also acknowledge the Lehigh University Faculty Innovation Grant for seed funding to assist with peptide development and analysis. In addition, we would like to thank Dr. Alfred Miller for providing the XPS spectra, Dr. JuttaMarzillier for qRT- PCR assistance, and Drs. Himanshu Jain and Matthias Falk for useful discussion during the project. In addition, the qRT-PCR instrument was provided through an in-kind donation to Lehigh University from BD.

REFERENCES

1. Hattar, S., Asselin, A., Greenspan, D., Oboeuf, M., Berdal, A. and Sautier, J.-M. (2005) Potential of Biomimetic Surfaces to Promote in Vitro Osteoblast-Like Cell Differentiation. Biomaterials, 26, 839-848.

2. Hench, L.L., Xynos, I.D., Edgar, A.J., Buttery, L.D.K., Polak, J.M., Zhong, J.-P., et al. (2002) Gene Activating Glasses. Journal of Inorganic Materials, 17, 897-909.

3. Xynos, I.D., Edgar, A.J., Buttery, L.D.K., Hench, L.L. and Polak, J.M. (2000) Ionic Products of Bioactive Glass Dissolution Increase Proliferation of Human Osteoblasts and Induce Insulin-Like Growth Factor II mRNA Expression and Protein Synthesis. Biochemical and Biophysical Research Communications, 276, 461-465.

4. Barboza, E.P., Caúla, A.L., de Oliveira Caúla, F., de Souza, R.O., Neto, L.G., Sorensen, R.G., et al. (2004) Effect of Recombinant Human Bone Morphogenetic Protein-2 in an Absorbable Collagen Sponge with Space-Providing Biomaterials on the Augmentation of Chronic Alveolar Ridge Defects. Journal of Periodontology, 75, 702-708.

5. Bergeron, E., Marquis, M.E., Chrétien, I. and Faucheux, N. (2007) Differentiation of Preosteoblasts Using a Delivery System with BMPs and Bioactive Glass Microspheres. Journal of Materials Science: Materials in Medicine, 18, 255- 263.

6. Best, S.M., Porter, A.E., Thian, E.S. and Huang, J. (2008) Bioceramics: Past, Present and for the Future. Journal of the European Ceramic Society, 28, 1319-1327.

7. Blaker, J.J., Gough, J.E., Maquet, V., Notingher, I. and Boccaccini, A.R. (2003) In Vitro Evaluation of Novel Bioactive Composites Based on Bioglass (R)-Filled Polylactide Foams for Bone Tissue Engineering Scaffolds. Journal of Biomedical Materials Research Part A, 67A, 1401-1411.

8. Blaker, J.J., Nazhat, S.N. and Boccaccini, A.R. (2004) Development and Characterisation of Silver-Doped Bioactive Glasscoated Sutures for Tissue Engineering and Wound Healing Applications. Biomaterials, 25, 1319-1329.

9. Chen, Q.Z.Z., Thompson, I.D. and Boccaccini, A.R. (2006) 45S5 Bioglass®—Derived Glass—Ceramic Scaffolds for Bone Tissue Engineering. Biomaterials, 27, 2414-2425.

10. Hench, L.L. (2009) Genetic Design of Bioactive Glass. Journal of the European Ceramic Society, 29, 1257-1265.

11. Hench, L.L. (1998) Bioceramics. Journal of the American Ceramic Society, 81, 1705-1728.

12. Hench, L.L., Xynos, I.D. and Polak, J.M. (2004) Bioactive Glasses for in Situ Tissue Regeneration. Journal of Biomaterials Science-Polymer Edition, 15, 543-562.

13. Knabe, C., Stiller, M., Berger, G., Reif, D., Gildenhaar, R., Howlett, C.R. and Zreiqat, H. (2005) The Effect of Bioactive Glass Ceramics on the Expression of Bone-Related Genes and Proteins in Vitro. Clinical Oral Implants Research, 16, 119-127.

14. Xynos, I.D., Hukkanen, M.V.J., Batten, J.J., Buttery, L.D., Hench, L.L. and Polak, J.M. (2000) Bioglass® 45S5 Stimulates Osteoblast Turnover and Enhances Bone Formation in Vitro: Implications and Applications for Bone Tissue Engineering. Calcified Tissue International, 67, 321-329.

15. Neo, M., Nakamura, T., Ohtsuki, C., Kokubo, T. and Yamamuro, T. (1993) Apatite Formation on 3 Kinds of Bioactive Material at an Early Stage in Vivo—A Comparative Study by Transmission Electron Microscopy. Journal of Biomedical Materials Research, 27, 999-1006.

16. Maquet, V., Boccaccini, A.R., Pravata, L., Notingher, I. and Jérôme, R. (2003) Preparation, Characterization, and in Vitro Degradation of Bio-Resorbable and Bioactive Composites Based on Bioglass®-Filled Polylactide Foams. Journal of Biomedical Materials Research Part A, 66A, 335-346.

17. Marques, A.C., Almeidaa, R.M., Thiema, A., Wang, S.J., Falk, M. and Jain, H. (2009) Sol-Gel-Derived Glass Scaffold with High Pore Interconnectivity and Enhanced Bioactivity. Journal of Materials Research, 24, 3495-3502.

18. Garcia, A.J., Ducheyne, P. and Boettiger, D. (1998) Effect of Surface Reaction Stage on Fibronectin-Mediated Adhesion of Osteoblast-Like Cells to Bioactive Glass. Journal of Biomedical Materials Research, 40, 48-56.

19. Zhang, K., Wang, Y.B., Hillmyer, M.A. and Francis, L.F. (2004) Processing and Properties of Porous Poly(L-lactide)/ Bioactive Glass Composites. Biomaterials, 25, 2489-2500.

20. Arcos, D., Ragel, C.V. and Vallet-Regi, M. (2001) Bioactivity in Glass/PMMA Composites Used as Drug Delivery System. Biomaterials, 22, 701-708.

21. Boccaccini, A.R. and Maquet, V. (2003) Bioresorbable and Bioactive Polymer/Bioglass®Composites with Tailored Pore Structure for Tissue Engineering Applications. Composites Science and Technology, 63, 2417-2429.

22. Yao, J., Radina, S., Leboy, P.S. and Ducheyne, P. (2005) The Effect of Bioactive Glass Content on Synthesis and Bioactivity of Composite Poly (Lactic-Co-Glycolic Acid)/Bioactive Glass Substrate for Tissue Engineering. Biomaterials, 26, 1935-1943.

23. Kontonasaki, E., Sivropoulou, A., Papadopoulou, L., Garefis, P., Paraskevopoulos, K.M. and Koidis, P. (2006) Attachment and Proliferation of Human Periodontal Ligament Fibroblasts on Fibronectin-Coated Bioactive Glass Modified Ceramics. Bioceramics, 18, 727-730.

24. Kaufmann, E., Ducheyne, P. and Shapiro, I.M. (2000) Evaluation of Osteoblast Response to Porous Bioactive Glass (45S5) Substrates by RT-PCR Analysis. Tissue Engineering, 6, 19-28.

25. Lu, H.H., El-Amin, S.F., Scott, K.D. and Laurencin, C.T. (2003) Three-Dimensional, Bioactive, Biodegradable, Polymer- Bioactive Glass Composite Scaffolds with Improved Mechanical Properties Support Collagen Synthesis and Mineralization of Human Osteoblast-Like Cells in Vitro. Journal of Biomedical Materials Research Part A, 64A, 465-474.

26. Moura, J., Teixeira, L.N., Ravagnani, C., Peitl, O., Zanotto, E.D., Beloti, M.M., et al. (2007) In Vitro Osteogenesis on a Highly Bioactive Glass-Ceramic (Biosilicate®). Journal of Biomedical Materials Research Part A, 82A, 545-557.

27. Oonishi, H., Kushitani, S., Yasukawa, E., Iwaki, H., Hench, L.L., Wilson, J., et al. (1997) Particulate Bioglass Compared with Hydroxyapatite as a Bone Graft Substitute. Clinical Orthopaedics and Related Research, 334, 316-325.http://www.ncbi.nlm.nih.gov/pubmed/9005929

28. Sepulveda, P., Jones, J.R. and Hench, L.L. (2001) Characterization of Melt-Derived 45S5 and Sol-Gel-Derived 58S Bioactive Glasses. Journal of Biomedical Materials Research, 58, 734-740.

29. Silver, I.A., Deas, J. and Erecinska, M. (2001) Interactions of Bioactive Glasses with Osteoblasts in Vitro: Effects of 45S5 Bioglass®, and 58S and 77S Bioactive Glasses on Metabolism, Intracellular Ion Concentrations and Cell Viability. Biomaterials, 22, 175-185.

30. Varanasi, V.G., Saizb, E., Loomer, P.M., Ancheta, B., Uritani, N., Ho, S.P., et al. (2009) Enhanced Osteocalcin Expression by Osteoblast-Like Cells (MC3T3-E1) Exposed to Bioactive Coating Glass (SiO_2-CaO-P_2O_5-MgO-K_2O-Na_2O System) Ions. Acta Biomaterialia, 5, 3536-3547.

31. Verrier, S., Blakera, J.J., Maquet, V., Hench, L.L. and Boccaccini, A.R. (2004) PDLLA/Bioglass® Composites for Soft-Tissue and Hard-Tissue Engineering: An in Vitro Cell Biology Assessment. Biomaterials, 25, 3013-3021.

32. Vogel, M., Voigta, C., Gross, U.M. and Müller-Mai, C.M. (2001) In Vivo Comparison of Bioactive Glass Particles in Rabbits. Biomaterials, 22, 357-362.

33. Yuan, H.P., de Bruijn, J.D., Zhang, X.D., van Blitterswijk, C.A. and de Groot, K. (2001) Bone Induction by Porous Glass Ceramic Made from Bioglass® (45S5). Journal of Biomedical Materials Research, 58, 270- 276.http://www.ncbi.nlm.nih.gov/pubmed/11319740

34. Gao, T.J., Arob, H.T., Ylänen, H. and Vuorio, E. (2001) Silica-Based Bioactive Glasses Modulate Expression of Bone Morphogenetic Protein-2 mRNA in Saos-2 Osteoblasts in Vitro. Biomaterials, 22, 1475-1483.

35. Kim, H.W., Lee, H.H. and Knowles, J.C. (2008) Nanofibrous Glass Tailored with Apatite-Fibronectin Interface for Bone Cell Stimulation. Journal of Nanoscience and Nanotechnology, 8, 3013-3019. http://www.ncbi.nlm.nih.gov/pubmed/18681040

36. Leach, J.K., Kaiglerb, D., Wang, Z., Krebsbach, P.H. and Mooney, D.J. (2006) Coating of VEGF-Releasing Scaffolds with Bioactive Glass for Angiogenesis and Bone Regeneration. Biomaterials, 27, 3249-3255.

37. Bacakova, L., Filova, E., Kubies, D., Machova, L., Proks, V., Malinova, V., et al. (2007) Adhesion and Growth of Vascular Smooth Muscle Cells in Cultures on Bioactive RGD Peptide-Carrying Polylactides. Journal of Materials Science-Materials in Medicine, 18, 1317-1323.

38. De Giglio, E., Sabbatinib, L., Colucci, S. and Zambonin, G. (2000) Synthesis, Analytical Characterization, and Osteoblast Adhesion Properties on RGD-Grafted Polypyrrole Coatings on Titanium Substrates. Journal of Biomaterials Science, Polymer Edition, 11, 1073-1083.

39. Morgan, A.W., Roskova, K.E., Lin-Gibson, S., Kaplan, D.L., Becker, M.L. and Simon Jr., C.G. (2008) Characterization and Optimization of RGD-Containing Silk Blends to Support Osteoblastic Differentiation. Biomaterials, 29, 2556- 2563.

40. Itoh, D., Yoneda, S., Kuroda, S., Kondo, H., Umezawa, A., Ohya, K., et al. (2002) Enhancement of Osteogenesis on Hydroxyapatite Surface Coated with Synthetic Peptide (EEEEEEEPRGDT) in Vitro. Journal of Biomedical Materials Research, 62, 292-298.

41. Moursi, A.M., Damsky, C.H., Lull, J., Zimmerman, D., Doty, S.B., Aota, S., et al. (1996) Fibronectin Regulates Calvarialosteoblast Differentiation. Journal of Cell Science, 109, 1369-1380. http://www.ncbi.nlm.nih.gov/pubmed/8799825

42. Yang, X.B., Roacha, H.I., Clarke, N.M.P., Howdle, S.M., Quirk, R., Shakesheff, K.M., et al. (2001) Human Osteoprogenitor Growth and Differentiation on Synthetic Biodegradable Structures after Surface Modification. Bone, 29, 523- 531.

43. Massia, S.P. and Hubbell, J.A. (1991) An RGD Spacing of 440 nm Is Sufficient for Integrin Alpha V Beta 3-Mediated Fibroblast Spreading and 140 nm for Focal Contact and Stress Fiber Formation. Journal of Cell Biology, 114, 1089-1100.http://www.ncbi.nlm.nih.gov/pubmed/1714913

44. Chen, D., Zhao, M. and Mundy, G.R. (2004) Bone Morphogenetic Proteins. Growth Factors, 22, 233-241. http://www.ncbi.nlm.nih.gov/pubmed/15621726

45. Cheng, H.W., Jiang, W., Phillips, F.M., Haydon, R.C., Peng, Y., Zhou, L., et al. (2003) Osteogenic Activity of the Fourteen Types of Human Bone Morphogenetic Proteins (BMPs). Journal of Bone and Joint Surgery, American Volume, 85-A, 1544-1552.http://www.ncbi.nlm.nih.gov/pubmed/12925636

46. Lee, K.S., Kim, H.J., Li, Q.L., Chi, X.Z., Ueta, C., Komori, T., et al. (2000) Runx2 Is a Common Target of Transforming Growth Factor Beta1 and Bone Morphogenetic Protein 2, and Cooperation between Runx2 and Smad5 Induces Osteoblast-Specific Gene Expression in the Pluripotentmesenchymal Precursor Cell Line C2C12. Molecular and Cellular Biology, 20, 8783-8792.http://www.ncbi.nlm.nih.gov/pubmed/11073979

47. Wozney, J.M. (1992) The Bone Morphogenetic Protein Family and Osteogenesis. Molecular Reproduction and Deve- lopment, 32, 160-167.

48. Lee, J., Lee, J. and Murphy, W. (2010) Modular Peptides Promote Human Mesenchymal Stem Cell Differentiation on Biomaterial Surfaces. Acta Biomateriala, 6, 21-28.

49. Jedlicka, S.S., Rickus, J.L. and Zemyanov, D.Y. (2007) Surface Analysis by X-Ray Photoelectron Spectroscopy of Sol-Gel Silica Modified with Covalently Bound Peptides. Journal of Physical Chemistry B, 111, 11850-11857.

50. Jedlicka, S.S., Little, K.M., Nivens, D.E., Zemlyanov, D. and Ri, J.L. (2007) Peptide Ormosils as Cellular Substrates. Journal of Materials Chemistry, 17, 5058-5067.

51. Vueva, Y., Gama, A., Teixeira, A.V., Almeida, R.M., Wang, S.J., Falk, M.M., et al. (2010) Monolithic Glass Scaffolds with Dual Porosity Prepared by Polymer-Induced Phase Separation and Sol-Gel. Journal of the American Ceramic Society, 93, 1945-1949.

52. Wang, S., Falk, M.M., Rashad, A., Saad, M.M., Marques, A.C., Almeida, R.M., et al. (2011) Evaluation of 3D Nano- Macro Porous Bioactive Glass Scaffold for Hard Tissue Engineering. Journal of Materials Science-Materials in Medicine, 22, 1195-1203.http://www.ncbi.nlm.nih.gov/pubmed/21445655

53. Wang, S.J. and Jain, H. (2010) High Surface Area Nano-Macroporous Bioactive Glass Scaffold for Hard Tissue Engineering. Journal of the American Ceramic Society, 93, 3002-3005.

54. Hamid, R., Rotshteyn, Y., Rabadi, L., Parikh, R. and Bullock, P. (2004) Comparison of Alamar Blue and MTT Assays for High Through-Put Screening. Toxicology in Vitro, 18, 703-710.

55. Benesch, J., Mano, J.F. and Reis, R.L. (2008) Proteins and Their Peptide Motifs in Acellular Apatite Mineralization of Scaffolds for Tissue Engineering. Tissue Engineering Part B-Reviews, 14, 433-445.

56. Takeuchi, A., Ohtsuki, C., Kamitakahara, M., Ogata, S.-I., Miyazaki, T. and Tanihara, M. (2008) Biomimetic Deposition of Hydroxyapatite on a Synthetic Polypeptide with Beta Sheet Structure in a Solution Mimicking Body Fluid. Journal of Materials Science: Materials in Medicine, 19, 387-393. http://www.ncbi.nlm.nih.gov/pubmed/17607510

57. Lian, J.B., Javed, A., Zaidi, S.K., Lengner, C., Montecino, M., van Wijnen, A.J., et al. (2004) Regulatory Controls for Osteoblast Growth and Differentiation: Role of Runx/Cbfa/AML Factors. Critical Reviews in Eukaryotic Gene Expression, 14, 1-41.http://www.ncbi.nlm.nih.gov/pubmed/15104525

58. Ducy, P., Geoffroy, V. and Karsenty, G. (1996) Study of Osteoblast-Specific Expression of one Mouse Osteocalcin Gene: Characterization of the Factor Binding to OSE2. Connective Tissue Research, 35, 7-14. http://www.ncbi.nlm.nih.gov/pubmed/9084638

59. Gaur, T., Lengner, C.J., Hovhannisyan, H., Bhat, R.A., Bodine, P.V.N., Komm, B.S., et al. (2005) Canonical WNT Signaling Promotes Osteogenesis by Directly Stimulating Runx2 Gene Expression. Journal of Biological Chemistry, 280, 33132-33140.http://www.ncbi.nlm.nih.gov/pubmed/16043491

60. Hanada, K., Dennis, J.E. and Caplan, A.I. (1997) Stimulatory Effects of Basic Fibroblast Growth Factor and Bone Morphogenetic Protein-2 on Osteogenic Differentiation of Rat Bone Marrow-Derived Mesenchymal Stem Cells. Jour- nal of Bone and Mineral Research, 12, 1606-1614. http://www.ncbi.nlm.nih.gov/pubmed/9333121

61. Banerjee, C., Javed, A., Choi, J.Y., Green, J., Rosen, V., van Wijnen, A.J., et al. (2001) Differential Regulation of the Two Principal Runx2/Cbfa1 N-Terminal Isoforms in Response to Bone Morphogenetic Protein-2 during Development of the Osteoblast Phenotype. Endocrinology, 142, 4026-4039. http://www.ncbi.nlm.nih.gov/pubmed/11517182

62. Ito, Y. and Miyazono, K. (2003) RUNX Transcription Factors as Key Targets of TGF-Beta Superfamily Signaling. Current Opinion in Genetics & Development, 13, 43-47.

63. Hoffmann, A., Preobrazhenska, O., Wodarczyk, C., Medler, Y., Winkel, A., Shahab, S., et al. (2005) Transforming Growth Factor-Beta-Activated Kinase-1 (TAK1), a MAP3K, Interacts with Smad Proteins and Interferes with Osteogenesis in Murinemesenchymal Progenitors. Journal of Biological Chemistry, 280, 27271-27283.

64. Li, J., Tsuji, K., Komori, T., Miyazono, K., Wrana, J.L., Ito, Y., et al. (1998) Smad2 Overexpression Enhances Smad4 Gene Expression and Suppresses CBFA1 Gene Expression in Osteoblasticosteosarcoma ROS17/2.8 Cells and Primary Rat Calvaria Cells. Journal of Biological Chemistry, 273, 31009-31015.

65. Stein, G.S. and Lian, J.B. (1993) Molecular Mechanisms Mediating Proliferation Differentiation Interrelationships during Progressive Development of the Osteoblast Phenotype. Endocrine Reviews, 14, 424-442.http://www.ncbi.nlm.nih.gov/pubmed/8223340

66. Stein, G.S. and Lian, J.B. (1993) Cellular and Molecular Biology of Bone. Academic Press, San Diego.

9

LIPOSOMES IN TISSUE ENGINEERING AND REGENERATIVE MEDICINE

Nelson Monteiro[1,2], Albino Martins[1,2], Rui L. Reis[1,2] and Nuno M. Neves[1,2]

[1]3B's Research Group—Biomaterials, Biodegradables and Biomimetics Department of Polymer Engineering, University of Minho, Headquarters of the European Institute of Excellence on Tissue Engineering and Regenerative Medicine, AvePark, Zona Industrial da Gandra S. Cla´udio do Barco, 4806-909, Caldas das Taipas, Guimara˜es, Portugal

[2]ICVS/3B's, PT Government Associate Laboratory, Braga/Guimara˜es, Portugal

ABSTRACT

Liposomes are vesicular structures made of lipids that are formed in aqueous solutions. Structurally, they resemble the lipid membrane of living cells. Therefore, they have been widely investigated, since the 1960s, as models to study the cell membrane, and as carriers for protection and/or delivery of bioactive agents. They have been used in different areas of research including vaccines, imaging, applications in cosmetics and tissue engineering. Tissue engineering is defined as a strategy for promoting the regeneration of tissues for the human body. This strategy may involve the coordinated application of defined cell types with structured biomaterial scaffolds to produce living structures. To create a new tissue, based on this strategy, a controlled stimulation of cultured cells is needed, through a systematic combination of bioactive agents and mechanical signals. In this review, we highlight the potential role of liposomes as a platform for the sustained and local delivery of bioactive agents for tissue engineering and regenerative medicine approaches.

1. INTRODUCTION

Lipids are hydrophobic or amphiphilic small molecules [1]. Therefore, they can be classified as fatty acyls, glycerolipids, glycerophospholipids, sphingolipids, sterol lipids, prenol lipids, saccharolipids and polyketides [2,3]. The amphiphilic nature of some lipids allows them to form organized structures such as vesicles or membranes, when immersed in an aqueous environment. They can be extracted from plant- or animal-derived tissues by low-polarity solvents such as chloroform.

Lipids play a vital role in physiological and pathophysiological events of living systems [4]. It is believed that life started when nucleic acids were enclosed within a membrane. The biological membrane separates nucleic acids from the external environment and controls the transfer of information and the transport of ions and molecules between the inside and outside of the cellular membrane. A cell membrane is a complex and dynamic system which consists of two lipid molecules held together by hydrophobic interactions, and self-assembled as a continuous bilayer with proteins embedded within the membrane or transiently associated with it (figure 1) [6].

This dynamic system is obviously necessary for life. The inner part of the cell membrane aggregates the hydrophobic moieties of the lipids, and the outside is hydrophilic. Integral proteins are usually in the hydrophobic core of the bilayer [5]. Peripheral proteins are bound to the surface of the membrane. The fluid mosaic model hypothesizes that the plasma membrane and organelle membranes consist of proteins embedded in a fluid phospholipid bilayer. The position of proteins is not static. Like the phospholipids in the bilayer, membrane proteins are in constant motion [5]. Therefore, the cell membrane allows the chemical reactions to occur much more efficiently in an enclosed area and protects the genetic information. The main lipid biosynthetic organelle is the endoplasmic reticulum which produces the bulk of the structural phospholipids and cholesterol [7]. The most important lipid function in the organism is their role in the plasma membrane.

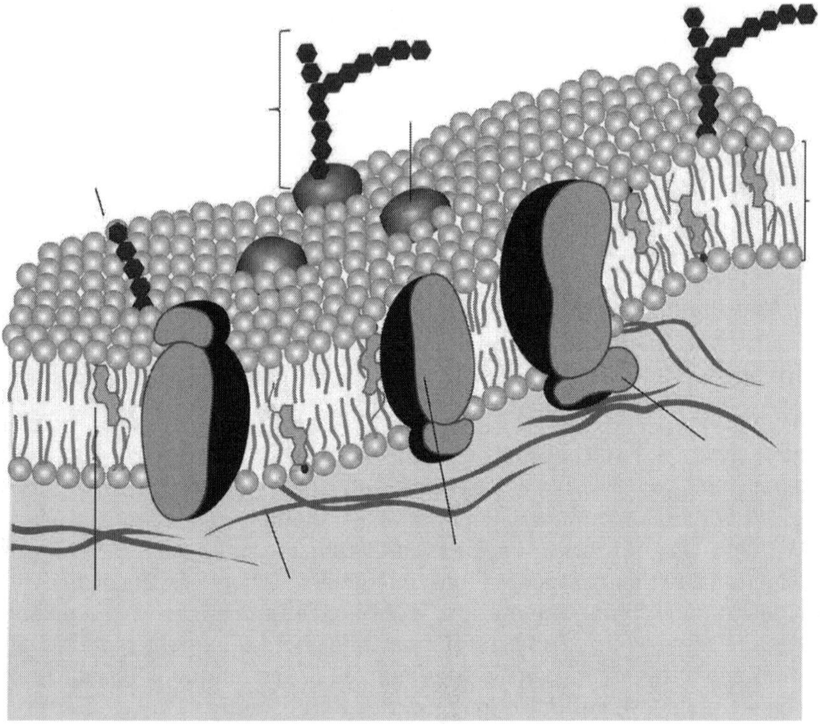

Figure 1. A cell membrane is a fluid with various proteins attached to the lipid bilayer. Adapted from [5]. (Online version in colour.)

Glycerophospholipids, or phospholipids, are key components of the lipid bilayer of cells [8–10]. Phospholipids, together with membrane proteins and cholesterol, are involved in many cell functions such as in the control of cell shape, compartmentalization, the storage of compounds, ion transport, metabolism, cell signalling processes and cell fusion processes [10]. Phospholipids consist of a glycerol which is linked to a phosphate group (PO_4^{2-}) and to two fatty acids. In some cases, the phosphate group is bonded to another small organic molecule, such as a choline. Figure 2 shows an example of a phospholipid and the main parts of its composition. Further details of lipids and their properties will be discussed in §2.1.

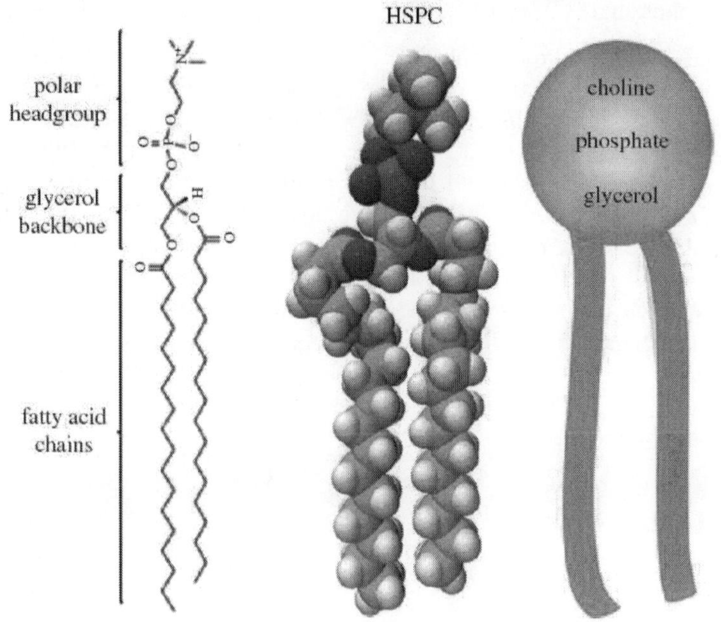

Figure 2. Chemical, three-dimensional and schematic structure of L-α-phosphatidylcholine, hydrogenated (soy) (HSPC) composed of fatty acid chains, glycerol backbone and the headgroup (choline). Adapted from [11]. (Online version in colour.)

Phospholipids may be classified as natural or synthetic. Natural phospholipids may be obtained from various sources such as soya bean or egg yolk. In terms of the polar headgroups, phospholipids are classified as phosphatidylcholine (PC), phosphatidylethanolamine (PE), phosphatidylserine (PS), phosphatidylinositol (PI), phosphatidylglycerol (PG) and phosphatidic acid (PA). PC and PE are the most abundant phosphatides in plants and animals and are also the most used to produce liposomes [12]. However, natural phospholipids are less stable than the synthetic phospholipids [13]. Synthetic phospholipids can be produced from natural lipids. The modification of the non-polar and polar regions of phospholipid molecules allows the creation of an unlimited variety of well-defined and characterized phospholipids [13]. Examples of synthetic lipids are dipalmitoyl phosphatidylcholine (DPPC), dimyristoyl phosphatidylcholine (DMPC), distearoyl phosphatidylcholine (DSPC) and hydrogenated soy phosphatidylcholine (HSPC), 1-palmitoyl-2-oleoyl-sn-glycero-3-phosphocholine (POPC), (1-palmitoyl-2-stearoyl(5-DOXYL)-sn-glycero-3phosphocholine (SLPC), 1,2-dioleoyl-sn-glycero-3-phosphocholine (DOPC), 1,2-distearoyl-sn-glycero-3-phospho-(1′-rac-glycerol) (DSPG), 1,2-

dipalmitoyl-sn-glycero-3-phosphoglycerol (DPPG) and 1,2-distearoyl-sn-glycero-3-phosphate (DSPA).

Sterol lipids, such as cholesterol (Chol) and its derivatives, are hydrophobic lipids that have an important role in the animal cell membrane [7]. Steroids are a family of lipids distinguished by the four-ring structure shown in figure 3. The various steroids differ from one another by the functional groups or side groups attached to those rings. Chol is distinguished by a hydrocarbon 'tail' formed of isoprene subunits. It is an important component of plasma membranes in many organisms. In mammals, it is also used as the starting point for the synthesis of several of the signalling molecules called hormones. Oestrogen, progesterone and testosterone are examples of hormones derived from Chol [14].

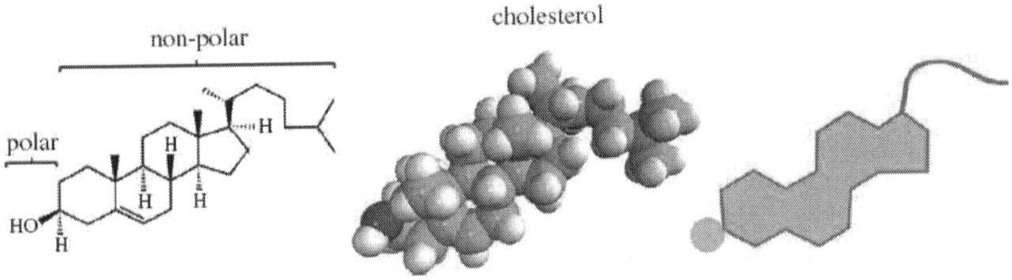

Figure 3. Chemical, three-dimensional and schematic structure of cholesterol (Chol). Adapted from [14]. (Online version in colour.)

Compounds that contain both hydrophilic and hydrophobic elements are referred as amphiphilic. The amphiphilic nature of phospholipids is of utmost importance for their biological function [14]. Specifically, this structure is responsible for their presence and functional role in cellular membranes. Phospholipids spontaneously self-assemble into ordered lyotropic liquid-crystalline phases in the presence of water [10]. The formation of these lipid-based structures depends on the phospholipid intrinsic factors, such as the nature and size of the lipid headgroup and the length and degree of unsaturation of the acyl chains, and on the extrinsic factors, such as temperature, pH, concentration and the presence of solutes and other lipids [7,10]. Examples of lipid-based structures are monolayers, lipid bilayers, micelles, liposomes and tubules [7,15]. Herein, we will review the use of lipids to form liposomes and discuss some of their applications, mainly focusing in tissue engineering (TE) and regenerative medicine.

2. LIPOSOMES

Liposomes are self-assembled vesicles that have the ability to encapsulate aqueous solutions and hydrophobic compounds. They are considered the oldest nanocarrier systems, which were first discovered in the mid-1960s by A. D. Bangham [16]. Bangham performed experiments to determine how lipids behave when immersed in water. When transmission electron microscopes became available, Bangham was able to obtain high-resolution images of the lipid–water mixtures. The images showed that lipids form water-filled vesicles, resembling cells, which he called liposomes [14]. For further details about the development of liposomes, the reader is directed to a recent review that cites the significant contributions of the early pioneers in the liposome field [17]. Since their discovery, a large number of studies have been carried out to understand their biophysical and biochemical properties, and possible applications. They have been used as: model membrane systems to study the basic nature of cell membranes [6], in biochemistry and molecular biology [6,18], in analytical methods [18], in microfluidic technologies [19], as a template for the production of nanogels [20], in cosmetics and food technology [21,22], in imaging [23], as drug delivery systems in pharmacology [24–26], and in TE [27]. In this section, we will revisit the liposome properties, their formulation/functionalization, preparation methods and stability.

2.1. Liposome properties

Liposomes are spherical lipid bilayers with diameters ranging from the nanometre to the micrometre scale [28]. Liposomes have different advantages when compared with other alternative carrier systems [29,30]. One of the main advantages is the fact that they are made of natural materials, i.e. lipids, and they can be easily synthesized in the laboratory.

The properties of the liposomes are mainly dependent on the characteristics of the lipids. A phospholipid has a headgroup, a glycerol backbone and two fatty acid chains (the so-called tails), as described above and depicted in figure 2. One of the oxygen groups of phosphoric acid may be esterified to a variety of organic molecules including glycerol, choline, ethanolamine, serine and inositol. Examples of negative lipids are PG, PS, PI and PA. Other lipids, such as 1,2-dioleoyl-3-dimethylammonium-propane (DODAP) and 1,2-dioleoyl-3-trimethylammonium-propane (DOTAP), are mixed with neutral phospholipids to produce positively charged liposomes [3]. The nature of the phospholipid is also related to the length of the fatty acid chains. Therefore, the fatty acids differ in the number of carbon atoms and the degree of unsaturation [3]. When a double bond exists between two carbon atoms (C = C) in a hydrocarbon chain, the chain is said to be unsaturated, whereas hydrocarbon chains without double bonds (i.e. C – C) are said to be saturated. The length and degree of saturation of the lipid chain influence the gel liquid-crystalline phase transition temperature, T_c (table 1). Considering the number of carbon atoms, the fatty acid can be named as lauric (C12), myristic (C14), palmitic (C16) and stearic (C18). In general, unsaturated fatty acids occur in natural phospholipids such as PC. DPPC, DMPC, DSPC and HSPC are the most common synthetic phospholipids.

Table 1. Crystalline phase transition (T_c) of some phospholipids. Adapted from [31].

phospholipid	acyl chain length, no. unsaturation	T_c (°C)
DSPC	18 : 0, 18 : 0	55
HSPC	16 – 18 (mixture)	52
DPPC	16 : 0, 16 : 0	42
POPC	16 : 0, 18 : 1	−7
SLPC	18 : 0, 18 : 2	−16.7
DOPC	18 : 1, 18 : 1	−21
DSPG	18 : 0, 18 : 0	53
DPPG	16 : 0, 16 : 0	41.1
DSPA	18 : 0, 18 : 0	58

In aqueous environments, the lipids tend to self-assemble into vesicles. Interactions between themselves, hydrophilic interactions between polar headgroups, van der Waals interactions between hydrocarbon chains and also with water (hydrophilic interactions and hydrophobic effect) lead to the formation of lipid-based structures such as liposomes and micelles (figure 4) [32]. Micelles are tiny droplets created when the hydrophilic heads of phospholipids face the water, and the hydrophobic tails cluster together to hide from the water molecules. Phospholipids with compact and short tails tend to form micelles. Phospholipids with longer tails tend to form liposomes (lipid bilayer with two sheets of phospholipid molecules). The head of the phospholipid could contain highly polar covalent bonds, as well as positive and negative charges. Typically, these charges and polar bonds of the head interact with water molecules, when a phospholipid is placed in a solution, whereas the fatty acid tails of a phospholipid do not interact with water (non-polar), which means that they do not form hydrogen bonds with the hydrocarbon tail [14].

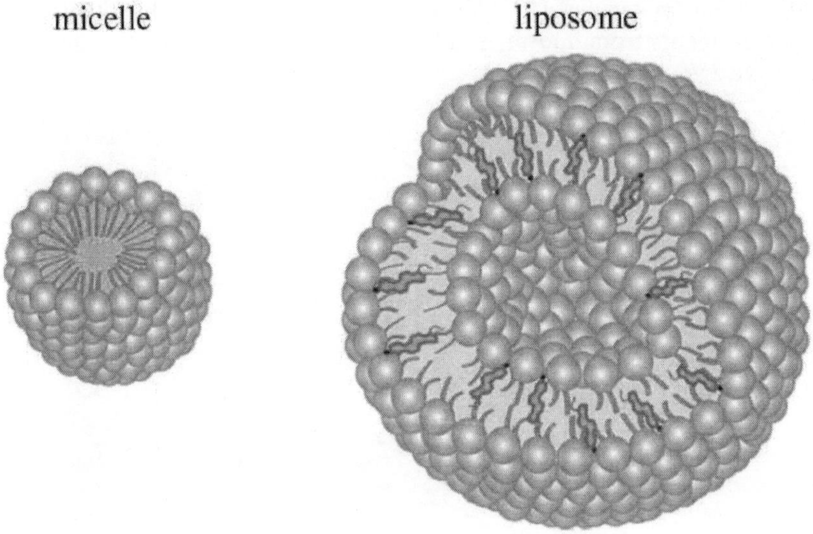

Figure 4. Alternative lipid-based particles: micelle and a liposome. (Online version in colour.)

The relative fluidity and the mobility of each lipid molecule within the bilayer constitute important properties of the liposomes. The lipid bilayer has the tendency to allow a given substance to pass across it—the so-called selective permeability [32]. This means that the internal environment of a liposome can become different from the outer space. Indeed, this capacity to keep different environments between external and internal space is also one of the main characteristics of cells [14,15].

As previously mentioned, the liposomes comprise highly selective membranes (figure 5). Small non-polar molecules move across the lipid bilayer quickly, whereas large molecules and charged substances cross the membrane slowly [14,33]. Additionally, water molecules and ions are also capable of moving across the lipid bilayer. Basically, water moves across lipid bilayers from regions of high concentration to regions of low concentration by osmosis. The solutes move by diffusion from a region of high concentration to a region of low concentration [14]. A comparison between glucose and sucrose permeability indicates that the smaller molecules (glucose) diffuses faster than larger molecules (sucrose) by two orders of magnitudes [33]. Comparing the diffusion of glucose and ions, glucose being a larger molecule, its permeability coefficient is approximately 6.40 to 250 times higher than Cl^-, K^+ and Na^+. The permeability of the lipid bilayer is very sensitive to the charge of the ion, being larger in the case of Cl^- by more than one order of magnitude compared with monovalent cations [33]. Moreover, liposomes have low permeability to hydrophilic molecules and high permeability to hydrophobic molecules [17]. More details about the release kinetics of bioactive agents from liposomes will be discussed in the section 'Release of bioactive agents from liposomes'.

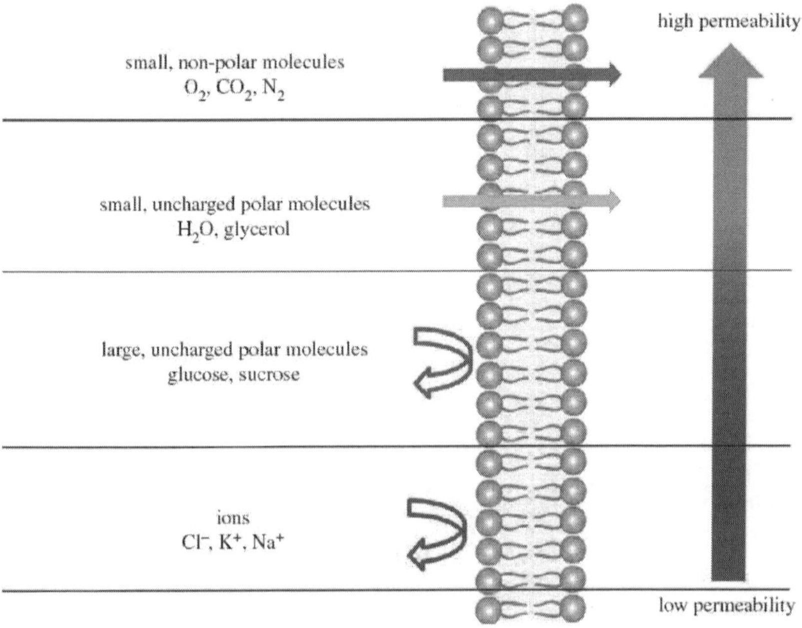

Figure 5. Selective permeability of lipid bilayers. Adapted from [14]. (Online version in colour.)

The degree of fatty acid saturation also affects the permeability of the lipid bilayers to specific molecules (figure 6) [15]. Because C – H bonds have more free energy than C = C bonds, saturated fats have higher chemical energy than unsaturated fats [14]. The double bonds create spaces among the tightly packed tails. Consequently, lipid bilayers containing many unsaturated fatty acids have more gaps and may be more permeable than bilayers with fewer unsaturated fatty acids [14].

Another parameter that can affect the mobility of the lipids within the bilayer is the temperature [34]. At a given temperature, a lipid bilayer can exist in a gel- or fluid-phase (figure 7). Depending on the lipid T_c, membranes composed of different lipids can exhibit different fluidity levels at the same temperature. The T_c of phospholipids depends on the following: (i) the length of the acyl chain in the lipid; (ii) the degree of saturation of the hydrocarbon chains in the lipid; (iii) the ionic strength of the suspension medium; and (iv) the type of the polar headgroup [36].

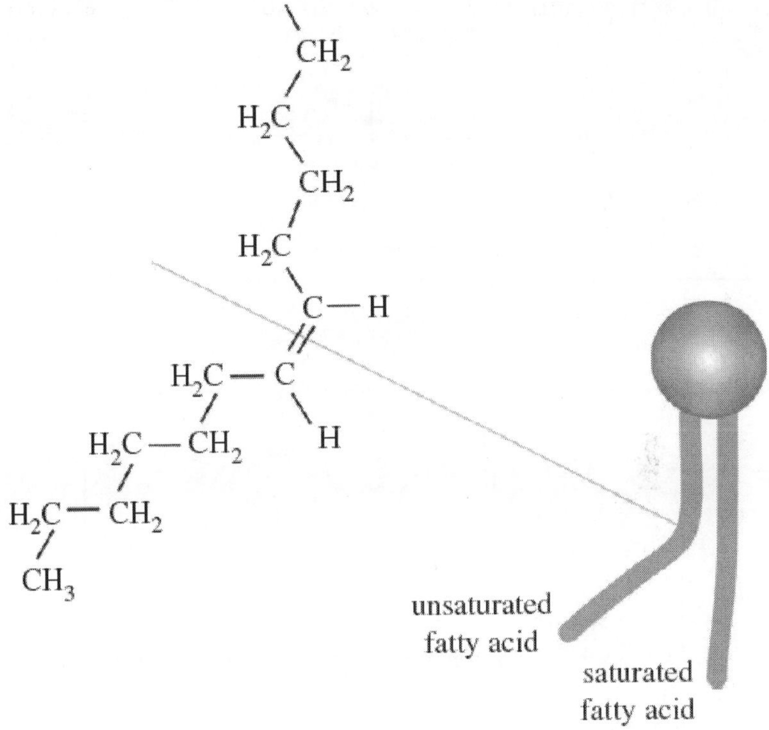

Figure 6. Unsaturated and saturated hydrocarbons. A double bond in a hydrocarbon chain creates spaces in the double layer. Adapted from [14]. (Online version in colour.)

Lipid bilayers dominated by phospholipids with long and saturated hydrocarbon tails should be stiffer and less permeable, because the interactions among the tails are stronger, leading to a high T_c. Indeed, hydrophobic interactions become stronger as saturated hydrocarbon tails increase in length [37]. Additionally, longer tail lipids have more area to interact [31,35,38]. Conversely, unsaturated lipids have a significantly lower T_c than saturated lipids [32].

After the formation of the liposomes, the movement of the molecules within and across the lipid bilayer is influenced by the temperature and the structure of the hydrocarbon tails [14]. As illustrated in figure 7, at a temperature below the T_c, phospholipids are in the gel phase, presenting low fluidity and permeability to encapsulated monovalent and divalent cations [15]. At this temperature, individual molecules within the bilayer move slowly. As a result, the hydrophobic tails in the interior of the lipid bilayer pack together more tightly [14]. At a temperature above the T_c, the

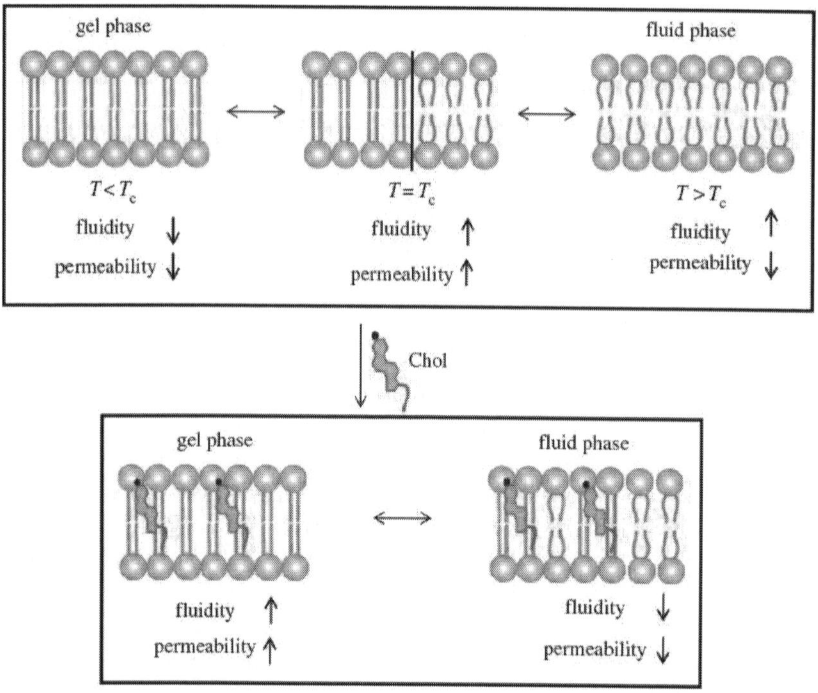

Figure 7. Effect of temperature and Chol on phospholipid bilayer permeability. Chol eliminates the transition phase of the lipid bilayer. The Chol increases the lipid bilayer permeability in the gel phase and decreases it in the fluid phase. Adapted from [15,35 ...

phospholipids are in a fluid phase and have high fluidity, but also relatively low permeability [15]. Therefore, the individual molecules within the lipid bilayer move rapidly. At a temperature equal to T_c, the lipid bilayer increases the permeability by several orders of magnitude. This phenomenon is attributed to the presence of highly permeable interfacial regions between coexisting gel and fluid bilayer domains [15,39].

As previously described, cholesterol and its derivatives are typically included in the liposome preparation. Therefore, the presence of Chol has a great influence on the properties of the lipid bilayers [31,33]. Specifically, the addition of Chol to a lipid bilayer decreases its fluidity and permeability to water within the fluid phase. The Chol molecule orients itself among the phospholipid molecules, with its hydroxyl group facing the water phase, the tricyclic ring sandwiched between the first few carbons of the fatty acyl chains, into the hydrocarbon core of the lipid bilayer [3]. Because the steroid rings of Chol are dense, adding Chol to a lipid bilayer should increase the density of the hydrophobic section. This decreases the flexibility of the surrounding lipid chains, increases the mechanical rigidity of the fluid bilayers and decreases their lateral diffusion (figure 7). Additionally, Chol can inhibit the crystallization of the hydrocarbon chains of saturated lipids to form a gel-state system [33,36]. It was observed that the effect of Chol in decreasing the permeability of Na^+, K^+, Cl^- and glucose at 36°C is independent of the surface charge and headgroups [33]. The permeability decreased by a factor of 4–18 for the cations, and by only 2 for glucose and Cl^-. The authors concluded that Chol affects the process of dissolution more than the process of diffusion [33]. Therefore, the addition of Chol to the lipid bilayer modifies its molecular packing. The lipid bilayer becomes more condensed when compared with pure phospholipids above the T_c, and more fluid when compared with the pure phospholipids below the T_c [33].

2.2. Formulation and functionalization

One of the main concerns in designing a liposome formulation is the stability of the bioactive agent and of the lipid components, which can be affected by the lipid concentration, the environment, pH and temperature, and also their susceptibility to enzyme degradation. Additionally, the liposome formulation depends on the bioactive agent to be encapsulated, the preparation method, the phospholipid composition and the intended application. Liposome formulations composed of phospholipids and Chol are defined as 'conventional liposomes' (figure 8*a*). Conventional liposome pharmacology and tissue distribution depend on the size, surface charge and membrane packing density [31]. For encapsulation of hydrophilic bioactive agents, Chol and saturated phospholipids are the most important factors that allow the membrane permeability to be reduced [31]. Cholesterol is commonly used in combination with phospholipids, because Chol can make liposome membranes stronger, as previously explained. The mole percentage of Chol within the liposome composition is commonly not more than 50% [12]. The hydrophobic interactions of Chol with membrane lipids forces phospholipid headgroups to shield Chol from water [40]. Cholesterol is necessary for the stabilization and maintenance of the bioactive agent in the core of the liposome, and this effect decreases with the increase in the temperature [33].

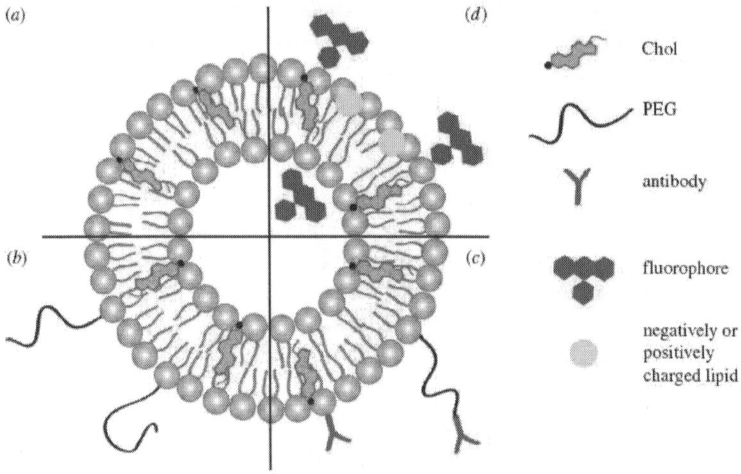

Figure 8. Cross section of a liposome: (*a*) conventional liposome; (*b*) SSL; (*c*) ligand-targeted liposome; (*d*) fluorescent liposome and charged liposomes. (Online version in colour.)

The discovery in 1990 of the steric stabilization was one of the main advances in the development of liposomes [17]. Coating the liposomes with hydrophilic polymers such as polyethylene glycol (PEG) changes their surface properties [41]. The liposome formulations composed of phospholipids, Chol and PEG are defined as 'sterically stabilized liposomes' (SSLs), also called 'stealth' liposomes, or PEGylated liposomes (figure 8*b*) [31,42]. It is known that PEG chains grafted onto lipid membranes act like polymeric chains grafted onto solid surfaces, stretching into brushes with increasing surface density. Based on this concept, it is believed that the PEGylated surface provides a steric barrier that prevents the adsorption of proteins onto the liposome surface [31,42]. Numerous formulations of SSLs have been described for the systemic delivery of bioactive agents [25]. However, this surface modification revealed some limitations due to its degradation under mechanical stress as a result of its ether structure and its non-biodegradability, as well as the resulting possible accumulation in the body [43]. Initially, it was believed that SSLs could be non-immunogenic, but an immune response can be elicited after subsequent administrations of PEGylated liposomes, leading to a rapid blood clearance or undesirable side effects [25,44]. Therefore, SSLs should be designed in a way that circumvents those PEG-associated limitations, mainly affecting the blood circulation half-life and the intracellular bioavailability [25,43,45]. A variety of synthetic and natural polymers such as poly(amino acid)s, heparin, dextran and chitosan were also proposed to replace PEG [43].

The effective delivery of bioactive agents towards target cells still represents an enormous challenge. One of the most promising strategies involves the covalent attachment of a ligand at the extremity of PEG chains grafted onto the liposome surface, which is intended to interact with antigens or receptors overexpressed at the surface of the target cells [46,47]. Antibodies and antibody fragments are the most widely used moieties/ligands due to the high specificity for their target antigens [28]. This has led to a new class of nanocarriers called ligand-targeted liposomes [48,49]. The conjugation of antibodies can be done directly either to the lipid bilayer of the liposomes in the presence or absence of PEG chains or to the distal end of the PEG chain (figure 8*c*). The former, antibody-targeted liposomes are rapidly cleared from the circulation, which limits their *in vivo* bio-distribution. PEG could eventually be used to overcome this limitation. However, when antibodies are attached at the liposome surface, their antigen binding may be masked by the presence of PEG in the same liposome, especially when longer chain PEG molecules are used. Thus, the latter strategy, coupling of ligands to the terminus of PEG molecules engrafted into the liposome surface, is the most used [17]. Trojan horse liposomes are brain transport vectors that include endogenous peptides, modified proteins and peptidomimetic monoclonal antibodies [49]. These liposomes target specific receptor/transport systems of the brain capillary endothelium and undergo receptor-mediated transcytosis through the blood–brain barrier.

Fluorescent lipids are also used in the liposome formulations (figure 8*d*). Liposomes carrying encapsulated fluorophores have been synthesized as novel fluorescent markers to image flow profiles in micro-fabricated structures [19]. Fluorescent-labelled lipids are incorporated within the lipid bilayer, and they have the following function: trigger rapid membrane fusion, enable fluorescence imaging of cell membranes and membrane traffic processes [50]. Moreover, negatively and positively charged lipids are also added to the liposome formulation to help in the liposome stabilization and avoid agglomeration [51].

Liposomes of DSPC/Chol (3 : 2) extruded through 400 nm filters were cleared 7.5 times faster than liposomes extruded through 200 nm filters, which in turn are cleared five times faster than small unilamellar vesicles (ULVs) [52]. The clearance of liposomes containing PEG-PE of different sizes (less than 70 nm, 150–200 nm and more than 300 nm) was investigated [53]. It was found that larger liposomes (more than 300 nm) and small liposomes (less than 70 nm) were more rapidly cleared from the circulation than the liposomes of size 150–200 nm [53]. In another study, the clearance was unaffected by the Chol content of the liposomes [54]. Liposomes of different composition with a particle size of about 90 nm were prepared using DSPC, Chol and cholesten-5-yloxy-*N*-(4-((1-imino-2-D-thiogalactosylethyl) amino) butyl) formamide (Gal-C4-Chol), and labelled with [^{3}H] cholesterol hexadecyl ether [55]. DSPC/Chol/Gal-C4-Chol (60 : 35 : 5) liposomes exhibit extensive hepatic uptake when compared with DSPC/Chol (60 : 40) liposomes [55]. Kawakami *et al.* [56,57] synthesized a mannosylated cholesterol derivative (Man-C4-Chol) for mannose receptor-mediated gene transfection to macrophages, and they studied the effect of the lipid composition of mannosylated cationic liposomes on *in vivo*transfection of plasmid DNA (pDNA). pDNA complexed with Man-C4-Chol liposomes showed higher transfection activity than that complexed with conventional cationic liposomes using mouse peritoneal macrophages. Therefore, the transfection efficiency of pDNA complexed with Man-C4-Chol liposomes was inhibited in the presence of mannose, suggesting that the complexes of pDNA and mannosylated cationic liposomes are recognized and taken up by the mannose receptors on macrophages. The liposome formulations, Man-C4-Chol (1/0.5/0.5), Man-C4-Chol/DOPE (3/2) and DOTMA/Chol (1/1), complexed with pDNA-encoding luciferase gene (pCMV-Luc) were compared by intravenous and intra-portal injections in mice. The highest gene expression was observed in the lung using the control cationic DOPE/Chol liposomes with both routes. Man-C4-Chol/DOPE liposome/DNA complexes showed the highest gene expression in the liver after intravenous and intra-portal injection. DOTMA/Chol/Man-C4-Chol liposome showed the highest gene expression in the liver by intravenous injection, but intra-portal injection showed high expression in the lung [56,57].

2.3. Classification

Liposomes could be classified based on the method of their preparation, by the number of bilayers present in the vesicle, or by their size [3]. However, the classification of liposomes by the number of bilayers and size are the most commonly used, rather than by the method of their preparation. Based on the number of bilayers and vesicles, the liposomes are classified as ULVs (25 nm to 1 μm), or multi-lamellar vesicles (MLVs, 0.1–15 μm), or multi-vesicular vesicles (MVVs, 1.6–10.5 μm). Furthermore, based on their size, unilamellar liposomes are classified as large unilamellar vesicles (LUVs, 100 nm to 1 μm) and small unilamellar vesicles (SUVs, 25–50 nm) (figure 9) [18].

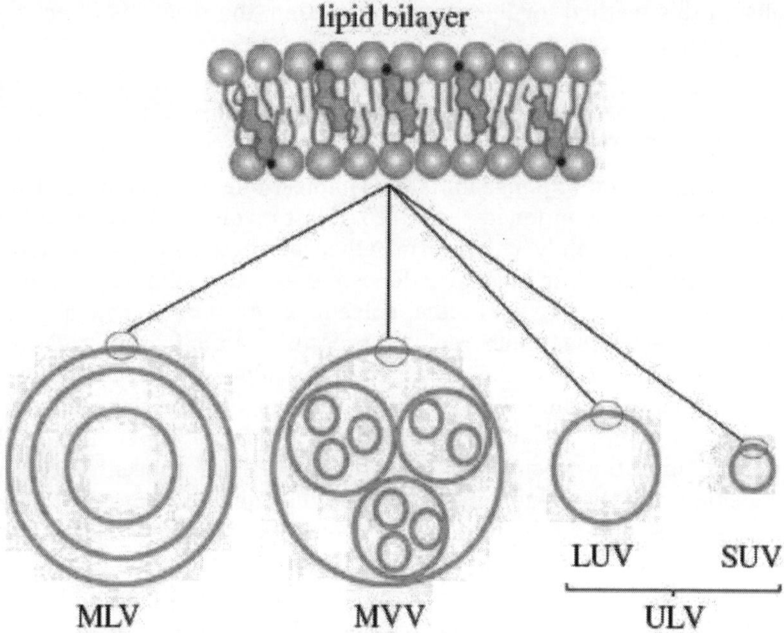

Figure 9. Lipid bilayer structure and types of liposomes: MLVs, MVVs, ULVs. Additionally, ULVs can be sub-classified as LUVs and SUVs. Adapted from [18]. (Online version in colour.)

2.4. Preparation methods

Many reports about the production of liposomes can be found in the literature [3,16,37,58–62]. Common liposome production methods include: thin-film hydration, reverse-phase evaporation, ethanol injection, polyol dilution, freeze–thaw, double emulsions, proliposome method, French press extrusion, detergent removal and high-pressure homogenization [12,37,51]. These methods typically produce LUVs or MLVs, depending on the selected method. Although all these methods can be used to manufacture liposomes, just three of them are usually used [63]: thin-film hydration, reverse-phase evaporation and the ethanol injection method, which are described below. One of the main concerns in liposome manufacturing is the toxicity related to the organic solvents used. Several techniques have been suggested for the removal of detergent and solvent traces from liposomes. These techniques include gel filtration, vacuum, centrifugation and dialysis [51]. A new method for the fast production of liposomes without the use of any hazardous chemicals or processes has been described. This method involves the hydration of the liposome components in an aqueous medium, followed by the heating (up to 120°C) of these components in the presence of glycerol (3% v/v) [64,65]. For more details on this method for liposome preparation, the reader is directed to the review in [51].

2.4.1. Thin-film hydration

The original method of Bangham is the simplest one when compared with the other methods mentioned above [16,61]. It involves the use of volatile organic solvents, mainly chloroform, ether or methanol, to dissolve or solubilize the lipids. The lipids are deposited as a thin film on the bottom wall of a round flask, while the solvent is evaporated by a rotary evaporation technique, under reduced pressure or a nitrogen stream. An aqueous buffer is added to the deposited lipids, allowing their hydration, at a temperature above the T_c of the lipid or of the highest melting component of the mixture [66]. Different MLVs can be produced depending on the hydration time, the re-suspension method, the lipid concentration and composition, and the volume of suspending the aqueous phase [12]. When comparing the thin-film hydration, the reverse-phase evaporation and the ethanol injection methods to prepare cationic liposomes, the results showed that liposomes prepared by the thin-film method were of the best quality and stability [63]. This liposome preparation method also presents some limitations such as low encapsulation ability and the difficulty in producing nano-size liposomes. However, sonication or extrusion through polycarbonate membranes can be employed to obtain ULVs [12,18,51].

2.4.2. Reverse-phase evaporation

In this method, the liposomes are formed from water-in-oil emulsions of phospholipids and buffer, in an excess of an organic phase. The phospholipids are firstly dissolved in organic solvents to form a film, then the solvents are removed by evaporation. The thin film is re-suspended in diethyl ether, followed by the addition of water. The preparation is then sonicated during a brief time period, forming a homogeneous emulsion. The organic solvents are removed under reduced pressure by continued rotary evaporation, resulting in the formation of a viscous gel-like intermediate phase characterized by a LUV dispersion. This method can encapsulate large macromolecules, with high encapsulation efficiency (EE) (20–68%). The main disadvantage of this method is the exposure of the material to be encapsulated to organic solvents and to sonication conditions, which may result in denaturation of sensitive molecules [12,66].

2.4.3. Ethanol injection

In this method, lipids dissolved in ethanol are rapidly injected into a buffer solution, where they spontaneously form SUVs, with a diameter of 30 nm

[67]. However, the size of the liposomes can be increased by increasing the lipid concentration [66]. This liposome preparation method has the advantage of avoiding chemical or physical treatment of lipids. However, the concentration of vesicles produced is low and it involves an extra step to remove ethanol [12].

2.5. Liposome characterization techniques

Methods of characterization of liposomes after production and upon storage are required for an adequate quality control [3]. Additionally, the chemical and physical characteristics of the liposomes allow their *in vitro* and *in vivo* behaviour to be predicted [31,68,69]. The main properties used to characterize the liposomes include the average diameter and the polydispersity index, the EE, the ratio of phospholipids to drug concentration and the lamellarity determination [69]. The average size and size distribution of liposomes are important parameters especially when the liposomes are intended for therapeutic use by the inhalation or parenteral routes [69]. For instance, small liposomes can pass through the fenestrae of the liver sinusoids and can circulate in the body for long time periods. Conversely, large liposomes are quickly cleared by macrophages [31]. Therefore, the potential therapeutic application of a given liposome is highly influenced by its average size and, consequently, by the possible liposome–cell interactions at the target sites of action.

Table 2 lists some techniques used to physically characterize the liposomes, in terms of the size distribution and zeta potential. Dynamic light scattering (DLS), sometimes referred to as photon correlation spectroscopy or quasi-elastic light scattering, is a technique enabling the measurement of the size of particles in the sub-micrometre range [72]. Typically, particle size measurements are taken between 2 and 5 min, which allows DLS to be classified as a rapid technique. To proceed with DLS analysis, the particles (up to tens of thousands) are suspended in a solution and illuminated by light in order for the particles to scatter light in a given index of refraction, different from that of the suspending solvent [3]. Therefore, liposomes are measured in their natural state, without the need for dehydration or staining.

Microscopic techniques such as atomic force microscopy (AFM), environmental scanning electron microscopy (ESEM), transmission electron microscopy (TEM) and confocal laser scanning microscopy (CLSM) can give more information about the nanoscale structures of liposomes. They can provide information regarding shape and morphology (i.e. by AFM and TEM), dimensions (by AFM, ESEM, TEM and CLSM), surface properties (just by AFM) and internal structure (only by CLSM) [70]. When comparing all these microscopy-based techniques, electron microscopy (i.e. ESEM or TEM) requires the specialized freeze fracture technique, which is prone to introduce distortion and artefacts from sample preparation, and the need to have hundreds of photographs in order to provide comparable statistics, constituting time-consuming techniques [69].

Table 2. Liposome physical characterization techniques.

physical characterization	technique	references
size distribution	DLS	[68–70]
	static light scattering	
	gel exclusion chromatography	
	light microscopy	
	laser diffraction	
	TEM	
	AFM	
	CLSM	
	small-angle X-ray scattering	
	flow cytometry	
	field-flow fractionation	
zeta potential	LDE	[68,71]
	capillary electrophoresis	

The zeta (ζ)-potential is used to measure the intensity of the repulsive electrostatic interaction between naturally charged colloidal particles [73]. Indeed, measurements of zeta potential are commonly used to predict the stability of colloidal systems. If all the particles in a suspension have a large negative or positive zeta potential, they will tend to repel each other and, therefore, there will be no tendency to aggregate. Typically, particles with zeta potentials more positive than +30 mV or more negative than −30 mV are considered stable [69]. However, if the particles have low zeta potential values, then there will be no force to prevent the particles from flocculating [69].

Laser Doppler electrophoresis (LDE) is the most widely used technique to measure the zeta potential. In this technique, a laser is used to provide a light source, illuminating particles within the samples. The incident laser beam passes through the centre of the sample cell and the light is scattered at a determined angle, which is detected. When an electric field is applied to the cell, any particles moving through the measurement volume will lead to fluctuation of the detected light with a frequency proportional to the particle speed. This information is passed to a digital signal processor and, then, the zeta potential is calculated [69]. This technique has been used to investigate how particle size changes as a function of any parameter of the preparation. It is used to monitor the effect of medium changes, e.g. pH, temperature, surfactant, blood serum, adsorption of proteins, and to develop formulations that resist aggregation/flocculation. It also enables the thickness of coatings at the surface of the liposomes to be measured. It has been widely used to predict the effectiveness of the liposomes' coating against opsonization *in vivo* [74]. Moreover, it gives information about whether the active agent used is encapsulated or adsorbed at the surface of liposomes. For instance, one can obtain information about DNA–liposome interactions, DNA–peptide interactions and DNA condensation during liposomal encapsulation, tracking the changes in size of liposome–drug complexes [74–76]. However, there are several factors that can affect the zeta-potential, such as pH, conductivity and concentration of a formulation component.

2.6. Liposome limitations

Liposomes have been widely used as carriers to encapsulate and to protect bioactive agents from the surrounding environments. Indeed, liposome drug products were the first type of therapeutic nanoparticles being introduced in the market [77]. For instance, Doxil, a doxorubicin-loaded liposome, was approved by the US Food and Drug Administration in 1995 for treatment of Kaposi's sarcoma [77]. In spite of the successful clinical application and of the advantages of this drug carrier, liposomes also present limitations, namely their low stability, low solubility, short circulation half-life, susceptibility to lipid oxidation and hydrolysis, leakage and fusion, reproducibility, difficulties in scaling up, high production costs and sterilization issues [62,78].

Stability is the main concern in all steps of liposome production, storage and administration, and includes the examination of: (i) the chemical stability of the lipids; (ii) the maintenance of the vesicle size and structure; (iii) the retention of entrapped contents; and (iv) the influence of biological fluids on the integrity and permeability properties of the liposomes [66]. Chemical and physical stability are important parameters that affect the biological performance of liposomes [12]. The physical stability of liposomes is mainly related to possible aggregation/agglomeration, fusion and their content leakage to the surrounding environment [31]. Liposomes stored under sterile conditions, in phosphate-buffered saline under nitrogen, can retain their entrapped bioactive agent for extended periods of time [66].The relative retention of a bioactive agent depends on the liposome type (i.e. MLV > reverse-phase evaporation vesicle > SUV), the temperature (i.e. 4°C > 25°C > 35°C) and lipid composition (i.e. saturated phospholipids > saturated phospholipids with equimolar cholesterol > unsaturated phospholipids with equimolar cholesterol > unsaturated phospholipids) [66]. Lipids can suffer auto-oxidation, which is usually induced by light, metal ions or temperature. Additionally, the hydrolysis of the phospholipids to fatty acids and 1- and 2-acyl-lysophospholipids leads to the production of glycerol phosphor compounds and causes chemical degradation of the liposomes during storage [79]. Antioxidants, complexing agents (e.g. EDTA) and inert atmosphere (e.g. nitrogen) are commonly used to avoid hydrolysis of the phospholipids. Other ways to overcome problems concerning the chemical decomposition of liposomes is their storage in a dry state (i.e. freeze dried), using a cryoprotectant. For example, addition of trehalose, which replaces water during the freeze drying, is effective in preventing fusion and dehydration damage in phospholipid vesicles [80].

Initially, it was assumed that, as liposomes were primarily made of natural lipids such as PC, they would avoid the mononuclear phagocyte system clearance and would not be recognized as an antigen. However, it was realized that liposomes would be accumulated in organs such as liver and spleen, and cleared by macrophages (immune system), and the clearance was related to the composition and the size of the liposomes [31]. Therefore, this could be an advantage when those organs or cell populations are the intended target [36]. However, when other organs and cell populations are the target, this represents a main obstacle. Since this discovery, the goal has been to design new liposome formulations to bypass the immune system. Therefore, the physiological conditions of the body should be taken into consideration when designing a liposome formulation for a specific cell or organ target when the liposomes are injected intravenously [29].

Currently, there are numerous methods available for laboratory-scale production (§2.4), but only a few large-scale manufacturing techniques are available [62]. Liposome manufacturing involves numerous unit operations which are not easy to scale up to commercial production levels [78]. Therefore, all the large-scale manufacturing techniques have serious limitations in terms of entrapment of sensitive molecules, exposure to mechanical and/or chemical stress, temperature and sterilization conditions. A detailed review of liposome technology aimed at industrial-scale production can be found in [62]. Currently, there is no established consensus regarding the liposome manufacturing methods. Solvent extraction systems require significant investment and are very expensive to operate and maintain. The most used method to achieve sterility for pharmaceutical products is sterile filtration using a polycarbonate membrane [78]. However, this method has disadvantages; for example, it needs to be performed under aseptic conditions, it is relatively expensive as it operates under high pressure, it is time consuming and it is not effective in removing virus [78].

3. BIOACTIVE AGENT DELIVERY

3.1. Loading liposomes with bioactive agents

The main advantages of liposomes as a bioactive agent delivery system are: (i) they have a versatile structure which can be tailored for each application; (ii) they can accommodate any type of bioactive agents either in their inner compartment (i.e. hydrophilic molecules) or within the lipid bilayer (i.e. lipophilic molecules) or both (amphiphilic molecules); (iii) their nanoscale properties determine the solubility, diffusivity, bio-distribution and biological fatc; (iv) they are minimally toxic, non-immunogenic and fully biodegradable; (v) they are flexible to link with site-specific ligands to achieve active targeting; and (vi) they increase the efficacy and therapeutic index of bioactive agents [12,37]. Therefore, the retention of bioactive agents within liposomes is especially important, not only during storage but also during *in vivo* administration [17].

The assembling of phospholipids into a hydrated bilayer structure is coupled with the entrapment of a portion of the aqueous medium within the continuous closed bilayer. Therefore, it may seem that the encapsulation of bioactive agents into liposomes should be a trivial process. However, there are some limitations to the encapsulation of molecules into liposomes, such as the type of bioactive agent (i.e. molecular weight and chemical properties), the liposome (i.e. size and lipid composition) and the manufacturing method. Selecting bioactive agents with physical characteristics that make them susceptible to retention into the liposomes is another approach to control the loading and release rate [17]. There are two ways to encapsulate bioactive agents into liposomes: passively, when the bioactive agent is encapsulated during the liposome formation; or actively, when the bioactive agent is encapsulated after liposome formation [37]. The advantage of active encapsulation is that the bioactive agent loading can be performed independently of the time and site of liposome production [3,17]. However, they can only be applied to a small number of bioactive agents with specific physico-chemical properties [12].

Pay load (PL) and EE are the terms used to determine the amount of bioactive agent incorporated into the liposomes. PL is the bioactive agent–lipid (mol mol^{-1}) ratio and EE is the PL in the final liposome formulation compared with the initial PL used for liposome preparation. Sometimes, the EE is used to indicate the percentage of bioactive agent encapsulated in relation to the amount of bioactive agent offered for encapsulation during liposome preparation, but the amount of lipid is not quantified. This quantification method is misleading, as it highly depends on the initial amount of bioactive agent offered to a specific lipid quantity [12] (figure 10).

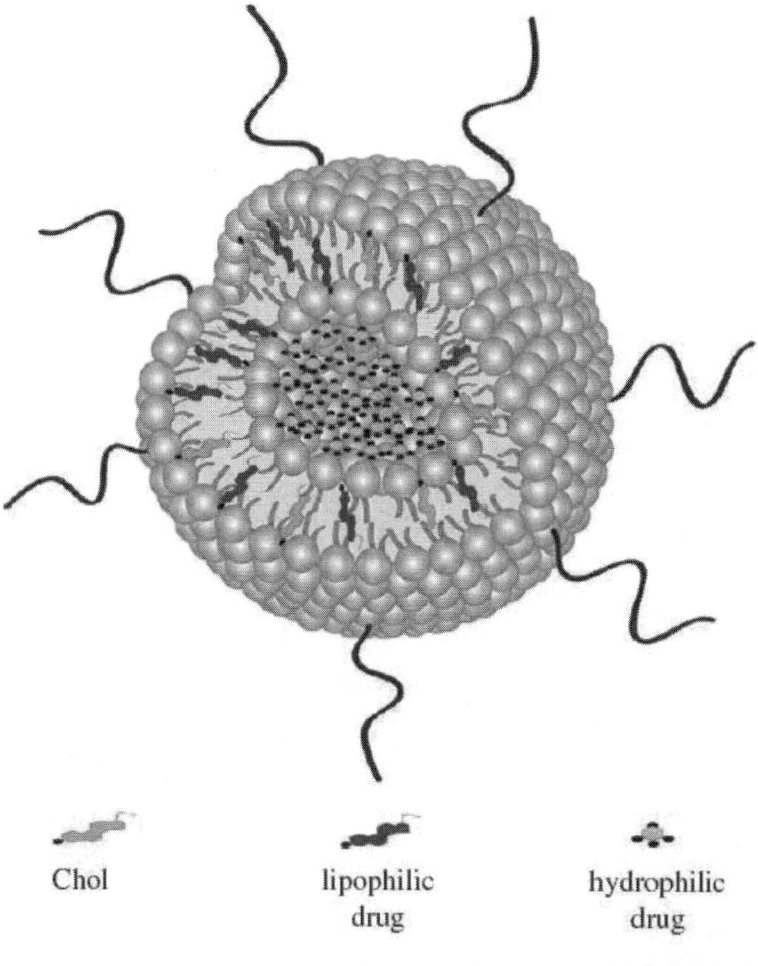

Figure 10. Representation of hydrophilic and lipophilic drug encapsulation into the liposome. Hydrophilic drug is encapsulated in the inner core of the liposome. Lipophilic drug is encapsulated into the lipid bilayer. Chol and the lipophilic drug compete for the ...

3.1.1. Hydrophilic bioactive agents

These are dissolved in the external aqueous phase, during liposome preparation, and become entrapped within the inner compartment of the liposome after solvent evaporation [81]. Thin-film hydration is the simplest method for the encapsulation of hydrophilic bioactive agents. Encapsulation becomes more difficult and inefficient as the size of the bioactive agent increases, e.g. large plasmids [36]. Additionally, the EE and bioactive agent-to-lipid ratios (i.e. PL) achieved by thin-film hydration are low [36]. The percentage of encapsulation of hydrophilic bioactive agent depends on the size of liposomes: MLVs > LUVs > SUVs [82]. Also, the liposome charge affects the entrapment of hydrophilic bioactive agent: positively charged > negatively charged > neutral liposomes [83]. The preparation of liposomes using unsaturated phospholipids (e.g. DOPC) demonstrated a higher EE than those liposomes using saturated phospholipids (e.g. DPPC) [84]. The EE was more dependent on the number of unsaturated bonds than on the alkyl chain length of the phospholipid molecule [84].

To improve the EE of hydrophilic bioactive agents, it is necessary to use different methods of liposome preparation, such as reverse-phase evaporation, dehydration–rehydration of empty liposomes and freeze–thaw cycling [36]. There are other strategies to increase the EE of hydrophilic bioactive agents into liposomes: by changing the pH and the ions and creating ammonium sulfate gradients (also called active loading) [3,17,85]. The development of these strategies for improving encapsulation and retention of bioactive agents into liposomes relies on experimental studies aimed at understanding the molecular factors governing the barrier properties [85]. For example, bioactive agents that are weak base/acid can diffuse through the lipid membrane and accumulate into the inner compartment of the liposome [85,86]. The ammonium sulfate gradient approach differs from most other chemical approaches as it does not require liposomes with an acidic interior or an alkaline extra-liposome phase [86]. This approach has been used to encapsulate bioactive agents inside the liposomes at high efficiency (more than 90%) [86]. Moreover, to solve the problem of stabilization of bioactive agents, one may increase the rigidity of the liposome membrane, while concurrently decreasing their tendency for aggregation by properly selecting the liposome lipid components [12]. The use of a thin polymeric coating made of chitosan or alginate may be a strategy to increase the stability of bioactive agents into the liposomes [87–89]. However, the polymeric coatings may increase the size of the liposomes and may also interact with the phospholipid bilayer. Another strategy is to covalently cross-link inter-lipid bilayers. For instance, cross-linked MLVs can reduce systemic toxicity and improve therapeutic efficacy [90].

3.1.2. Lipophilic bioactive agents

In order to maximize the association of lipophilic bioactive agents with liposomes, the most common practice is to mix the lipids with the lipophilic bioactive agent and evaporate the solvent to form a thin-lipid film. This mixture is re-hydrated in the buffer and the liposome-associated bioactive agent is separated from the free bioactive agent [66]. The interaction of such bioactive agents within the lipid bilayer depends on the amount and structure of the lipophilic bioactive agent, which results in alterations in the vesicle properties such as permeability, size and stability of the bilayer [91]. The encapsulation of lipophilic bioactive agents decreases as the lipid chain length increases, and may destabilize the liposomes [92]. Conversely, the addition of Chol increases the rigidity and stability of liposome membranes, as was mentioned in the section on 'Liposome properties'. However, if the bioactive agent is lipophilic, it might be displaced by adding Chol in the lipid bilayer. Therefore, the presence of Chol decreases the encapsulation of lipophilic bioactive agents [91,92]. A system has been proposed to enhance the EE of lipophilic bioactive agents [93]. This consists of combining liposomes and cyclodextrin–bioactive agent complexes by forming bioactive agent-in-cyclodextrins-in-liposomes [93]. Cyclodextrins are hydrophobic inside and hydrophilic outside, cavity-forming, water-soluble oligosaccharides that can accommodate water-insoluble bioactive agents into their cavities, increasing their water solubility. Therefore, the water-soluble bioactive agent–cyclodextrin complex is capable of being encapsulated within the aqueous compartments of the liposomes. An EE of 32.3 ± 11.9% for dehydroepiandrosterone and 31.9 ± 11.8% for hydroxypropyl-β-cyclodextrin was achieved by using this method [12,93].

3.2. Release of bioactive agents from liposomes

Another critical issue when designing liposomes as bioactive agent carriers is the control of the bioactive agents' release, which may dictate their therapeutic safety and efficacy. The release kinetics is affected by the nature of the bioactive agent, the liposome composition, the method of encapsulation (i.e. passive or active) and the intended application [12,94]. To understand the release of bioactive agents from liposomes, we have to understand the structure of the liposome and the possible interactions between the bioactive agent, the lipid bilayer and the surrounding environment. For example, encapsulation of dexamethasone (Dex) is higher in DSPC liposomes than in PC liposomes, and Dex is displaced from liposomes as the Chol content of liposome membranes increases [55,91]. Also, a kinetic release study performed in buffer or serum medium showed that the release of Dex was provoked by the dilution of liposome [91]. In another study, the release profile of Dex from the DSPE–PEG-coated liposomes was performed in phosphate-buffered saline using dialysis tubes for 21 days [55]. The release profile of Dex showed an initial burst release within 12 h. Following the initial release, a slower release was observed until day 6. Afterwards, Dex continued to be released at a slower but steady rate until day 21 [55].

Bioactive agents (i.e. molecules or ions) encapsulated into the liposomes tend to move spontaneously to the outer environment of the lipid bilayer. The movement of molecules and ions that results from their kinetic energy is known as diffusion [14]. Therefore, the bioactive agent tends to move from a region of high concentration to a region of low concentration. This process depends on the type of bioactive agent and the lipid composition [15,36,39]. For instance, when the initial concentration of Ca^{2+} is high inside the liposome, but low outside, heating to T_c results in immediate diffusion of the Ca^{2+} into the medium [15,39]. Therefore, phospholipids with high T_c may reduce the diffusion of ions and low molecular weight molecules. The influence of the liposomal composition on release of ibuprofen was investigated [95]. The long alkyl chain lipid enhanced ibuprofen EE and retention at 37°C: dilignoceroyl phosphatidylcholine (C_{24}PC) > DSPC > DMPC > PC. After 30 min of incubation, PC liposomes released 15.5% of the ibuprofen load compared with 1.5% for C_{24}PC liposomes, and major differences in release were evident after 24 h of incubation [95]. At 37°C, both PC (T_c < 0°C) and DMPC (T_c = 23°C) liposomes were in the fluid state resulting in an increasing amount of drug released compared with the higher transition temperature lipids DSPC (T_c = 55°C) and C_{24}PC (T_c= 80°C). Therefore, C_{24}PC liposomes released less ibuprofen than DSPC liposomes despite both systems being in the ordered gel phase. This behaviour may be explained by the increased van der Waals interactions between the longer lipid chains and the increased lipid phase area within the liposomes enhancing the drug binding and bilayer stability [95]. Moreover, the inclusion of charged lipids in the liposome formulations may influence bioactive agents' release. The addition of 2 mol of the anionic lipid dicetylphosphate (DCP) reduced the EE and increased the release of ibuprofen (69.0 ± 3.7%), and this was explained by the electrostatic repulsive forces between the carboxyl group of ibuprofen and the anionic headgroup of DCP [95]. The addition of 2 mol of the cationic lipid stearylamine (SA) to the liposome formulation (PC : Chol—16 mol : 4 mol) increased the EE by approximately 8–47% and also increased the release of ibuprofen (71.2 ± 2.8%) when compared with PC : Chol [95]. The authors suggested that the electrostatic attraction between the positively charged head group in SA and the carboxyl group present in dissociated ibuprofen appears to have little effect on ibuprofen association with MLVs. Moreover, addition of SA to the liposome formulation also resulted in a reversal of surface charge (due to the cationic headgroup of SA) and an increase in vesicle size of approximately 0.8 μm compared with the unmodified PC : Chol formulation. This could be the reason for the increase in the release of the drug. These results indicate that the presence of charged lipids (anionic or cationic) within the liposome bilayer increases the permeability of the lipid bilayer and lipid–drug binding.

A successful treatment using bioactive agent-loaded liposomes depends also on the route of administration (e.g. subcutaneously, orally and intravenously). Conventional liposomes, subcutaneously administrated, aim to target the lymphatic system for imaging, distribution of therapeutic agents or vaccination [12]. Moreover, conventional liposomes may be developed to circumvent the endoplasmic reticular system or to mask the toxic side effects of bioactive agents [12]. Liposomes administered intravenously face barriers such as the endothelial lining of the vasculature and the blood–brain barrier. In those particular situations, it is important to keep the bioactive agent in liposomes until they reach the target site [36]. If the bioactive agent leaks out of the liposome at a rapid rate, it will be lost before reaching the site of action, and no therapeutic benefit will be obtained. On the other hand, if the bioactive agent leaks out of the liposomes slowly, it will be able to reach the site of action, but the levels of released bioactive agent will never reach the desired therapeutic concentrations [36]. The low integrity of liposomes, after *in vivo*administration and in contact with the blood components, may result in removal of some lipid molecules and consequent opening of pores into the liposome bilayer, through which the loaded bioactive agent may leak out. Also, the physical instability of liposomes that results in aggregation and fusion may release the bioactive agent from liposomes [36]. For example, neutral liposomes tend to aggregate and increase the size of the liposomes [96].

The interaction of liposomes with cells is a critical aspect of bioactive agents' efficacy. In some clinical applications, the bioactive agent has to be released inside the cells (e.g. tumour cells) to have a beneficial therapeutic effect. *In vitro* and *in vivo* studies have shown that the main interactions of liposomes with cells are simple adsorption (by specific interactions with cell-surface components, electrostatic forces, or by non-specific weak hydrophobic forces) or following endocytosis [37]. Many attempts to activate the release of bioactive agents from liposomes in the vicinity of or inside the cells have been described [17,97]. Active release relies on developing mechanisms to increase the permeability or to destabilize the liposome bilayer once it reaches the target site. A variety of stimuli such as pH [46,98], temperature [39,99], ultrasonic waves [100], magnetic fields [101,102] and light [103,104] are currently being investigated to improve the target release of bioactive agent. Although the concept of triggered release is very promising, more studies have to be performed to prove its application in humans [17].

Go to:

4. APPLICATIONS OF LIPOSOMES IN TISSUE ENGINEERING

In living organisms, the lipids are considered basic building blocks, because they are formed of a polar headgroup, one or more hydrophobic tail regions and a backbone structure that connects the two [15]. Therefore, researchers have been taking advantage of the characteristics and versatility of lipids to find new applications [12,15]. Liposomes and micelles are the most widely used lipid-based nanoparticles [105]. Other lipid structures can be obtained such as cubic-, hexagonal- or sponge-phase structures [106]. They offer the advantage of stability and could be used as novel biomaterials mimicking biological membranes. It is believed that phospholipid-based materials may be increasingly used as tools for the manipulation of cell and tissue responses to biomaterials, namely for the controlled release of bioactive agents and for reconstructive surgery [15].

Nowadays, the most clinically used applications of liposome-based therapy are in the treatment of cancer and of systemic fungal infections [36]. However, the future of liposomes is not limited to those therapeutic applications. Liposomes, being a highly flexible platform, have been used in different areas of research including the production of vaccines, imaging, cosmetics and TE [27,36]. Herein, we will focus on lipids as a biomaterial and review some applications of the liposomes, mainly in TE and regenerative medicine. It is expected that therapeutic nanoparticles, such as liposomes, offer the potential to dramatically improve the effectiveness and side-effect profile of new and existing bioactive agents [77]. Moreover, many tissue-engineered scaffolds have already been approved for human use [27]. Therefore, combining liposomes with scaffolds has the potential for clinical translation in the near future.

4.1. Combining liposomes with scaffolds

The field of biomaterials has been advancing towards the nanoscale design of bioactive systems of drug release or scaffolding, aimed at TE and regenerative medicine strategies [107]. The combination of drug-loaded nanoparticles with scaffolds can be used to spatially and temporally control the release of bioactive agents, leading to a sustained and local delivery [108]. Over recent years, there have been an increasing number of studies aiming to direct stem cell fate through the delivery of growth/differentiation factors, DNA or interference RNA (RNAi). The encapsulation of the bioactive agents into nanoparticles such as liposomes has many advantages, as described in the section 'Loading liposomes with bioactive agents'. However, the single use of liposomes is limited due to the absence of a three-dimensional mechanical support frequently required to promote tissue regeneration [107].

Many types of scaffolds can be made of natural and/or synthetic polymers, using different processing techniques. For an extensive analysis of the topic, the reader is directed to reviews of TE scaffolding in [109–112]. TE scaffolds can be designed to physically and chemically control the release pattern of any incorporated bioactive agents [27]. Particularly, hydrogels are a set of scaffolds that present an enormous potential for application as smart and stimuli-responsive biomaterials [113]. The delivery of bioactive agents from scaffolds to promote regeneration is often challenging due to the complex fabrication processes. For instance, bioactive agents can be incorporated in the matrix or encapsulated into hollow nanofibres [114]. When the bioactive agent is incorporated in the matrix, there is a strong burst release within the first few hours [115]. Therefore, when the bioactive agent is encapsulated into hollow nanofibres, there is a better control of the bioactive agent release profile, but some of the complications during nanofibre preparation do not allow for easy large-scale production [116].

The combination of liposomes with scaffolds has been already reviewed elsewhere [15,27,108,117]. All these strategies aspire to combine the advantageous properties of liposomes and polymer matrices, aiming at developing materials that can sequester and maintain liposomes at a local tissue site. Stimuli-responsive liposomes have also been used as devices to control chemical reactions, which result in the rapid formation of a biomaterial such as the *in situ* formation of minerals, polymers or mineral/polymer composite biomaterials [15,118–121]. This exploits the release of entrapped substances from liposomes at temperatures near the lipid T_c (i.e. approx. 37°C) (see figure 7; see also the 'Liposome properties' section) [39]. The liposome structure is very sensitive to organic solvents, temperature and pH. Therefore, many ways to immobilize liposomes at the scaffold surface have been proposed [27,122–125]. There are two ways to immobilize liposomes at the surface of the scaffolds: (i) non-specific immobilization, which means that the liposomes are adsorbed at the surface of the scaffold and are easily removed during the cell culture at each medium exchange; (ii) specific immobilization, which means that the liposomes are covalently bound at the surface of the scaffold, increasing their stability. A scaffold system that uses naturally occurring interactions between liposomes and the fibrinogen was used to obviate the need for chemical conjugation [126]. To facilitate liposome adsorption, scaffold surfaces were coated with various extracellular matrix proteins, which were able to transfect a higher number of cells, while at the same time reducing the amount of DNA required [127]. Recently, a chemical modification of electrospun polycaprolactone (PCL) nanofibre meshes (NFMs) was reported enabling the immobilization of bioactive-loaded liposomes onto their surfaces [124,125]. To achieve this, initial UV-ozone irradiation was used to generate reactive free radicals that were immediately subjected to aminolysis. These modified surfaces were reacted with 2IT to generate sulfide (SH) pendant groups. Dexamethasone and pDNA-encoding RUNX2-loaded liposomes were covalently bonded to the SH groups present at the surface of electrospun NFMs [124,125]. The availability of the drug-release vehicle at the surface of the NFMs (where initial cellular contact occurs) enables a sustainable release of the Dex in the vicinity of the cells in culture and, consequently, increases its efficacy and bioavailability [125]. It was concluded that the amount of liposomes immobilized is specifically controlled by the amount of SH groups available at the nanofibres' surface [124,125]. Another strategy to combine liposomes and nanofibres is to use coaxial electrospinning. This technique enables the incorporation of liposomes into nanofibres [128]. Table 3 shows the applications of liposomes combined with scaffolds for TE approaches.

Table 3. Examples of lipid/liposome-based strategies for bioactive agent release in TE and regenerative medicine approaches. Signal transducers and activators of transcription (STAT), human bone marrow-derived mesenchymal stem cells (hBMSCs), luciferase (Luc), ...

material	bioactive agent	applications	comments	year/ references
cell culture substrate	pDNA-Luc-EGFP	gene delivery	absorbed lipoplex transfected primary cells with improved cellular viability relative to bolus delivery	2005 [127]
	microRNA and pre-miRNA	neural regeneration	cells transfected with miR-124 using lipofectamine played a key role in the differentiation of progenitor cells to mature neurons	2007 [129]
	siRNA knockdown chordin	bone regeneration	endogenously produced chordin constrains the osteogenic differentiation of human MSCs. MSCs were transfected with siRNA using lipofectamine	2008 [130]
	pDNA-Luc-EGFP	gene delivery	enhancing lipoplex delivery should target the material design to increase internalization	2009 [131]
	pDNA-βGal, pDNA-Luc-EGFP	gene delivery	the presence of the peptide in the lipoplex increased internalization efficiency	2009 [132]
	pDNA-BMP2-VEGF-BMP-2-VEGF	bone regeneration	BMSCs transfected with reconstructed plasmid p-hBMP-2-VEGF-loaded liposomes could secrete a high level of BMP-2 and hVEGF	2010 [133]
	siRNA-STAT3	bone regeneration	elimination of the STAT3 signalling pathway by the inhibitors, STAT3 siRNA, resulted in the acceleration and augmentation of BMP-induced osteogenic differentiation in MSCs	2010 [134]
	miR-335-5p	bone regeneration	miRNA activates Wnt signalling and promotes osteogenic differentiation by downregulating DKK1	2011 [135]
	Dex	bone regeneration	Dex-loaded liposomes do not have any cytotoxic effect, and they are able to promote an earlier induction of hBMSC differentiation into the osteogenic lineage	2012 [55]
chitosan – fibrin matrix	quinacrine	drug delivery system	liposome immobilized in fibrin chitosan matrix released quinacrine for 9 days for neutral liposomes, 5 days for PEGylated liposomes and 17 days after cross linking of the matrix	2006 [136]
chitosan scaffolds	pDNA-TGF-β1	cartilage regeneration	autologous MSCs transfected with the TGF-β1 gene using lipofectamine, and seeded into chitosan scaffolds filled a defect with regenerated hyaline-like cartilage tissue	2007 [137]
chitosan scaffold – fibrin gel	tirofiban	drug delivery system	the release periods of tirofiban from chitosan scaffold-fibrin gel loaded with liposomes can sustain 20% longer and significantly less burst release than with liposomes	2008 [138]
collagen scaffold	pDNA-IGF-1	cartilage regeneration	chondrocyte-seeded scaffolds prolonged and elevated IGF-1 expression by using the cross-linking method of plasmid incorporation along with the addition of the lipid transfection reagent	2007 [139]
collagen scaffold	pDNA-GDNF	neural regeneration	combination of scaffold and lipid transfection of MSCs is a promising approach for the long-term production of glial cell line-derived neurotrophic factor	2008 [140]
collagen/HA scaffold	DOX, LYZ, BMP2	bone regeneration	thiol bisphosphonate-decorated liposomes increased retention in the collagen/HA scaffolds after subcutaneous implantation in rats	2012 [122]
material	bioactive agent	applications	comments	year/ references
fibrin scaffold	tobramycin	treatment of *Pseudomonas* keratitis	tobramycin-loaded liposomes immobilized in fibrin scaffold inhibit colonies of *Pseudomonas* in albino rabbit corneas	1992 [141]
fibrin scaffold	protein horseradish peroxidase	drug delivery	protein-filled liposomes were absolutely stable within the fibrin scaffold	2000 [142]
fibrin clots	TGF-β1	cartilage regeneration	fibrin containing TGF-β1-loaded liposome released TGF-β1 at a slower rate than the one containing the free TGF-β1	2003 [143]
fibronectin-coated PLG	pDNA-NGF	neural regeneration	sustained lipoplex release was obtained for up to 50 days, with rates controlled by the fabrication conditions	2006 [144]
fibrin sealant	pDNA-VEGF	wound healing	topical fibrin-mediated administration of a VEGF-A plasmid using lipofectamine may serve as an alternative in treating ischaemic skin flaps	2007 [145]
fibrin hydrogel	pDNA-βGal	gene delivery	cell transfection depended strongly on the local cell-pDNA microenvironment (two dimensions versus three dimensions)	2009 [146]
fibrin scaffold	pDNA-βGal-EGFP	multiple gene delivery	the pDNA-loaded lipoplexes immobilized in fibrin showed significantly higher transfection efficiency for two reporter genes at 7 days than pDNA-loaded lipoplexes alone in a rabbit ear ulcer model	2009 [126]
fibrin scaffold – fibrin microsphere	pDNA	gene delivery	fibrin microspheres were shown to degrade and release DNA differentially compared with fibrin scaffold	2011 [147]
GAM	pDNA-encoding BMP4	bone regeneration	transfection efficiencies improved *in vitro* when liposomes where released from resorbable GAM	2005 [148]
PCL/PVA NFM	TGF-β, bFGF, IGF-I	cartilage regeneration	coaxial electrospinning enables the incorporation of liposomes into nanofibres and enhanced MSC proliferation	2012 [128]
PCL NFM	Dex	bone regeneration	Dex-loaded liposomes immobilized at the surface of electrospun PCL NFMs did not exhibit any cytotoxic effect, being able to successfully promote the osteogenic differentiation of hBMSCs	2013 [125]
PHEMA fibrous scaffolds	fetal bovine serum	cartilage regeneration	attached fetal bovine serum-loaded liposomes significantly improved both chondrocyte adhesion and proliferation	2012 [123]
PLG scaffold	pDNA-βGal	spinal cord injury	*in vivo*, lipoplexes immobilized at PLG scaffolds increased transgene expression levels for at least three weeks in a rat spinal cord hemisection model	2009 [149]
scaffold with tendon cell sheets	collagen V siRNA	tendon regeneration	downregulation of collagen V α1 or α2 chains by siRNAs using lipofectamine had different effects on collagen I and decorin gene expressions of tenocytes	2011 [150]
PCL NFM	DNA (RUNX2)	bone regeneration	RUNX2-loaded liposomes immobilized at the surface of electrospun PCL NFMs induce a long-term gene expression of eGFP and RUNX2 by cultured hBMSCs	2014 [124]

Tissue regeneration depends not only on the bioactive agent itself such as growth factors (GFs), but also on the various parameters associated with its presentation, including concentration, spatio-temporal gradients, combination with other GFs and the target cell type [151–154]. Bioactive agent-loaded liposomes combined with scaffolds offer various intrinsic benefits such as: (i) effective concentration; (ii) stable concentration gradients; (iii) multiple bioactive agent delivery; and (iv) spatial patterning [27]. The bioactive agent delivered by the liposome–scaffold device can be of two types: (i) growth/differentiation factor, by the incorporation of the proper bioactive agent-loaded liposomes into the scaffold; and (ii) nucleic acid delivery, by the incorporation of the DNA (or RNAi) into liposomes that encodes (or silences) a specific protein or by cell delivery, in which cells act as 'GF factories'. These types of local bioactive agent release system will be discussed in §4.2.

4.2. Growth/differentiation factor delivery

The tissue regeneration process involves complex cascades of bioactive agents such as GFs, cytokines and other molecules. GFs are endogenous polypeptides that act through the cell-surface receptors to regulate cellular activities such as proliferation, migration and differentiation [155]. The outcome of GF therapeutics mainly depends on their delivery mode due to their rapid clearance *in vivo* [156]. Furthermore, one of the main challenges in TE is to find the best way to induce the correct differentiation of stem cells. So far, the most common approach relies on the use of cocktails of growth/differentiation factors supplemented in the culture medium. For instance, bone morphogenetic protein 2 (BMP-2) is added to induce the osteogenic differentiation, whereas transforming growth factor β(TGF-β) is used to promote the chondrogenic differentiation. An adequate combination of signalling molecules should be provided by controlled release systems in order to promote the desired regenerative outcome [112,157–159]. Therefore, liposomes can be used as carriers for the spatio-temporal controlled delivery of GFs, improving stem cell proliferation and differentiation *in vitro* [160,161].

Liposomes were used with some success in an animal model for cartilage repair [143]. They were used as release systems of TGF-β1 over a period of some weeks, avoiding the typical side effects of systemic administration, due to its direct injection into the joint cavity. Moreover, the conjugation of TGF-β1-loaded liposomes with a scaffold improved its release kinetics and local efficacy [143]. Bisphosphonate-coated liposomes displayed a strong binding to a collagen/hydroxyapatite (HA) composite scaffold, increasing their retention in the collagen/HA scaffolds after their subcutaneous implantation in rats [122]. The bisphosphonate-coated liposomes were able to entrap BMP-2 and deliver it locally [122]. The immobilization of fetal bovine serum-loaded liposomes in a poly(2-hydroxyethyl methacrylate) fibrous scaffold significantly improved chondrocyte adhesion and proliferation [123]. The aim of this study was to investigate the interaction between liposomes and fibre scaffolds and to develop a novel drug delivery system. The release profile of Dex showed an initial burst release, although Dex continued to be released at a slower but steady rate until day 21 [55]. This time frame was selected in accordance with the culture time usually required to obtain a complete osteogenic differentiation of mesenchymal stem cells (MSCs) *in vitro*. Dex-loaded liposomes did not have any cytotoxic effect on human bone marrow-derived mesenchymal stem cells (hBMSCs). They were able to promote an earlier induction of differentiation of hBMSCs into the osteogenic lineage [55]. Biological assays showed that Dex-loaded liposomes immobilized at the surface of electrospun PCL NFMs did not exhibit any cytotoxic effect, being able to successfully promote the osteogenic differentiation of hBMSCs [125].

4.3. Therapeutic gene delivery

The use of growth/differentiation factors to induce stem cell differentiation *in vivo* has some limitations such as short half-lives, denaturation during the encapsulation processes, time-consuming, long time periods to obtain the differentiated cells, use of cocktails of GFs and difficulty to differentiate the cells into one specific lineage [108,162]. Therefore, gene therapy, encoding transcription factors or encoding for a specific or to a set of proteins may be a good approach to overcome these limitations and to control stem cell differentiation [124]. Transcription factors would ensure that expression of all natural splice variants occurs in a coordinated time and sequence, and may regulate a cascade of multiple different genes. Gene therapy was initially envisioned as the insertion of a functioning gene into the host cell genome to replace a hereditary genetic defect or, more recently, to provide a new function in a cell such as overexpressing GFs or even killing cancer cells [36,148]. When the pDNA enters into the nucleus, it is intercalated in the DNA of the host cell and, then, transcribed into messenger RNA (mRNA). Therefore, therapeutic proteins or GFs are produced using the cell machinery outside of the nucleus.

Another way to control stem cell differentiation relies on the delivery of RNAi, as previously mentioned [162]. RNAi acts by binding to nucleic acids, inhibiting gene transcription and translation; in other words, RNAi acts by silencing genes of interest through the eradication of target mRNAs through the introduction of small interfering RNAs (siRNAs), small hairpin RNAs (shRNAs) or micro-RNAs (miRNAs) [25,108,162]. The main advantage of this strategy is that the RNAi induces the silencing of targeted genes without integration into the host genome. Therefore, this field has grown considerably in the last decades, due to its huge potential to treat diseases by replacing defective or missing genes or silencing unwanted gene expression [163]. Notable progress has been achieved by the delivery of miRNAs to reprogramme somatic cells into pluripotent stem cells (iPSCs), thus obviating the need to introduce pDNA into donor cells [164].

The delivery of genes can be made by an *ex vivo* approach or directly into a target cell or tissue. The *ex vivo* method generally uses autologous cells that are recovered from the patient's body. The cells are generally transduced with viral vectors, containing recombinant genes, and re-inserted/transplanted into the target tissue [148]. Likewise, pDNA or RNAi is poorly taken up by the host cells, being subjected to degradation when exposed to blood proteins [165]. Viral and non-viral vectors are examples of carriers used to transfect cells [166–168]. The widely used method to transfer genes to stem cells is performed through viral vectors (i.e. lentivirus and retrovirus), because of their higher transgene expression and transduction efficiency. Thus, when stem cells are used to correct a genetic pathology and to express the therapeutic gene for the duration of a patient's life, viral vectors are preferred [169]. Conversely, when stem cells are used to treat non-inherited diseases and are only required to express the therapeutic gene for a short period of time, non-viral vectors are preferred [169]. Although viral vectors are more efficient, they possess some limitations such as high production cost, safety issues including mutagenesis, the immunogenicity of the virus proteins, lack of desired tissue selectivity and generation of infectious viruses due to recombination [170–172]. Therefore, there is a need for a delivery system that not only protects the pDNA/RNAi and facilitates its cellular uptake, but additionally enhances the potential for a targeted and effective delivery [165,169]. Non-viral delivery systems (i.e. liposomes, cationic lipids, polymers and proteins) have comparatively lower transfection efficiency, but they have been proposed due to their safety, easy production and higher pDNA size encapsulation [27,108,162,172–174]. Moreover, non-viral vectors provide flexibility in formulation design which can be tailored to interact with the DNA cargo. Also, by incorporating targeting ligands, it is possible to specify the route of vector administration and enhance the delivery to specific tissues or cells.

Liposomes are considered the first non-viral delivery system used in cell biology [27]. Cationic lipids have emerged as the primary option for gene delivery [77,175]. Various cationic lipids were synthesized and evaluated for the transfection of cells [175]. These cationic liposomes show high transfection efficiency, which can be attributed partly to their interactions with negatively charged cell membranes [27]. Lipofectamine, a leading commercial reagent, is a cationic liposome formulation commonly used to transfect cells [176]. However, their success *in vivo* as a gene therapy strategy has been limited due to the lack of colloidal stability, short duration of gene expression and cytotoxicity [175,177,178]. Cationic polymers have also been proposed as non-viral gene delivery systems, due to their flexible properties, facile synthesis, robustness and proven gene delivery efficiency. For further details about the most recent scientific advances in cationic polymers and their derivatives not only for gene delivery purposes but also for various alternative therapeutic applications, the reader is directed to the review in [179]. The use of liposomes combined with scaffolds may contribute to overcoming the issues of toxicity, long-term expression and silencing efficiency. Specifically, the topical delivery of gene-loaded liposomes via a biomaterial scaffold may reduce their exposure to immune cells, enhance cellular uptake and possibly allow for a sustained delivery [27,162].

Liposome–scaffold systems may be used as a biomaterial to deliver genes in an efficient, cell-controlled and spatially localized manner for TE applications. Bone and cartilage regeneration with a gene therapy strategy is one of the clinically relevant applications [108]. Those strategies have focused on the delivery of genes encoding BMPs and TGF-β that initiate bone and cartilage progenitor cell differentiation, and coordinate the pathways of newly formed bone ossification and cartilage maturation [133,137,148,180]. Tissue vascularization was exploited by the delivery of genes encoding VEGF using bone marrow stromal cells [181]. Liposomes loaded with DNA encoding the 165 amino acid form of VEGF were injected into rat skin stimulating wound healing [182]. The flap survival was enhanced by 14%, and the histological analysis showed new vessel formation [182]. A plasmid expression vector containing VEGF was constructed to be administered to the wound bed of rat abdominal skin flaps in a fibrin sealant [145]. The topical fibrin-mediated administration of a VEGF-A plasmid increased flap survival by 7 days. pDNA-encoding RUNX2-loaded liposomes were covalently immobilized at the surface of PCL nanofibres [124]. The biological result using hBMSCs showed that hBMSCs cultured on RUNX2-loaded liposomes immobilized at the surface of electrospun PCL NFM showed enhanced levels of metabolic activity and total protein synthesis. RUNX2-loaded liposomes immobilized at the surface of electrospun PCL NFMs induce a long-term gene expression of eGFP and RUNX2 by cultured hBMSCs. Furthermore, osteogenic differentiation of hBMSCs was also achieved by the overexpression of other osteogenic markers in medium free of osteogenic supplementation.

4.4. Magnetite cationic liposomes

The main properties of magnetic particles are that they are non-toxic and biocompatible, they can be injected and they can eventually accumulate in the target tissue or organ by the application of an external magnetic field [183]. They are attracted to high magnetic flux density, which is used for bioactive agent targeting and bio-separation including cell sorting. Because of their small size (i.e. approx. 10 nm), they can be encapsulated into the inner compartment of liposomes (figure 11). Magnetite cationic liposomes (MCLs) are magnetic nanoparticles, positively charged, that can interact with the negatively charged cell membranes. They find many applications such as in hyperthermic treatments and in TE and regenerative medicine strategies [183]. Table 4 compiles some of the studies in which MCLs were used in TE approaches. The application of magnetic nanoparticles and magnetic force for TE is termed 'magnetic force-based tissue engineering (Mag-TE)' [183]. For instance, MSC sheets created by magnetic nanoparticle-containing liposomes may represent a new modality for therapeutic angiogenesis and bone tissue regeneration [193]. MCLs successfully facilitated cell seeding into the interior space of the scaffolds [187].

Table 4. Applications of MCLs.

applications	achievements	year/ references
cell sheet with RGD-MCLs in TE	coating the culture surface with RGD-MCLs facilitated cell growth, cell sheet construction and cell sheet harvesting, using magnetic force without enzymatic treatment	2005 [184]
three-dimensional tissue-engineered tubular structures	tubular structures were constructed using magnetic force. Two types of tissue were used to create tubular structures: urinary tissue and vascular tissue	2005 [185]
cell sheet approach for choroidal neovascularization	MCLs were used to construct and deliver retinal pigment epithelium cell sheets *in vitro*. 15-layered cell sheets were formed after 24 h of culture	2005 [186]
enhancement of cell seeding into the deep internal space of the scaffolds (collagen sponges and polylactic acid sponge)	fibroblasts labelled with MCLs were seeded onto scaffolds. Cell-seeding efficiency increased significantly in all scaffolds when compared with those without magnetic force	2006 [187]
bone TE using bone marrow stromal cells and three-dimensional HA scaffolds	magnetically labelled BMSCs were successfully seeded into the internal space of scaffolds with a high cell density. The levels of alkaline phosphatase and osteocalcin were significantly higher than those by static seeding	2007 [188]
in vitro reconstruction of three-dimensional bone tissues without the use of scaffolds	MSCs magnetically labelled with MCLs formed multi-layered sheet-like structures after a 24 h culture period and maintained the ability to differentiate into osteoblasts, adipocytes or chondrocytes after a 21 day culture period. Transplantation of the MSC sheets into the bone defect in the crania of nude rats showed new bone formation	2007 [189]
incorporation of capillary-like structures into dermal cell sheets	human umbilical vein endothelial cells co-cultured with magnetically labelled normal human dermal fibroblast (NHDF) sheets showed tube-like formation of human aortic endothelial cells, resembling early capillaries, within or on the surface NHDF sheets	2007 [190]
effective cell seeding onto decellularized blood vessels for vascular TE	cells labelled with MCLs increased the cell seeding into porcine decellularized carotid artery scaffold. The scaffold was successfully constructed with two human cells, smooth muscle cells and dermal fibroblasts	2007 [191]
viral vector (VEGF) labelled with MCLs for fabrication of angiogenic cell sheets	a retroviral vector encoding VEGF was labelled with MCLs, to magnetically attract the particles onto a monolayer of mouse myoblast C2C12 cells. Subcutaneous transplantation of C2C12/VEGF cell sheets into nude mice produced thick tissues, with a high cell density, and promoted vascularization	2010 [192]
MSC sheet for therapeutic angiogenesis and tissue regeneration	human MSCs incubated with MCLs formed multi-layered cell sheets according to magnetic force. MSC sheets layered onto the ischaemic tissues of nude mice before skin closure showed greater angiogenesis than the control and MSC injected	2011 [193]
construction of three-dimensional artificial skeletal muscle tissues	MCLs were used to magnetically label C2C12 myoblast cells for the construction of artificial skeletal muscle tissues by an applied magnetic force. Elongated and multi-nucleated myotubes were observed within the tissue	2011 [194]
iPS cell sheets for reparative angiogenesis	mouse iPS cell-derived Flk-1(+) cells were incubated with MCLs. Implantation of the Flk-1(+) cell sheet accelerated revascularization of ischaemic hindlimbs relative to the contralateral limbs in nude mice and increased the expressions of VEGF and bFGF in ischaemic tissue	2013 [195]

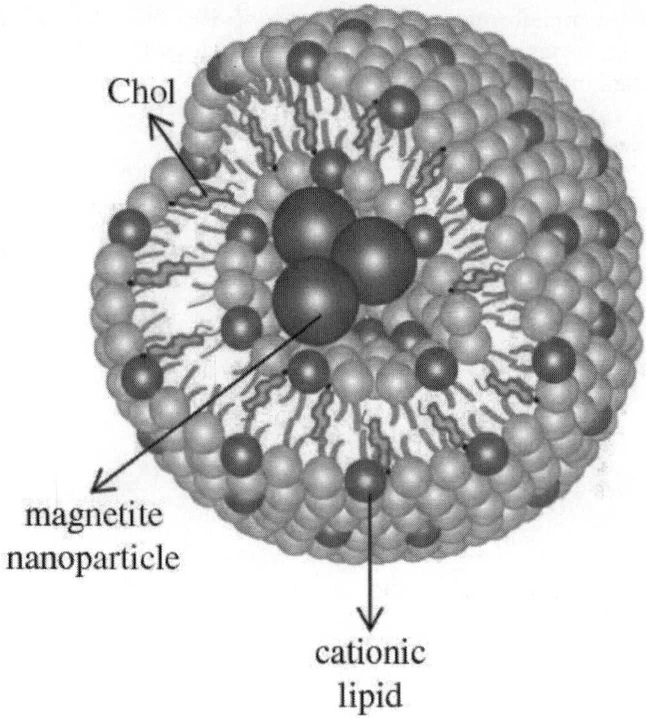

Figure 11. Magnetite cationic liposome. (Online version in colour.)

4.5. Liposomes as templates

We have focused herein on the use of liposomes as a device for the release of bioactive agents. However, liposomes can be used as templates to produce polymeric nanoparticles and nanogels (figure 12) [20,196,197]. In this approach, the inner compartment of the liposome can be used as the reaction vessel [15,196]. For example, the encapsulation of sodium alginate was performed by the thin-film method. Liposomes were extruded through a polycarbonate membrane of uniform pore size followed by incubation with high concentrations of calcium chloride. The diffusion of calcium into the liposome interior resulted in alginate gelation within the liposome of size 748 ± 280 nm [197]. Another study reported the production of alginate nanogels with a size distribution between 120 and 200 nm [196]. Liposomes were also used as a template to prepare monodisperse PEG hydrogel nanoparticles [20]. The procedure for the preparation of PEG nanoparticles using liposomes consisted of the encapsulation of a photo-polymerizable PEG hydrogel solution into the cavity of the liposomes, extrusion through a polycarbonate membrane and photopolymerization of the contents inside the liposomes by UV irradiation (figure 12*b*). The size distribution of the prepared particles was 94 ± 12 nm after extrusion through the membrane with pore size of 100 nm. This approach also enables the surface of the hydrogel nanoparticles with functional groups to be modified in a one-step procedure [20].

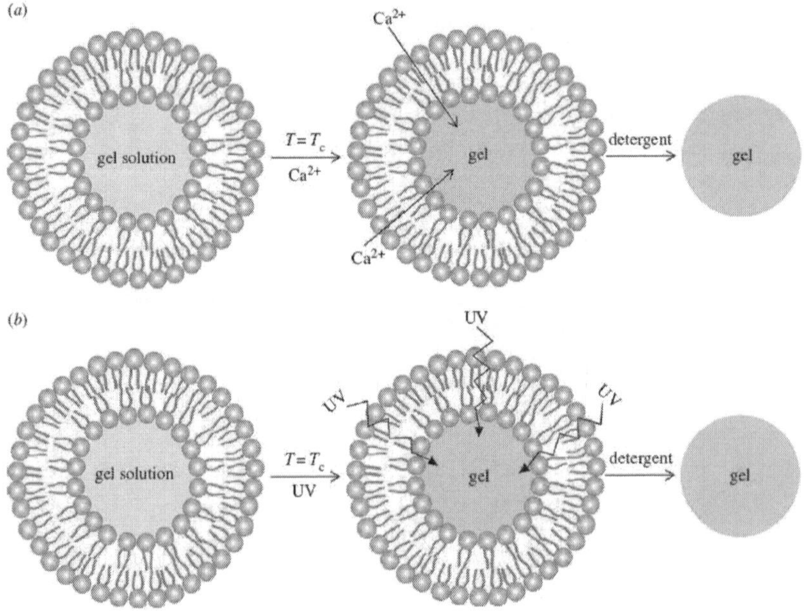

Figure 12. Strategies to produce nanogels using liposomes as a template. (*a*) Ca^{2+} calcium alginate gelation and (*b*) UV-photopolymerization. (Online version in colour.)

The ability to encapsulate a gel in the inner compartment of liposomes using specific polymers has particular implications beyond drug delivery [198]. Specifically, it can mimic a eukaryotic cell, which is considered to be a gel enclosed within a lipid bilayer membrane. The gel may be formed by the polymerization of cytoskeletal proteins such as actin, filament and tubulin. Therefore, a liposome with a gelled core might be a better study model of a biological cell [199]. Moreover, it could also be used as a container for single molecule fluorescence studies, e.g. for localizing a single DNA or protein molecule within the interior.

The lipid–polymer hybrid (LPH) nanoparticles have also been investigated to deliver therapeutic compounds in a controlled manner [200]. These hybrid nanoparticles combine the unique strengths of liposome and polymeric nanoparticles. They may be used to overcome their limitations in terms of bioactive agent EE, storage stability and release [200]. LPH nanoparticle synthesis requires the use of microfluidics technology to improve the mixing process, but is restricted by a low throughput. However, a pattern-tunable micro-vortex platform can allow mass production and size control of LPH nanoparticles, with superior reproducibility and homogeneity.

5. CONCLUDING REMARKS

Liposomes are vesicular structures made of lipids that are formed in aqueous solutions, and they resemble the lipid membrane of cells. Since their discovery, a lot of research has been carried out to achieve effective delivery of therapeutic bioactive agents, mainly in cancer research, with some products now coming onto the market. However, there are some limitations to overcome. For example, new liposome-based formulations have to be developed to overcome the rapid blood clearance of PEGylated liposomes. The concept of triggered release is very promising and more studies are needed to validate its applicability*in vivo*, in humans. Also, large-scale manufacturing methods and technologies are required for marketed products providing sterile, well-characterized and stable products.

In this review, we have highlighted some applications of liposomes in TE and regenerative medicine. Liposomes can be viewed as a platform to induce the differentiation of stem cells through the release of bioactive agents (i.e. GFs or nucleic acids). One of the main challenges in stem cell research is the successful differentiation of stem cells into a specific lineage. However, conventional methods to induce the overexpression of lineage-specific proteins relies on the use of growth/differentiation factor cocktails with sub-optimal outcomes. Liposomes can control growth/differentiation factor release and avoid their side effects. Another possible alternative relies on the delivery of nucleotides (i.e. pDNA and RNAi). The true potential of gene delivery is the possibility to guide the stem cell fate in the absence of inductive factors. The incorporation of pDNA into the host cells' genome raises biocompatibility issues in the context of TE, where differentiated cells are used for subsequent *in vivo* applications. RNAi may be a possible alternative to guide stem cells' fate without incorporation of pDNA into the host cells' genome. Cationic liposomes are a possible alternative to transfect cells, but they have a lower transfection efficiency [169]. The combination of liposomes with scaffolds represents a good approach to overcome this limitation.

The efficacy of using liposome–scaffolds in TE and regenerative medicine is clear from the literature. However, the potential of this strategy has to be investigated in order to optimize liposome formulations and the best material for the specific application. We believe that liposome–scaffolds may enhance stem cell differentiation and bring novelties in TE and regenerative medicine.

FUNDING STATEMENT

The authors thank the Portuguese Foundation for Science and Technology for the PhD grant to N.S.M. (SFRH/BD/62465/2009), the post-doctoral grants of A.M. (SFRH/BPD/73663/2010). This study was also partly supported by POLARIS (FP7-REGPOT-2012-2013-1), RL3—TECT—NORTE-01-0124-FEDER-000020, co-financed by the North Portugal Regional Operational Programme (ON.2—O Novo Norte), under the National Strategic Reference Framework (NSRF), through the European Regional Development Fund (ERDF), the OsteoGraphy (PTDC/EME-MFE/2008) and MaxBone (PTDC/SAU-ENB/115179/2009) projects.

REFERENCES

1. Fahy E, et al. 2009. Update of the LIPID MAPS comprehensive classification system for lipids. J. Lipid Res. 50, S9–S14
2. LIPID-MAPS. 2014. Lipidomics gateway See http://www.lipidmaps.org.
3. Vemuri S, Rhodes CT. 1995. Preparation and characterization of liposomes as therapeutic delivery systems: a review. Pharm. Acta Helv. 70, 95–111
4. Subramaniam S, Fahy E, Gupta S, Sud M, Byrnes RW, Cotter D, Dinasarapu AR, Maurya MR. 2011. Bioinformatics and systems biology of the lipidome. Chem. Rev. 111, 6452–6490
5. Mader S. 2010. Inquiry into life, 13, ilustrada edn New York, NY: McGraw-Hill Education.
6. Richter RP, Berat R, Brisson AR. 2006. Formation of solid-supported lipid bilayers: an integrated view. Langmuir 22, 3497–3505
7. van Meer G, Voelker DR, Feigenson GW. 2008. Membrane lipids: where they are and how they behave. Nat. Rev. Mol. Cell Biol. 9, 112–124
8. Singer SJ, Nicolson GL. 1972. The fluid mosaic model of the structure of cell membranes. Science 175, 720–731
9. Berridge MJ, Irvine RF. 1989. Inositol phosphates and cell signaling. Nature 341, 197–205
10. Messersmith PB, Vallabhaneni S, Nguyen V. 1998. Preparation of calcium-loaded liposomes and their use in calcium phosphate formation. Chem. Mat. 10, 109–116

11. Avanti-Polar-Lipids. See http://avantilipids.com/.

12. Antimisiaris SG, Kallinteri P, Fatouros DG. 2007. Liposomes and drug delivery. InPharmaceutical manufacturing handbook (ed. S Cox Gad), pp. 443–533 Hoboken, NJ: John Wiley and Sons, Inc.

13. Eibl H, Kaufmann-Kolle P. 1995. Medical application of synthetic phospholipids as liposomes and drugs. J. Liposome Res. 5, 131–148

14. Freeman S. 2011. Biological science—lipids, membranes, and the first cells. San Fransicso, CA: Benjamin Cummings.

15. Collier JH, Messersmith PB. 2001. Phospholipid strategies in biomineralization and biomaterials research. Annu. Rev. Mater. Res. 31, 237–263

16. Bangham A, Standish M, Watkins J. 1965. Diffusion of univalent ions across the lamellae of swollen phospholipids. J. Mol. Biol. 13, 238–252

17. Allen TM, Cullis PR. 2013. Liposomal drug delivery systems: from concept to clinical applications. Adv. Drug Deliv. Rev. 65, 36–48

18. Gómez-Hens A, Manuel Fernández-Romero J. 2005. The role of liposomes in analytical processes. Trends Anal. Chem. 24, 9–19

19. Singh AK, Cummings EB, Throckmorton DJ. 2001. Fluorescent liposome flow markers for microscale particle image velocimetry. Anal. Chem. 73, 1057–1061

20. An SY, et al. 2009. Preparation of monodisperse and size-controlled poly(ethylene glycol) hydrogel nanoparticles using liposome templates. J. Colloid Interface Sci. 331, 98–103

21. Reineccius Gary A. 1995. Liposomes for controlled release in the food industry. InEncapsulation and controlled release of food ingredients (eds SJ Risch, GA Reineccius), pp. 113–131 Washington, DC: American Chemical Society.

22. Xia S, Xu S. 2005. Ferrous sulfate liposomes: preparation, stability and application in fluid milk. Food Res. Int. 38, 289–296

23. Boerman OC, Laverman P, Oyen WJG, Corstens FHM, Storm G. 2000. Radiolabeled liposomes for scintigraphic imaging. Prog. Lipid Res. 39, 461–475

24. Voinea M, Simionescu M. 2002. Designing of 'intelligent' liposomes for efficient delivery of drugs. J. Cell. Mol. Med. 6, 465–474

25. Gomes-da-Silva LC, Fonseca NA, Moura V, de Lima MCP, Simoes S, Moreira JN. 2012. Lipid-based nanoparticles for siRNA delivery in cancer therapy: paradigms and challenges. Acc. Chem. Res. 45, 1163–1171

26. Ulbrich W, Lamprecht A. 2010. Targeted drug-delivery approaches by nanoparticulate carriers in the therapy of inflammatory diseases. J. R. Soc. Interface 7, S55–S66

27. Kulkarni M, Greiser U, O'Brien T, Pandit A. 2010. Liposomal gene delivery mediated by tissue-engineered scaffolds. Trends Biotechnol. 28, 28–36

28. Banerjee R. 2001. Liposomes: applications in medicine. J. Biomater. Appl. 16, 3–21

29. Fenske DB, Chonn A, Cullis PR. 2008. Liposomal nanomedicines: an emerging field.Toxicol. Pathol. 36, 21–29

30. Immordino ML, Dosio F, Cattel L. 2006. Stealth liposomes: review of the basic science, rationale, and clinical applications, existing and potential. Int. J. Nanomed. 1, 297–315

31. Drummond DC, Meyer O, Hong K, Kirpotin DB, Papahadjopoulos D. 1999.Optimizing liposomes for delivery of chemotherapeutic agents to solid tumors. Pharmacol. Rev. 51, 691–744.

32. Frézard F. 1999. Liposomes: from biophysics to the design of peptide vaccines. Braz. J. Med. Biol. Res. 32, 181–189

33. Papahadjopoulos D, Nir S, Ohki S. 1972. Permeability properties of phospholipid membranes: effect of cholesterol and temperature. BBA Biomembr. 266, 561–583

34. Koynova R, Caffrey M. 1998. Phases and phase transitions of the phosphatidylcholines. BBA Rev. Biomembr. 1376, 91–145.

35. Frézard F, Schettini DA, Rocha OGF, Demicheli C. 2005. Liposomes: physicochemical and pharmacological properties, applications in antimony-based chemotherapy. Quim. Nova 28, 511–518

36. Maurer N, Fenske DB, Cullis PR. 2001. Developments in liposomal drug delivery systems. Expert Opin. Biol. Ther. 1, 923–947

37. Akbarzadeh A, Rezaei-Sadabady R, Davaran S, Joo SW, Zarghami N, Hanifehpour Y, Samiei M, Kouhi M, Nejati-Koshki K. 2013. Liposome: classification, preparation, and applications. Nanoscale Res. Lett. 8, 102

38. Batista CM, Carvalho CMBd, Magalhães NSS. 2007. Liposomes and their therapeutic: state of art applications. Rev. Farm. Bioquim. Univ. Sao Paulo 43, 167–179.

39. Murphy WL, Messersmith PB. 2000. Compartmental control of mineral formation: adaptation of a biomineralization strategy for biomedical use. Polyhedron 19, 357–363

40. Huang JY, Feigenson GW. 1999. A microscopic interaction model of maximum solubility of cholesterol in lipid bilayers. Biophys. J. 76, 2142–2157

41. Woodle MC, Lasic DD. 1992. Sterically stabilized liposomes. Biochim. Biophys. Acta1113, 171–199

42. Tsukanova V, Salesse C. 2004. On the nature of conformational transition in poly(ethylene glycol) chains grafted onto phospholipid monolayers. J. Phys. Chem. 108, 10 754–10 764

43. Knop K, Hoogenboom R, Fischer D, Schubert US. 2010. Poly(ethylene glycol) in drug delivery: pros and cons as well as potential alternatives. Angew. Chem. Int. Ed. Engl.49, 6288–6308

44. Ishida T, Harada M, Wang XY, Ichihara M, Irimura K, Kiwada H. 2005. Accelerated blood clearance of PEGylated liposomes following preceding liposome injection: effects of lipid dose and PEG surface-density and chain length of the first-dose liposomes. J. Control Release 105, 305–317

45. Kale AA, Torchilin VP. 2007. 'Smart' drug carriers: PEGylated TATp-modified pH-sensitive liposomes. J. Liposome Res. 17, 197–203

46. Moura V, Lacerda M, Figueiredo P, Corvo ML, Cruz MEM, Soares R, de Lima MCP, Simoes S, Moreira JN. 2012. Targeted and intracellular triggered delivery of therapeutics to cancer cells and the tumor microenvironment: impact on the treatment of breast cancer.Breast Cancer Res. Treat. 133, 61–73

47. Shi N, Pardridge WM. 2000. Noninvasive gene targeting to the brain. Proc. Natl Acad. Sci. USA 97, 7567–7572

48. Paszko E, Senge MO. 2012. Immunoliposomes. Curr. Med. Chem. 19, 5239–5277

49. Pardridge WM. 2002. Drug and gene targeting to the brain with molecular Trojan horses. Nat. Rev. Drug Discov. 1, 131–139 (

50. Kleusch C, Hersch N, Hoffmann B, Merkel R, Csiszar A. 2012. Fluorescent lipids: functional parts of fusogenic liposomes and tools for cell membrane labeling and visualization. Molecules 17, 1055–1073

51. Mozafari MR. 2005. Liposomes: an overview of manufacturing techniques. Cell. Mol. Biol. Lett. 10, 711–719.

52. Senior J, Crawley JCW, Gregoriadis G. 1985. Tissue distribution of liposomes exhibiting long half-lives in the circulation after intravenous injection. Biochim. Biophys. Acta 839, 1–8

53. Litzinger DC, Buiting AMJ, van Rooijen N, Huang L. 1994. Effect of liposome size on the circulation time and intraorgan distribution of

amphipathic poly(ethylene glycol)-containing liposomes. BBA Biomembr. 1190, 99–107

54. Barza M, Stuart M, Szoka F. 1987. Effect of size and lipid-composition on the pharmacokinetics of intravitreal liposomes. Invest. Ophthalmol. Vis. Sci. 28, 893–900.

55. Murao A, Nishikawa M, Managit C, Wong J, Kawakami S, Yamashita F, Hashida M. 2002. Targeting efficiency of galactosylated liposomes to hepatocytes *in vivo*: effect of lipid composition. Pharm. Res. 19, 1808–1814

56. Kawakami S, Sato A, Yamada M, Yamashita F, Hashida M. 2001. The effect of lipid composition on receptor-mediated *in vivo* gene transfection using mannosylated cationic liposomes in mice. Stp Pharma. Sci. 11, 117–120.

57. Kawakami S, Sato A, Nishikawa M, Yamashita F, Hashida M. 2000. Mannose receptor-mediated gene transfer into macrophages using novel mannosylated cationic liposomes. Gene Ther. 7, 292–299

58. Lichtenberg D, Barenholz Y. 1988. Liposomes—preparation, characterization, and preservation. Methods Biochem. Anal. 33, 337–462

59. Olson F, Hunt CA, Szoka FC, Vail WJ, Papahadjopoulos D. 1979. Preparation of liposomes of defined size distribution by extrusion through polycarbonate membranes.Biochim. Biophys. Acta 557, 9–23

60. Vanrooijen N, Sanders A. 1994. Liposome-mediated depletion of macrophages—mechanism of action, preparation of liposomes and applications. J. Immunol. Methods174, 83–93

61. Bangham AD. 1993. Liposomes: the Babraham connection. Chem. Phys. Lipids 64, 275–285

62. Wagner A, Vorauer-Uhl K. 2011. Liposome technology for industrial purposes. J. Drug Deliv. 2011, 1–9

63. Yang SY, Chen JY, Zhao D, Han DE, Chen XJ. 2012. Comparative study on preparative methods of DC-Chol/DOPE liposomes and formulation optimization by determining encapsulation efficiency. Int. J. Pharm. 434, 155–160

64. Mozafari MR, Reed CJ, Rostron C. 2002. Development of non-toxic liposomal formulations for gene and drug delivery to the lung. Technol. Health Care 10, 342–344.

65. Mozafari MR, Reed CJ, Rostron C, Martin DS. 2004. Transfection of human airway epithelial cells using a lipid-based vector prepared by the heating method. J. Aerosol Med.17, 100.

66. Szoka F, Papahadjopoulos D. 1980. Comparative properties and methods of preparation of lipid vesicles (liposomes). Annu. Rev. Biophys. Bioeng. 9, 467–508

67. Brunner J, Skrabal P, Hauser H. 1976. Single bilayer vesicles prepared without sonication physicochemical properties. Biochim. Biophys. Acta 455, 322–331

68. Torchilin VP, Weissig V. 2003. Liposomes: a practical approach, 2nd edn Oxford, UK: Oxford University Press.

69. Laouini A, Jaafar-Maalej C, Limayem-Blouza I, Sfar S, Charcosset C, Fessi H. 2012.Preparation, characterization and applications of liposomes: state of the art. J. Colloid Sci.1, 147–168.

70. Ruozi B, Belletti D, Tombesi A, Tosi G, Bondioli L, Forni F, Vandelli MA. 2011.AFM, ESEM, TEM, and CLSM in liposomal characterization: a comparative study. Int. J. Nanomed. 6, 557–563

71. Franzen U, Vermehren C, Jensen H, Ostergaard J. 2011. Physicochemical characterization of a PEGylated liposomal drug formulation using capillary electrophoresis. Electrophoresis 32, 738–748

72. Chu B, Lin FL. 1974. Laser light-scattering study of a ternary liquid-mixture—ethanol water chloroform. J. Chem. Phys. 61, 5132–5146

73. Uskokovic V, Odsinada R, Djordjevic S, Habelitz S. 2011. Dynamic light scattering and zeta potential of colloidal mixtures of amelogenin and hydroxyapatite in calcium and phosphate rich ionic milieus. Arch. Oral Biol. 56, 521–532

74. Malvern-Instrument-Company. 2013. Dynamic light scattering and zeta potential Seehttp://www.malvern.com/.

75. Phillips NC, Heydari C. 1996. Modulation of cationic liposomal DNA zeta potential and liposome–protein interaction by amphiphilic poly (ethylene glycol)*. Pharm. Pharmacol. 2, 73–76.

76. Perrie Y, Gregoriadis G. 2000. Liposome-entrapped plasmid DNA: characterisation studies. Biochim. Biophys. Acta 1475, 125–132

77. Burgess P, Hutt PB, Farokhzad OC, Langer R, Minick S, Zale S. 2010. On firm ground: IP protection of therapeutic nanoparticles. Nat. Biotech. 28, 1267–1270

78. Toh M-R, Chiu GNC. 2013. Liposomes as sterile preparations and limitations of sterilisation techniques in liposomal manufacturing. Asian J. Pharm. Sci. 8, 88–95

79. Grit M, Crommelin JA. 1993. Chemical-stability of liposomes—implications for their physical stability. Chem. Phys. Lipids 64, 3–18

80. Cabral ECM, Zollner RL, Santana MHA. 2004. Preparation and characterization of liposomes entrapping allergenic proteins. Braz. J. Chem. Eng. 21, 137–146

81. Xu X, Khan MA, Burgess DJ. 2012. Predicting hydrophilic drug encapsulation inside unilamellar liposomes. Int. J. Pharm. 423, 410–418

82. Maestrelli F, González-Rodríguez ML, Rabasco AM, Mura P. 2006. Effect of preparation technique on the properties of liposomes encapsulating ketoprofen–cyclodextrin complexes aimed for transdermal delivery. Int. J. Pharm. 312, 53–60

83. Ma L, Ramachandran C, Weiner ND. 1991. Partitioning of a homologous series of alkyl paraaminobenzoates in dipalmitoylphosphatidylcholine liposomes—effect of liposome type. Int. J. Pharm. 77, 127–140

84. Sakai H, Gotoh T, Imura T, Sakai K, Otake K, Abe M. 2008. Preparation and properties of liposomes composed of various phospholipids with different hydrophobic chains using a supercritical reverse phase evaporation method. J. Oleo Sci. 57, 613–621

85. Joguparthi V, Anderson BD. 2008. Liposomal delivery of hydrophobic weak acids: enhancement of drug retention using a high intraliposomal pH. J. Pharm. Sci. 97, 433–454

86. Haran G, Cohen R, Bar LK, Barenholz Y. 1993. Transmembrane ammonium-sulfate gradients in liposomes produce efficient and stable entrapment of amphipathic weak bases.Biochim. Biophys. Acta 1151, 201–215

87. Gonçalves MCF, Mertins O, Pohlmann AR, Silveira NP, Guterres SS. 2012. Chitosan coated liposomes as an innovative nanocarrier for drugs. J. Biomed. Nanotechnol. 8, 240–250

88. Li N, Zhuang C-Y, Wang M, Sui C-G, Pan W-S. 2012. Low molecular weight chitosan-coated liposomes for ocular drug delivery: *in vitro* and *in vivo* studies. Drug Deliv. 19, 28–35

89. Haidar ZS. 2010. Bio-inspired/-functional colloidal core-shell polymeric-based nanosystems: technology promise in tissue engineering, bioimaging and nanomedicine.Polymers 2, 323–352

90. Joo K-I, Xiao L, Liu S, Liu Y, Lee C-L, Conti PS, Wong MK, Li Z, Wang P. 2013.Crosslinked multilamellar liposomes for controlled delivery of anticancer drugs.Biomaterials 34, 3098–3109

91. Tsotas VA, Mourtas S, Antimisiaris SG. 2007. Dexamethasone incorporating liposomes: effect of lipid composition on drug trapping efficiency and vesicle stability.Drug Deliv. 14, 441–445

92. Bhardwaj U, Burgess DJ. 2010. Physicochemical properties of extruded and non-extruded liposomes containing the hydrophobic drug dexamethasone. Int. J. Pharm. 388, 181–189

93. McCormack B, Gregoriadis G. 1994. Drugs-in-cyclodextrins-in liposomes: a novel concept in drug delivery. Int. J. Pharm. 112, 249–258

94. Lindner LH, Hossann M. 2010. Factors affecting drug release from liposomes. Curr. Opin. Drug Discov. Devel. 13, 111–123.

95. Mohammed AR, Weston N, Coombes AGA, Fitzgerald M, Perrie Y. 2004. Liposome formulation of poorly water soluble drugs: optimisation of drug loading and ESEM analysis of stability. Int. J. Pharm. 285, 23–34

96. Ahl PL, Bhatia SK, Meers P, Roberts P, Stevens R, Dause R, Perkins WR, Janoff AS. 1997. Enhancement of the *in vivo* circulation lifetime of l-α-distearoylphosphatidylcholine liposomes: importance of liposomal aggregation versus complement opsonization. BBA Biomembr. 1329, 370–382

97. Samoshin AV, Veselov IS, Chertkov VA, Yaroslavov AA, Grishina GV, Samoshina NM, Samoshin VV. 2013. Fliposomes: new amphiphiles based on trans-3,4-bis(acyloxy)-piperidine able to perform a pH-triggered conformational flip and cause an instant cargo release from liposomes. Tetrahedron Lett. 54, 5600–5604

98. Fonseca C, Moreira JN, Ciudad CJ, de Lima MCP, Simoes S. 2005. Targeting of sterically stabilised pH-sensitive liposomes to human T-leukaemia cells. Eur. J. Pharm. Biopharm. 59, 359–366

99. Sasaki Y, Iwamoto S, Mukai M, Kikuchi J-I. 2006. Photo- and thermo-responsive assembly of liposomal membranes triggered by a gemini peptide lipid as a molecular switch. J. Photochem. Photobiol. A Chem. 183, 309–314

100. Herbst SM, et al. 2009. Delivery of stem cells to porcine arterial wall with echogenic liposomes conjugated to antibodies against CD34 and intercellular adhesion molecule-1.Mol. Pharm. 7, 3–11

101. Tanaka H, Sugita T, Yasunaga Y, Shimose S, Deie M, Kubo T, Murakami T, Ochi M. 2005. Efficiency of magnetic liposomal transforming growth factor-beta 1 in the repair of articular cartilage defects in a rabbit model. J. Biomed. Mater. Res. A 73, 255–263

102. Matsuo T, Sugita T, Kubo T, Yasunaga Y, Ochi M, Murakami T. 2003. Injectable magnetic liposomes as a novel carrier of recombinant human BMP-2 for bone formation in a rat bone-defect model. J. Biomed. Mater. Res. A 66, 747–754

103. Dai Y-Q, Qin G, Geng S-Y, Yang B, Xu Q, Wang J-Y. 2012. Photo-responsive release of ascorbic acid and catalase in CDBA-liposome for commercial application as a sunscreen cosmetic. RSC Adv. 2, 3340–3346

104. Randles EG, Bergethon PR. 2013. A photodependent switch of liposome stability and permeability. Langmuir 29, 1490–1497

105. Buse J, El-Aneed A. 2010. Properties, engineering and applications of lipid-based nanoparticle drug-delivery systems: current research and advances. Nanomedicine 5, 1237–1260

106. Shanmugam T, Banerjee R. 2011. Nanostructured self assembled lipid materials for drug delivery and tissue engineering. Ther. Deliv. 2, 1485–1516

107. Santo VE, Gomes ME, Mano JF, Reis RL. 2012. From nano- to macro-scale: nanotechnology approaches for spatially controlled delivery of bioactive factors for bone and cartilage engineering. Nanomedicine 7, 1045–1066

108. Storrie H, Mooney DJ. 2006. Sustained delivery of plasmid DNA from polymeric scaffolds for tissue engineering. Adv. Drug Deliv. Rev. 58, 500–514

109. Dawson E, Mapili G, Erickson K, Taqvi S, Roy K. 2008. Biomaterials for stem cell differentiation. Adv. Drug Deliv. Rev. 60, 215–228

110. Burg KJL, Porter S, Kellam JF. 2000. Biomaterial developments for bone tissue engineering. Biomaterials 21, 2347–2359

111. Martins A, Araújo JV, Reis RL, Neves NM. 2007. Electrospun nanostructured scaffolds for tissue engineering applications. Nanomedicine 2, 929–942

112. Santo VE, Gomes ME, Mano JF, Reis RL. 2012. Controlled release strategies for bone, cartilage and osteochondral engineering. Part I: recapitulation of tissue healing and variables for the design of delivery systems. Tissue Eng. Part B Rev. 19, 308–326

113. Alexander C. 2008. Stimuli-responsive hydrogels—Drugs take control. Nat. Mater.7, 767–768

114. Yoo HS, Kim TG, Park TG. 2009. Surface-functionalized electrospun nanofibers for tissue engineering and drug delivery. Adv. Drug Deliv. Rev. 61, 1033–1042

115. Thakur RA, Florek CA, Kohn J, Michniak BB. 2008. Electrospun nanofibrous polymeric scaffold with targeted drug release profiles for potential application as wound dressing. Int. J. Pharm. 364, 87–93

116. Sirc J, et al. 2012. Controlled gentamicin release from multi-layered electrospun nanofibrous structures of various thicknesses. Int. J. Nanomed. 7, 5315–5325

117. Adler AF, Leong KW. 2010. Emerging links between surface nanotechnology and endocytosis: impact on nonviral gene delivery. Nano Today 5, 553–569

118. DiTizio V, Ferguson GW, Mittelman MW, Khoury AE, Bruce AW, DiCosmo F. 1998. A liposomal hydrogel for the prevention of bacterial adhesion to catheters.Biomaterials 19, 1877–1884

119. Sanborn TJ, Messersmith PB, Barron AE. 2002. In situ crosslinking of a biomimetic peptide-PEG hydrogel via thermally triggered activation of factor XIII. Biomaterials 23, 2703–2710

120. Pederson AW, Ruberti JW, Messersmith PB. 2003. Thermal assembly of a biomimetic mineral/collagen composite. Biomaterials 24, 4881–4890

121. Zhang ZY, Shum P, Yates M, Messersmith PB, Thompson DH. 2002. Formation of fibrinogen-based hydrogels using phototriggerable diplasmalogen liposomes. Bioconjug. Chem. 13, 640–646

122. Wang G, Mostafa NZ, Incani V, Kucharski C, Uludag H. 2012. Bisphosphonate-decorated lipid nanoparticles designed as drug carriers for bone diseases. J. Biomed. Mater. Res. Part A 100A, 684–693

123. Rampichova M, et al. 2012. A simple drug anchoring microfiber scaffold for chondrocyte seeding and proliferation. J. Mater. Sci. Mater. Med. 23, 555–563

124. Monteiro N, Ribeiro D, Martins A, Faria S, Fonseca NA, Moreira JN, Reis RL, Neves NM. 2014. Instructive nanofibrous scaffold comprising runt-related transcription factor 2 gene delivery for bone tissue engineering. ACS Nano 8, 8082–8094

125. Monteiro N, Martins A, Pires R, Faria S, Fonseca NA, Moreira JN, Reis RL, Neves NM. 2014. Immobilization of bioactive factor-loaded liposomes at the surface of electrospun nanofibers targeting tissue engineering. Biomater. Sci. 2, 1195–1209

126. Kulkarni M, Breen A, Greiser U, O'Brien T, Pandit A. 2009. Fibrin–lipoplex system for controlled topical delivery of multiple genes. Biomacromolecules 10, 1650–1654

127. Bengali Z, Pannier AK, Segura T, Anderson BC, Jang J-H, Mustoe TA, Shea LD. 2005. Gene delivery through cell culture substrate adsorbed DNA complexes. Biotechnol. Bioeng. 90, 290–302

128. Mickova A, Buzgo M, Benada O, Rampichova M, Fisar Z, Filova E, Tesarova M, Lukas D, Amler E. 2012. Core/shell nanofibers with embedded liposomes as a drug delivery system. Biomacromolecules 13, 952–962

129. Makeyev EV, Zhang J, Carrasco MA, Maniatis T. 2007. The microRNA miR-124 promotes neuronal differentiation by triggering brain-specific alternative Pre-mRNA splicing. Mol. Cell 27, 435–448

130. Kwong FNK, Richardson SM, Evans CH. 2008. Chordin knockdown enhances the osteogenic differentiation of human mesenchymal stem cells. Arthritis Res. Ther. 10, R65

131. Bengali Z, Rea JC, Gibly RF, Shea LD. 2009. Efficacy of immobilized polyplexes and lipoplexes for substrate-mediated gene delivery. Biotechnol. Bioeng. 102, 1679–1691

132. Rea JC, Gibly RF, Barron AE, Shea LD. 2009. Self-assembling peptide–lipoplexes for substrate-mediated gene delivery. Acta Biomater. 5, 903–912 (

133. Guo-ping W, Xiao-chuan H, Zhi-hui Y, Li G. 2010. Influence on the osteogenic activity of the human bone marrow mesenchymal stem cells transfected by liposome-mediated recombinant plasmid pIRES-hBMP2-hVEGF165 *in vitro*. Ann. Plast. Surg. 65, 80–84 (

134. Levy O, Ruvinov E, Reem T, Granot Y, Cohen S. 2010. Highly efficient osteogenic differentiation of human mesenchymal stem cells by eradication of STAT3 signaling. Int. J. Biochem. Cell Biol. 42, 1823–1830

135. Zhang J, Tu Q, Bonewald LF, He X, Stein G, Lian J, Chen J. 2011. Effects of miR-335-5p in modulating osteogenic differentiation by specifically downregulating Wnt antagonist DKK1. J. Bone Miner. Res. 26, 1953–1963 (

136. Chung TW, Yang MC, Tsai WJ. 2006. A fibrin encapsulated liposomes-in-chitosan matrix (FLCM) for delivering water-soluble drugs—influences of the surface properties of liposomes and the crosslinked fibrin network. Int. J. Pharm. 311, 122–129

137. Guo C-A, Liu X-G, Huo J-Z, Jiang C, Wen X-J, Chen Z-R. 2007. Novel gene-modified-tissue engineering of cartilage using stable transforming growth factor-beta 1-transfected mesenchymal stem cells grown on chitosan scaffolds. J. Biosci. Bioeng. 103, 547–556

138. Wang SS, Yang MC, Chung TW. 2008. Liposomes/chitosan scaffold/human fibrin gel composite systems for delivering hydrophilic drugs—release behaviors of tirofiban *in vitro*. Drug Deliv. 15, 149–157

139. Capito RM, Spector M. 2007. Collagen scaffolds for nonviral IGF-1 gene delivery in articular cartilage tissue engineering. Gene Ther. 14, 721–732

140. Bolliet C, Bohn MC, Spector M. 2008. Non-viral delivery of the gene for glial cell line-derived neurotrophic factor to mesenchymal stem cells *in vitro* via a collagen scaffold.Tissue Eng. Part C Methods 14, 207–219

141. Fruchtperry J, Assil KK, Ziegler E, Douglas H, Brown SI, Schanzlin DJ, Weinreb RN. 1992. Fibrin-enmeshed tobramycin liposomes—single application topical therapy of pseudomonas keratitis. Cornea 11, 393–397

142. Meyenburg S, Lilie H, Panzner S, Rudolph R. 2000. Fibrin encapsulated liposomes as protein delivery system—studies on the *in vitro* release behavior. J. Control. Release69, 159–168

143. Giannoni P, Hunziker EB. 2003. Release kinetics of transforming growth factor-beta 1 from fibrin clots. Biotechnol. Bioeng. 83, 121–123

144. Whittlesey KJ, Shea LD. 2006. Nerve growth factor expression by PLG-mediated lipofection. Biomaterials 27, 2477–2486

145. Michlits W, Mittermayr R, Schaefer R, Redl H, Aharinejad S. 2007. Fibrin-embedded administration of VEGF plasmid enhances skin flap survival. Wound Repair Regen. 15, 360–367

146. Lei P, Padmashali RM, Andreadis ST. 2009. Cell-controlled and spatially arrayed gene delivery from fibrin hydrogels. Biomaterials 30, 3790–3799

147. Kulkarni MM, Greiser U, O'Brien T, Pandit A. 2011. A temporal gene delivery system based on fibrin microspheres. Mol. Pharm. 8, 439–446

148. Winn SR, Chen JC, Gong X, Bartholomew SV, Shreenivas S, Ozaki W. 2005. Non-viral-mediated gene therapy approaches for bone repair. Orthod. Craniofac. Res. 8, 183–190

149. De Laporte L, Yan AL, Shea LD. 2009. Local gene delivery from ECM-coated poly(lactide-co-glycolide) multiple channel bridges after spinal cord injury. Biomaterials30, 2361–2368

150. Lu P, Zhang GR, Song XH, Zou XH, Wang LL, Ouyang HW. 2011. Col V siRNA engineered tenocytes for tendon tissue engineering. PLoS ONE 6, e21154

151. Tayalia P, Mooney DJ. 2009. Controlled growth factor delivery for tissue engineering. Adv. Mater. 21, 3269–3285

152. Schroeder JE, Mosheiff R. 2011. Tissue engineering approaches for bone repair: concepts and evidence. Injury 42, 609–613

153. Kock L, van Donkelaar CC, Ito K. 2012. Tissue engineering of functional articular cartilage: the current status. Cell Tissue Res. 347, 613–627

154. Balmayor ER, Pashkuleva I, Frias AM, Azevedo HS, Reis RL. 2011. Synthesis and functionalization of superparamagnetic poly-epsilon-caprolactone microparticles for the selective isolation of subpopulations of human adipose-derived stem cells. J. R. Soc. Interface 8, 896–908

155. Quaglia F. 2008. Bioinspired tissue engineering: the great promise of protein delivery technologies. Int. J. Pharm. 364, 281–297

156. Zhang SF, Uludag H. 2009. Nanoparticulate systems for growth factor delivery.Pharm. Res. 26, 1561–1580

157. Santo VE, Gomes ME, Mano JF, Reis RL. 2012. Controlled release strategies for bone, cartilage and osteochondral engineering. Part II: challenges on the evolution from single towards multiple bioactive factor delivery. Tissue Eng. Part B Rev. 19, 327–352

158. Kelleher CM, Vacanti JP. 2010. Engineering extracellular matrix through nanotechnology. J. R. Soc. Interface 7, S717–S729

159. Lee K, Silva EA, Mooney DJ. 2011. Growth factor delivery-based tissue engineering: general approaches and a review of recent developments. J. R. Soc. Interface8, 153–170

160. Tehrany EA, Kahn CJF, Baravian C, Maherani B, Belhaj N, Wang X, Linder M. 2012. Elaboration and characterization of nanoliposome made of soya; rapeseed and salmon lecithins: application to cell culture. Colloid Surf. B Biointerfaces 95, 75–81

161. Monteiro N, Martins A, Ribeiro D, Faria S, Fonseca NA, Moreira JN, Reis RL, Neves NM. 2013. On the use of dexamethasone-loaded liposomes to induce the osteogenic differentiation of human mesenchymal stem cells. J. Tissue Eng. Regen. Med.7

162. Yau WWY, Rujitanaroj P-O, Lam L, Chew SY. 2012. Directing stem cell fate by controlled RNA interference. Biomaterials 33, 2608–2628

163. Mao S, Sun W, Kissel T. 2010. Chitosan-based formulations for delivery of DNA and siRNA. Adv. Drug Deliv. Rev. 62, 12–27

164. Sun X, Fu X, Han W, Zhao Y, Liu H. 2010. Can controlled cellular reprogramming be achieved using microRNAs? Ageing Res. Rev. 9, 475–483

165. Podesta JE, Kostarelos K. 2009. Engineering cationic liposome: sirna complexes for*in vitro* and *in vivo* delivery. Methods Enzymol. 464, 343–354

166. von Groll A, Levin Y, Barbosa MC, Ravazzolo AP. 2006. Linear DNA low efficiency transfection by liposome can be improved by the use of cationic lipid as charge neutralizer. Biotechnol. Prog. 22, 1220–1224

167. Jang JH, Houchin TL, Shea LD. 2004. Gene delivery from polymer scaffolds for tissue engineering. Expert Rev. Med. Devices 1, 127–138

168. Pornpattananangkul D, Olson S, Aryal S, Sartor M, Huang C-M, Vecchio K, Zhang L. 2010. Stimuli-responsive liposome fusion mediated by gold nanoparticles. ACS Nano4, 1935–1942

169. Santos JL, Pandita D, Rodrigues J, Pego AP, Granja PL, Tomas H. 2011. Non-viral gene delivery to mesenchymal stem cells: methods, strategies and application in bone tissue engineering and regeneration. Curr. Gene Ther. 11, 46–57

170. Vo TN, Kasper FK, Mikos AG. 2012. Strategies for controlled delivery of growth factors and cells for bone regeneration. Adv. Drug Deliv. Rev. 64, 1292–1309

171. Barua S, Ramos J, Potta T, Taylor D, Huang HC, Montanez G, Rege K. 2011.Discovery of cationic polymers for non-viral gene delivery using combinatorial approaches. Comb. Chem. High Throughput Screen. 14, 908–924

172. Kawakami S, Higuchi Y, Hashida M. 2008. Nonviral approaches for targeted delivery of plasmid DNA and oligonucleotide. J. Pharm. Sci. 97, 726–745

173. Pack DW, Hoffman AS, Pun S, Stayton PS. 2005. Design and development of polymers for gene delivery. Nat. Rev. Drug Discov. 4, 581–593

174. Newland B, Dowd E, Pandit A. 2013. Biomaterial approaches to gene therapies for neurodegenerative disorders of the CNS. Biomater Sci 1, 556–576

175. Woodle MC, Scaria P. 2001. Cationic liposomes and nucleic acids. Curr. Opin. Colloid Interface Sci. 6, 78–84

176. Dalby B, Cates S, Harris A, Ohki EC, Tilkins ML, Price PJ, Ciccarone VC. 2004.Advanced transfection with Lipofectamine 2000 reagent: primary neurons, siRNA, and high-throughput applications. Methods 33, 95–103

177. Newland B, Moloney TC, Fontana G, Browne S, Abu-Rub MT, Dowd E, Pandit AS. 2013. The neurotoxicity of gene vectors and its amelioration by packaging with collagen hollow spheres. Biomaterials 34, 2130–2141

178. Yang F, et al. 2009. Genetic engineering of human stem cells for enhanced angiogenesis using biodegradable polymeric nanoparticles. Proc. Natl Acad. Sci. USA107, 3317–3322

179. Samal SK, Dash M, Van Vlierberghe S, Kaplan DL, Chiellini E, van Blitterswijk C, Moroni L, Dubruel P. 2012. Cationic polymers and their therapeutic potential. Chem. Soc. Rev. 41, 7147–7194

180. Lei L, Liao W, Sheng P, Huang G, Jiang L. 2006. Self-induction of rabbit marrow stromal stem cells into chondrocytes by transfection with reconstructed PGL3-transforming growth factor beta1 gene *in vitro*. Zhongguo Xiu Fu Chong Jian Wai Ke Za Zhi 20, 134–138.

181. Bai X-F, Tian W-D, Chen X-Z, Li Z-Y. 2005. Expression of human VEGF165 gene in rat bone marrow stromal cells *in vitro*. J. Sichuan Univ. 36, 468–470.

182. Liu PY, Tong WJ, Liu K, Han SH, Wang XT, Badiavas E, Rieger-Christ K, Summerhayes I. 2004. Liposome-mediated transfer of vascular endothelial growth factor cDNA augments survival of random-pattern skin flaps in the rat. Wound Repair Regen.12, 80–85

183. Ito A, Shinkai M, Honda H, Kobayashi T. 2005. Medical application of functionalized magnetic nanoparticles. J. Biosci. Bioeng. 100, 1–11

184. Ito A, Ino K, Kobayashi T, Honda H. 2005. The effect of RGD peptide-conjugated magnetite cationic liposomes on cell growth and cell sheet harvesting. Biomaterials 26, 6185–6193

185. Ito A, Ino K, Hayashida M, Kobayashi T, Matsunuma H, Kagami H, Ueda M, Honda H. 2005. Novel methodology for fabrication of tissue-engineered tubular constructs using magnetite nanoparticles and magnetic force. Tissue Eng. 11, 1553–1561

186. Ito A, Hibino E, Kobayashi C, Terasaki H, Kagami H, Ueda M, Kobayashi T, Honda H. 2005. Construction and delivery of tissue-engineered human retinal pigment epithelial cell sheets, using magnetite nanoparticles and magnetic force. Tissue Eng. 11, 489–496

187. Shimizu K, Ito A, Honda H. 2006. Enhanced cell-seeding into 3D porous scaffolds by use of magnetite nanoparticles. J. Biomed. Mater. Res. Part B 77B, 265–272

188. Shimizu K, Ito A, Honda H. 2007. Mag-seeding of rat bone marrow stromal cells into porous hydroxyapatite scaffolds for bone tissue engineering. J. Biosci. Bioeng. 104, 171–177

189. Shimizu K, Ito A, Yoshida T, Yamada Y, Ueda M, Honda H. 2007. Bone tissue engineering with human mesenchymal stem cell sheets constructed using magnetite nanoparticles and magnetic force. J. Biomed. Mater. Res. Part B 82B, 471–480

190. Ino K, Ito A, Kumazawa H, Kagami H, Ueda M, Honda H. 2007. Incorporation of capillary-like structures into dermal cell sheets constructed by magnetic force-based tissue engineering. J. Chem. Chem. Eng. Jpn 40, 51–58

191. Shimizu K, Ito A, Arinobe M, Murase Y, Iwata Y, Narita Y, Kagami H, Ueda M, Honda H. 2007. Effective cell-seeding technique using magnetite nanoparticles and magnetic force onto decellularized blood vessels for vascular tissue engineering. J. Biosci. Bioeng. 103, 472–478

192. Akiyama H, Ito A, Kawabe Y, Kamihira M. 2010. Genetically engineered angiogenic cell sheets using magnetic force-based gene delivery and tissue fabrication techniques. Biomaterials 31, 1251–1259

193. Ishii M, Shibata R, Numaguchi Y, Kito T, Suzuki H, Shimizu K, Ito A, Honda H, Murohara T. 2011. Enhanced angiogenesis by transplantation of mesenchymal stem cell sheet created by a novel magnetic tissue engineering method. Arterioscler. Thromb. Vasc. Biol. 31, 2210–2215

194. Yamamoto Y, Ito A, Fujita H, Nagamori E, Kawabe Y, Kamihira M. 2011.Functional evaluation of artificial skeletal muscle tissue constructs fabricated by a magnetic force-based tissue engineering technique. Tissue Eng. Part A 17, 107–114

195. Kito T, et al. 2013. iPS cell sheets created by a novel magnetite tissue engineering method for reparative angiogenesis. Sci. Rep. 3, 1418

196. Hong JS, Vreeland WN, DePaoli Lacerda SH, Locascio LE, Gaitan M, Raghavan SR. 2008. Liposome-templated supramolecular assembly of responsive alginate nanogels.Langmuir 24, 4092–4096

197. Monshipouri M, Rudolph AS. 1995. Liposome-encapsulated alginate—controlled hydrogel particle formation and release. J. Microencapsul. 12, 117–127

198. Joo V, Ramasamy T, Haidar ZS. 2011. A novel self-assembled liposome-based polymeric hydrogel for cranio-maxillofacial applications: preliminary findings. Polymers3, 967–974

199. Viallat A, Dalous J, Abkarian M. 2004. Giant lipid vesicles filled with a gel: shape instability induced by osmotic shrinkage. Biophys. J. 86, 2179–2187

200. Kim Y, Lee Chung B, Ma M, Mulder WJM, Fayad ZA, Farokhzad OC, Langer R. 2012. Mass production and size control of lipid–polymer hybrid nanoparticles through controlled microvortices. Nano Lett. 12, 3587–3591

10

HEMOCOMPATIBILITY OF ALBUMIN NANOPARTICLES AS A DRUG DELIVERY SYSTEM—AN IN VITRO STUDY

Mohamed A. Elblbesy[1,2]

[1]Department of Medical Laboratory Technology, Faculty of Applied Medical Science, University of Tabuk, Tabuk, Kingdom of Saudi Arabia

[2]Department of Medical Biophysics, Medical Research Institute, Alexandria University, Alexandria, Egypt

ABSTRACT

As a major plasma protein, albumin has a distinct advantage compared with other materials for nanoparticle preparation. It is cheap and easily available. The present work aimed to prepare bovine albumin nanoparticles (BAN) with a simple coacervation method and to test their hemocompatibility. The albumin nanoparticles obtained by this method had a range of sizes from 250 - 350 nm at pH = 7.4. In vitro hemocompatibility tests of the prepared (BAN) were conducted after the incubation of BAN with normal blood for 2 h at 37°C. Hemocompatibility tests showed that the reduction in the hemolysis percentage of erythrocytes was due to exposure to BAN. The other blood parameters such as hemoglobin (HG), mean corpuscle hemoglobin (MCH), and mean corpuscle hemoglobin concentration (MCHC) were in the normal range. The prothrombin time (PT) and erythrocyte sedimentation rate (ESR) decreased as the concentration of BAN increased. The results obtained in this study demonstrated that BAN could be used safely and without abnormal effect when interacted with blood through many biomedical applications.

Keywords: Albumin, Nanoparticles, Hemocompatibility

1. INTRODUCTION

Nanoparticles, liposomes, and micelles have received a considerable amount of attention by researchers as drug delivery carriers for site-specific drug delivery. The recent advances in nanoparticles in drug targeting have resulted in a new focus on the role of nanoparticles in disease management. Nanoparticles were first prepared using non-biodegradable polymers such as polyacrylamide, polymethyl methacrylate and polystyrene [1] . However, they could not be considered for systemic use due to the toxicity caused by the non-biodegradable property of the polymer. Hence, biodegradable polymers such as polycyanoacrylate have been used for the preparation of nanoparticles [2] . Nanoparticles have recently received much attention due to their versatile nature, particularly in the field of disease diagnosis, prevention and treatment at the cellular and molecular levels. Nanoparticles can be administered by oral and intravenous routes. In the oral route, they are generally used to increase the oral bioavailability of proteins, peptides and vaccines [3] . In the intravenous route, they are used for the site-specific delivery of a wide range of drugs such as antibiotics to treat intracellular infections and oligonucleotides to protect them from enzymatic degradation and to enhance their penetration across biological membranes. Nanoparticles are useful to deliver several drugs into the brain, including anti-cancer and anti-Alzheimer's drugs [4] . Hence, the usage of nanoparticles in disease management provides tremendous hope in the treatment of many deadly diseases such as cancer, AIDS, intracellular infections and cardiovascular diseases.

Albumin is the most abundant protein in plasma (35 - 50 g/l human serum) with a molecular weight of 66.5 kDa. It is stable in the pH range of 4 - 9, soluble in 40% ethanol, and it can be heated at 60°C for up to 10 h without deleterious effects. These properties, as well as its preferential uptake in tumors and inflamed tissue, widespread availability, biodegradability, and lack of toxicity and immunogenicity make serum albumin an ideal candidate for drug delivery. Recent applications of serum albumin (human and bovine) have demonstrated certain of its advantages as a natural and therefore biocompatible and biodegradable carrier for drugs [5] [6] . The use of albumin for preparing biodegradable nanoparticles has gained great interest in therapeutics, particularly in cancer treatments. Because the carriers deliver the drugs in a controlled manner to the targeted cancer cells, a lower amount of drug would be sufficient to obtain the therapeutic effect, which consequently minimizes the side effects. Albumin contains amino and carboxylic groups, and these groups can be used for surface modifications. Moreover, human serum albumin-based nanoparticles have been well tolerated without any serious side effects in clinical trial studies [7] [8] . Nanoparticles of minimal size can enhance the efficacy of drug metabolism with minimal side effects because they allow for the possibility of site-specific targeted delivery [9] [10] .

Albumin-based nanoparticle carrier systems represent an attractive strategy because a significant amount of drug can be incorporated into the particle matrix due to the different drug binding sites that are present in the albumin molecule [11] . Because of the defined albumin primary structure and high content of charged amino acids (e.g., lysine), albumin-based nanoparticles could allow the electrostatic adsorption of positively (e.g., ganciclovir) or negatively charged (e.g., oligonucleotide) molecules without the requirement of other compounds [12] [13] . In addition, albumin nanoparticles can be easily prepared under soft conditions by coacervation, controlled desolvation or emulsion formation. They have a smaller size (50 to 300 nm) compared with microparticles and, in general, they have better controlled release properties than liposomes which may improve patient acceptance and compliance. Commercially, albumins are obtained in significant quantities from egg white (ovalbumin), bovine serum (bovine serum albumin, BSA), and human serum (human serum albumin, HSA) and also available from soybeans, milk, and grains [14] .

Hemocompatibility is a very important factor to consider when deciding on the application of implantable biomaterials such as artificial blood vessels and orthopedic implants. With the development of blood-contacting materials or implantable devices, it is necessary to improve the hemocompatibility by surface modification or re-design. When blood contacts a foreign biomaterial surface, the first process that occurs is the competitive adsorption of plasma proteins [15] . Many studies have aimed to improve hemocompatibility of biomaterials by surface modification [16] [17] . The improvement of hemocompatibility on a biomaterial surface aims at reducing protein adsorption with the eventual goal of decreasing platelet adhesion

According to the variety of the applications of albumin nanoparticles, the present work aims to prepare bovine albumin nanoparticles (BAN) using the simple coacervation method. To evaluate the interaction between synthetic BAN and human blood, hemocompatibility tests were performed in vitro.

2. MATERIALS AND METHODS

2.1. Chemicals

Bovine albumin lyophilized powder (A2153 SIGMA), 0.5 M sodium hydroxide, and 4% glutaraldehyde. The chemicals used were analytical grade.

2.2. Albumin Nanoparticle Preparation

Bovine albumin was dissolved in distilled water (0.02 g/ml). The bovine albumin solution was filtered using a Millipore filter. The filtrate was placed in a beaker and adjusted to pH7.4 using 0.5 M sodium hydroxide solution. The solution was stirred on a magnetic stirrer, and the required amount of acetone was added dropwise from a syringe until the solution became turbid. BAN were cross-linked by adding 100 µl of a 4% glutaraldehyde-ethanol solution, and the solution was mixed continuously for 3 h at room temperature. After the cross-linking stage, BAN were again filtered through the Millipore membrane filter. The filtrate was centrifuged at 20,000 rpm at 20°C for 30 min. After centrifugation, the supernatant was removed, and the suspension was washed four times with acetone. Finally, BAN were suspended in an acetone-water mixture.

2.3. Albumin Nanoparticle Characterization

The sizes, morphology of the BAN were examined using scanning electron microscopy (SEM) (JEOL, JSM- 6360 LA Japan).

2.4. BAN Hemocompatibility Tests

Blood samples were drawn by venipuncture from 20 healthy adult human donors of the same age and gender. Blood from each donor was collected in two different evacuated tubes: one contained EDTA (1.5 mg/l) as an anti-coagulant, and the other contained sodium citrate, and both tubes contained 5 ml blood. After the gentle mixing, each blood sample was divided into six tubes. One of the tubes was used as a control without the addition of BAN.

The prepared BAN was added to blood in each tube with at the following concentrations: 0.34 mg/ml, 0.67 mg/ml, 1 mg/ml, 1.34 mg/ml, and 1.67 mg/ml. Control blood samples contained no BAN, and blood samples with BAN were incubated for 2 h at 37°C.

2.5. Complete Blood Count (CBC)

2 ml of anti-coagulated blood incubated with BAN were sent to a clinical lab to perform a CBC. CBC analysis was also conducted on the control samples without BAN.

2.6. Hemolysis Assay

Anti-coagulated blood incubated with BAN was centrifuged and washed with phosphate buffer saline (PBS) five times in a centrifuge according to the procedure reported in literature [16] [17] to completely remove plasma and obtain erythrocytes. Erythrocytes were suspended in PBS. A 0.2 ml aliquot of the diluted suspension was transferred to a 5 ml tube that was filled with 0.8 ml of water (as a positive control) and PBS buffer (as a negative control), respectively. The supernatant was analyzed using a spectrophotometer to record the absorbance at 540 nm. The hemolytic percentage (HP) can be calculated using the following equation:

$$HP\% = \frac{(D_t - D_{nc})}{(D_{pc} - D_{nc})} \times 100$$

where the D_t is the absorbance of a test sample, D_{nc} is the absorbance of the negative control and D_{pc} is the absorbance of the positive control.

2.7. Prothrombin Time (PT)

200 µl of thromboplastin with calcium was incubated at 37°C for 15 min. Plasma was separated from the blood incubated with BAN and with sodium citrate as an anti-coagulant. 100 µl of plasma at 37°C was added to thromboplastin with calcium. The clotting time was monitored using a stopwatch.

2.8. Erythrocyte Sedimentation Rate (ESR)

Blood samples collected on sodium citrate and incubated with BAN with different concentrations were checked for clots using two applicator sticks. Using a capillary pipette, the Wintrobe tube was filled to the 0 mark with a blood sample. The tube was left in an exactly vertical position in the rack. After 60 min, the level of the erythrocyte column was recorded. This result was the erythrocyte sedimentation rate in mm/h.

2.9. Statistical Analysis

The data from the hemocompatibility tests are represented as the mean ± SD in all graphs. Chart bars diagram of the values obtained were drawn. The Pearson's correlation coefficient (R^2) was determined.

3. RESULTS AND DISCUSSION

Human serum albumin nanoparticles containing noscapine were prepared by the coacervation method, in which ethanol was used as a dissolving agent, followed by crosslinking using glutaraldehyde [18] . The pH-coacervation method has widely been used in the encapsulation of water-soluble and insoluble drugs. With the modified pH-coacervation technique reported in this study, Safaa Sebak et al. obtained noscapine-loaded human serum albumin nanoparticles with diameters between 150 and 300 nm with a narrow size distribution [19] . In the present study, the coacervation method was used to produce BAN at pH 7.4 at 37°C. BAN obtained were in the size range of 250 to 350 nm with a mostly uniform shape as shown in Figure 1.

Many nanoparticle products have been designed for intravenous administration such as targeted and controlled drug delivery, as contrast agents for imaging or as vaccine carriers. However, the mechanism of interaction of nanoparticles with blood is not well understood. The evaluation of the compatibility of nanoparticles in blood can be critical in the design of the nanoparticle for in vivo applications. Various assays can be used to evaluate hemocompatibility including erythrocyte composure (hemolysis and hemagglutination), thrombogenicity, cytokine assays, complement activation (protein adsorption), phagocytosis, and toxicity [20] .

Erythrocytes are perfect osmometers where disturbances in osmotic and physical changes in blood can cause their lysis. Erythrocytes are also critical in maintaining proper flow characteristics in microcirculation, thus preventing vessel occlusion. In general, erythrocyte destruction can be caused by several factors such as bacterial infection, intrinsic erythrocytes membrane defects, mechanical shear due to blood flow, osmotic and pH changes in blood, drug-induced hemolysis, excessive complement, and platelet activation [21] . The exposure of erythrocytes to nanoparticles could result in hemolysis and/or hemagglutination. Because hemolysis results in the release of red-colored hemoglobin, colorimetric assays can be used to estimate the extent of damage. Nanoparticles also function as antigens in bridging erythrocytes, causing hemagglutination, which results in clumping or agglutination to form a diffuse lattice. This effect has serious consequences in humans leading to the obstruction of blood vessels, imbalance in osmolarity of blood, and the inability of erythrocytes to carry out normal functions [20] . Accordingly, the effect of BAN on erythrocytes hemolysis was analyzed. My results showed that the percentage of HP decreased as the BAN concentration increased. The correlation between BAN concentration and HP % was strong ($R^2 = 0.92$), and they were negatively correlated. The values of HP% were in the normal acceptable range between 0% - 1% for all of the examined concentrations of BAN as indicated in Figure 2. It was shown that albumin has a preventative effect against erythrocytes hemolysis [22] [23] . The result was in accordance with the previous finding that BAN reduced the percentage of HP. As shown in Figure 3, the HG

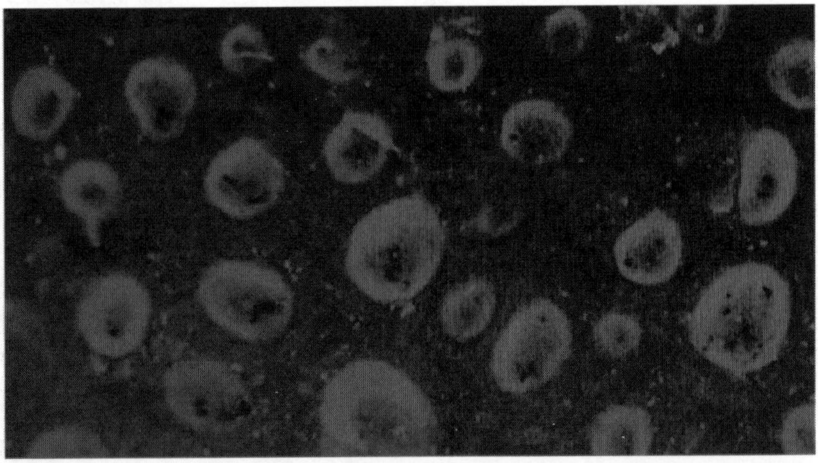

Figure 1. Scanning electron micrographs (SEMs) of BAN prepared pH 7.4, 60 k resolution. BAN were found to be of uniform size.

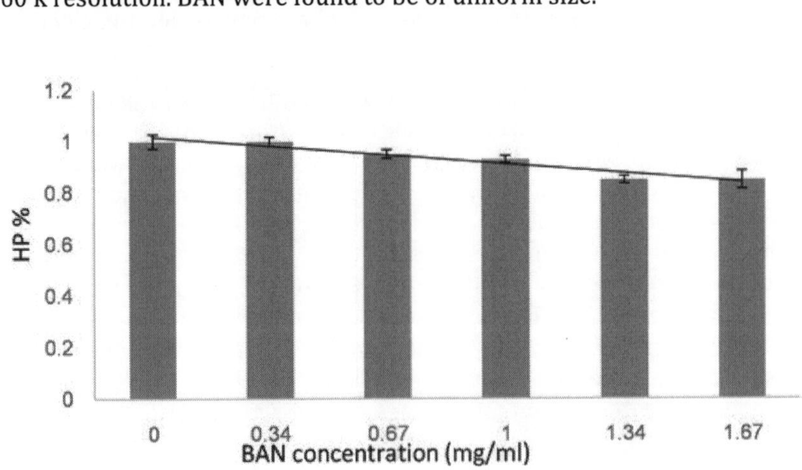

Figure 2. HP% is slightly changed with BAN concentrations.

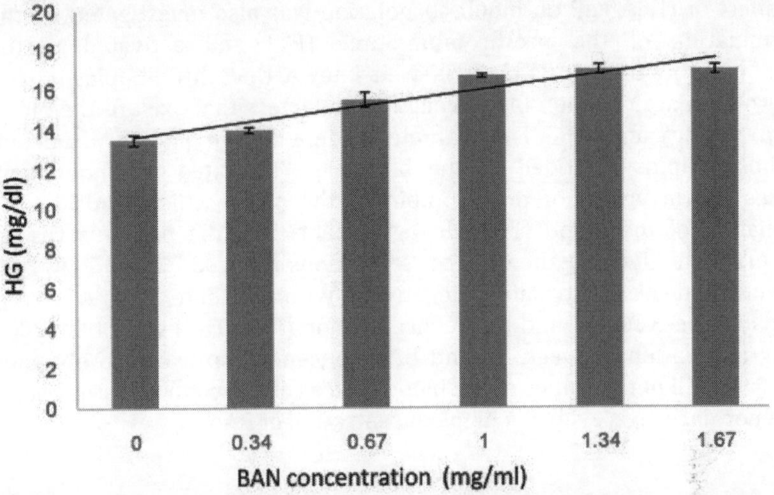

Figure 3. Ata BAN concentration of 0.67 mg/ml, the HG concentration begins to increase as the concentration of BAN increases.

(mg/dl) concentration was in the normal range, and no abnormal effect was observed due to the increase in BAN concentration. The relationship between the HG and BAN concentration was strongly positively correlated ($R^2 = 0.9$). MCH and MCHC were strongly correlated with the BAN concentration. Increases in MCH and MCHC were obtained as the concentration of BAN was increased (Figure 4, Figure 5). For MCH and MCHC, the relationship with BAN concentration was strongly correlated ($R^2 = 0.98$). Incubation and storage of RBCs led to an echinocytic shape transformation, which was reversible upon incubation in albumin solutions. This process was time-, concentration- and hematocrit-dependent. Treating RBC units at the end of their shelf-life by adding 20% albumin or washing them in 0.2% albumin reversed all degrees of echinocytosis towards discocytosis. In conclusion, albumin has the capacity to reverse echinocytosis generated by RBC storage [24] . Yubin Huang et al. indicated that the coexistence of a 10 - 44 vol% of rHSA-FeP did not cause any influence on the number of WBCs and RBCs during the 6-h observation period. This was further proven by the microscopic observation of the morphological change of the RBC. No deformation or aggregation of RBC was observed for the blood suspension with the 10, 20, and 44 vol% of rHSA-FeP after 6 h at 37°C [25] . MPV was linearly related to BAN concentrations. The relationship between them was strongly positively correlated ($R^2 = 0.93$). In accordance with previous studies, Figure 6 showed that there was no noticeable change in MPV due to the increase in BAN concentrations.

The effect of rHSA-FeP on blood coagulation was also investigated by the determination of the prothrombin time (PT) and activated partial thromboplastin time (APTT). It is well known that thromboplastin and prothrombin are common blood coagulation factors that exist in the blood serum, platelet, and WBC. For example, when a certain amount of calcium or phospholipids is added to the blood, the activated thromboplastin induces the inversion of prothrombin to thrombin, which leads to the coagulation of the blood [26] . In agreement with the previous results in Figure 7and Figure 8, the BAN concentrations reduced PT and ESR. BAN concentration was correlated negatively with PT and ESR. A strong correlation between PT and BAN concentration (R^2 = 0.88) was obtained. A strong correlation between PT and BAN concentration was also obtained (R^2 = 0.92). All of the values of the blood indices obtained in this study were in the normal range, even at a high concentration of BAN.

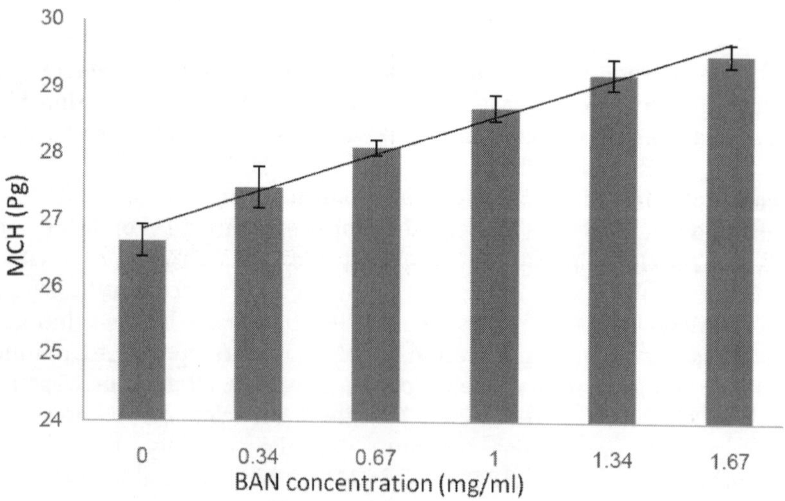

Figure 4. MCH correlate positively to BAN concentration.

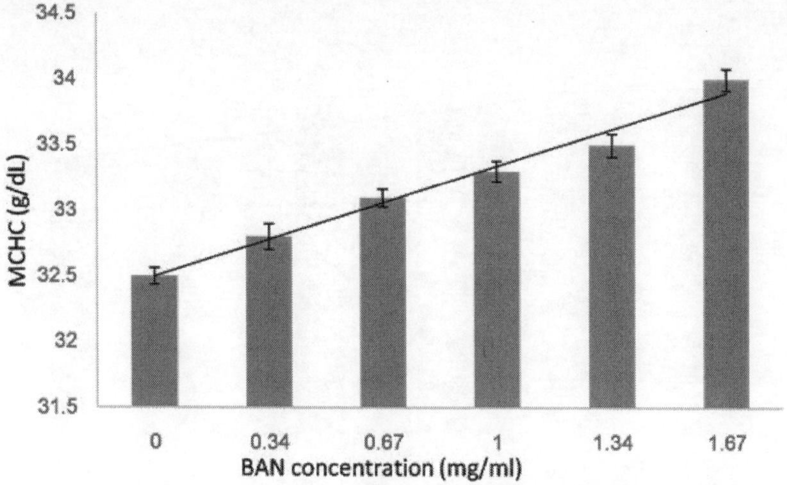

Figure 5. Variation in the values of MCHC is monitored as the BAN concentrations increase.

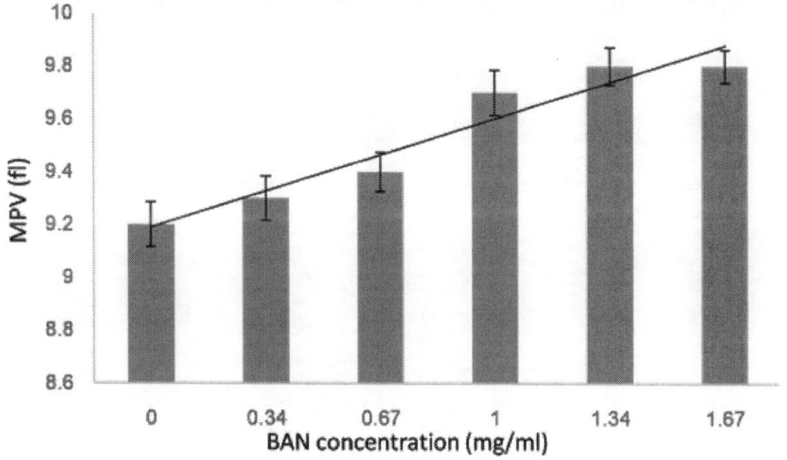

Figure 6. MPV increases linearly with BAN concentration. The plateau in MPV values at high concentrations of BAN is indicated.

4. CONCLUSION

It can be concluded that BAN has no harmful effect when it interacts directly with blood. The results obtained in this study emphasize that BAN has a high degree of biodegradability and biocompatibility. BAN can be used as a promising tool in many biological and medical applications.

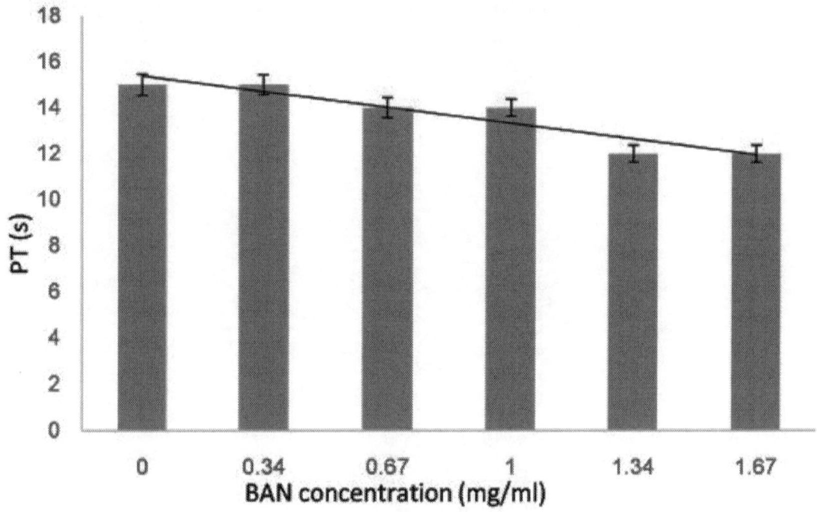

Figure 7. Higher concentrations of the BAN were associated with faster clotting formation, but the PT remained within the normal range.

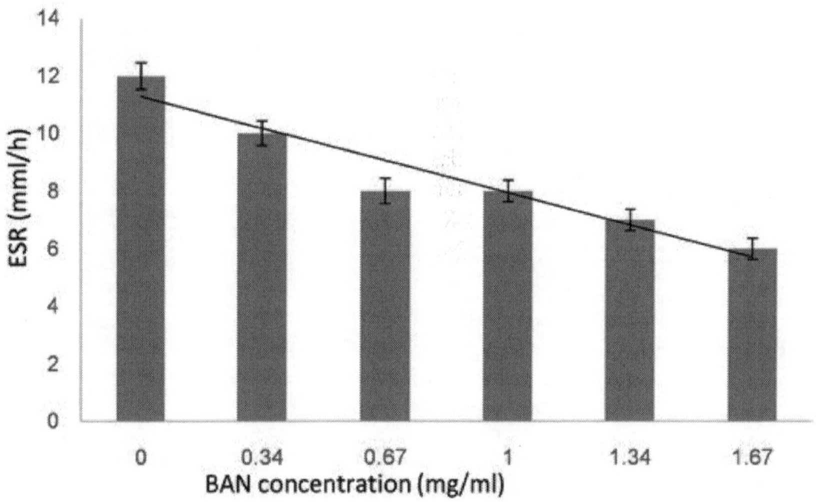

Figure 8. A notable decrease in ESR is shown at higher concentrations of BAN.

ACKNOWLEDGEMENTS

The author would like to acknowledge financial support of this work from the Deanship of Scientific Research (DSR), University of Tabuk (Tabuk, Saudi Arabia, under grant no. (S-1436-0145)).

REFERENCES

1. Kreuter, J. and Speiser, P.P. (1976) Studies of Poly (Methyl Methacrylate) Adjuvants. Journal of Pharmaceutical Sciences, 65, 1624-1627.

2. Couvreur, P., Kante, B., Grislain, L., Roland, M. and Speiser, P.P. (1982) Toxicity of Polyalkylcyanoacrylate Nanoparticles. 2. Doxorubicinloaded Nanoparticles. Journal of Pharmaceutical Sciences, 71, 790-792.

3. Yamamoto, T.H. and Kawashima, Y. (2001) Mucoadhesive Nanoparticulate Systems for Peptide Drug Delivery. Advanced Drug Delivery Reviews, 47, 39-54.

4. Wilson, B. (2009) Brain Targeting PBCA Nanoparticles and the Blood-Brain Barrier. Nanomedicine, 4, 499-502.

5. Temming, K., Meyer, D.L., Zabinski, R., Dijkers, E.C., Poelstra, K., Molema, G. and Kok, R.J. (2006) Evaluation of RGD-Targeted Albumin Carriers for Specific Delivery of Auristatin E to Tumor Blood Vessels. Bioconjugate Chemistry, 17, 1385-1394.

6. Rahimnejad, M., Jahanshahi, M. and Najafpour, G.D. (2006) Production of Biological Nanoparticles from Bovine Serum Albumin for Drug Delivery. African Journal of Biotechnology, 5, 1918-1923.

7. Geny, B., Mettauer, B., Muan, B., Bischoff, P., Epailly, E., Piquard, F., Eisenmann, B. and Haberey, P. (1993) Safety and Efficacy of a New Transpulmonary Echo Contrast Agent in Echocardiography Studies in Patients. Journal of the American College of Cardiology, 22, 1193-1198.

8. Gradishar, W.J., Tjulandin, S., Davidson, N., Shaw, H., Desai, N., Bhar, P., Hawkins, M. and O'Shaughnessy, J. (2005) Phase III Trial of Nanoparticle Albumin-Bound Paclitaxel Compared with Polyethylated Castor Oil-Based Paclitaxel in Women with Breast Cancer. Journal of Clinical Oncology, 23, 7794-7803.

9. Hawkins, M.J., Soon-Shiong, P. and Desai, N. (2008) Protein Nanoparticles as Drug Carriers in Clinical Medicine. Advanced Drug Delivery Reviews, 60, 876-885.

10. Brigger, I., Dubernet, C. and Couvreur, P. (2002) Nanoparticles in Cancer Therapy and Diagnosis. Advanced Drug Delivery Reviews, 54, 631-651.

11. Patil, G.V. (2003) Biopolymer Albumin for Diagnosis and in Drug Delivery. Drug Development Research, 58, 219-247.

12. Irache, J.M., Merodio, M., Arnedo, A., Camapanero, M.A., Mirshahi, M. and Espuelas, S. (2005) Albumin Nanoparticles for the Intravitreal Delivery of Anticytomegaloviral Drugs. Mini-Reviews in Medicinal Chemistry, 5, 293-305.

13. Weber, C., Coester, C., Kreuter, J. and Langer, K. (2000) Desolvation Process and Surface Characteristics of Protein Nanoparticles. International Journal of Pharmaceutics, 194, 91-102.

14. Arshady, R. (1989) Preparation of Microspheres and Microcapsules by Interfacial Polycondensation Techniques. Journal of Microencapsulation, 6, 13-28.

15. Chen, H.L., Luo, P., Huang, Z.Y., Chen, H.P., Chen, M. and Chen, D.H. (2013) Preparation and Blood Compatibility of Carbon/TiO2 Nanocomposite. Diamond and Related Materials, 38, 52-58.

16. Chen, Z., Zhang, R., Kodama, M. and Nakaya, T. (2000) Anticoagulant Surface Prepared by the Heparinization of Ionic Polyurethane Film. Journal of Applied Polymer Science, 76, 382-390.

17. Vallar, L. and Rivat, C. (1996) Regenerated Cellulose-Based Hemodialyzers with Immobilized Proteins as Potential Devices for Extracorporeal Immunoadsorption Procedures—An Assessment of Protein Coupling Capacity and In-Vitro Dialysis Performances. Artificial Organs, 20, 8-16.

18. Lin, W., Coombes, A.G., Davies, M.C., Davis, S.S. and Illum, L. (1993) Preparation of Sub-100 nm Human Serum Albumin Nanospheres Using a pH-Coacervation Method. Journal of Drug Targeting, 1, 237-243.

19. Sebak, S., Mirzaei, M., Malhotra, M., Kulamarva, A. and Prakash, S. (2010) Human Serum Albumin Nanoparticles as an Efficient Noscapine Drug Delivery System for Potential Use in Breast Cancer: Preparation and in Vitro Analysis. International Journal of Nanomedicine, 5, 525-532.

20. Evani, S.J. and Anand, K (2011) Hemocompatibility of Nanoparticles. In: Sitharaman, B., Eds., Nanobiomaterials Handbook, Taylor and Francis Group, London, 31.

21. Sowemimo-Coker, S.O. (2002) Red Blood Cell Hemolysis during Processing. Transfusion Medicine Reviews, 16, 46- 60.

22. Parsi, K., Exner, T., Connor, D.E., Herbert, A., Ma, J.E. and Joseph, D.D.F. (2008) The Lytic Effects of Detergent Sclerosants on Erythrocytes, Platelets, Endothelial Cells and Microparticles Are

Attenuated by Albumin and Other Plasma Components in Vitro. Journal of Vascular and Endovascular Surgery, 36, 216-223.

23. Rojas-Mancilla, E., Oyarce, A., Verdugo, V., Zheng, Z.P. and Ramírez-Tagle, R. (2015) The Cluster [Re6Se8I6]3-Induces Low Hemolysis of Human Erythrocytes in Vitro: Protective Effect of Albumin. The International Journal of Molecular Sciences, 16, 1728-1735.

24. Reinhart, S.A., Schulzki, T. and Reinhart, W.H. (2015) Albumin Reverses the Echinocytic Shape Transformation of Stored Erythrocytes. Clinical Hemorheology and Microcirculation, 60, 437-449.

25. Huang, Y.B., Komatsu, T., Nakagawa, A., Tsuchida, E. and Kobayashi, S. (2003) Compatibility in Vitro of Albumin-Heme (O2 Carrier) with Blood Cell Components. Journal of Biomedical Materials Research Part A, 66A, 292-297.

26. Owen, W., Esmon, C.T. and Jackson, C.M. (1974) The Conversion of Prothrombin to Thrombin. Journal of Biological Chemistry, 249, 594-605.

11

GOLD NANOPARTICLE AND BERBERINE ENTRAPPED INTO HYDROGEL MATRIX AS DRUG DELIVERY SYSTEM

Camila Rufino Souza[1], Henrique R. Oliveira[1], Wagner M. Pinheiro[2], Lubhandwa S. Biswaro[2], Ricardo B. Azevedo[2], Anderson J. Gomes[1*], Claure N. Lunardi[1]

[1]Laboratory of Nanobiotechnology, Faculty of Ceilandia, University of Brasilia, Brasilia, Brazil

[2]Institute of Biological Sciences, University of Brasilia, Brasilia, Brazil

ABSTRACT

In this study the novel hydrogel loaded with gold nanoparticle (AuNP) enhanced the berberine (BS) release when compared with other formulations of hydrogel. Hydrogels are hydrophilic polymer networks having the capacity to absorb water, ranging from about twenty to thousand times their dry weight. BS is a natural product, a quaternary ammonium salt from the group of isoquinoline alkaloids found in medicinal plants as Berberis Vulgaris. BS has some activity against dysentery, hypertension, inflammation, and liver disease in China and Japan. In this work, BS was used as a model drug to study its association with different types of hydrogel composites of polyvinyl alcohol (BS-PVA 10%); gellan gum (BS-GG 2%), gellan gum-PVA crosslinked with cysteine (cys) (BSGG2%PVA2%cys) and gellan gum-PVA cosslinked with cysteine associated with gold nanoparticles (AuNP-BSGG2%PVA2%cys). Several parameters such as fraction of retained water (Wf), hydration percentage (%H), Swelling (DSw) and time course profile (t = 100%) (TC) were evaluated for all preparations. The results showed that the AuNP-BS-GG2%PVA2%cys was able to retain more water and swelling than the other preparations. The time course of release of the BS to the medium was greater for AuNP-BS-GG2%PVA2%cys making it a candidate to drug delivery studies in biological assays. Also Scanning Electron Microscopy (SEM) images of the surface of these hydrogel were performed. Furthermore, crosslink of the resulting hydrogels were investigated by Fourier Transform Infrared Spectroscopy (FTIR) and differential scanning calorimetry (DSC) and thermogravimetric analysis (TGA). Thus, briefly, the aim of this work was to study three composition of hydrogel loaded with BS and its composition in relation to addition to AuNP and evaluate its profile for further drug delivery application using the Surface Plasmonic Resonance (SPR) as a tool improving the drug release in the new hydrogel.

Keywords: Hydrogel, Berberine, Drug Delivery, AuNPs

1. INTRODUCTION

Berberine (BS) is a natural product, a quaternary ammonium salt from the group of isoquinoline alkaloids, belongsto structural class of protoberberine, which positively-charged ion turns it hydro soluble. It is found in medicinal plants as Berberis Vulgaris [1] -[3] and the salts are derived of ammoniac form, solid, nonvolatile, crystallizable and inodorous when it is oxygenized. Due to its yellow color and characteristics the BS is applied with basic nature pigment [4] . The most interesting aspect of BS is related to its pharmacological properties been reported in the treatment of illness, such as: dysentery, hypertension, diabetes, inflammation, and also cancer [5] - [12] .

Hydrogel is report itself as polymeric network (reticulated structure), which has the feature to swell with water or biological fluids. Moreover, hydrogel storage a large amount of these fluids in its structure without dissolves itself [13] [14] . Besides, it exhibits some particular characteristics, such as hydrophilic behavior, insoluble in water, floppy, elastic and in contact with water, its volume increases considerably, without lose its shape, until achieve physic-chemical equilibrium. Hydrogels are able to mimic nature tissue, being biocompatibility [15] [16] . This feature leads to the amount of storage a large amount of water and special surface property. Hence, in this background, synthetics hydrogels are awaken interest at tissue engineering, cell encapsulate and drug delivery

[17] . Another type of nanomaterial that have attracted increasing attention due to their unique properties in multidisciplinary research fields are gold nanoparticles (AuNPs) [2] [3] . The gold core is inert and essentially non-toxic to cells [18] . The particles absorb and resonantly scatter visible and near-infrared light upon excitation of their surface plasmon oscillation (SPR).

Gold nanoparticles associated to hydrogels are based on the behavior of AuNPs properties like the SPR band leading to evaluation of parameters like size among others [19] . Another important characteristic of AuNPs is

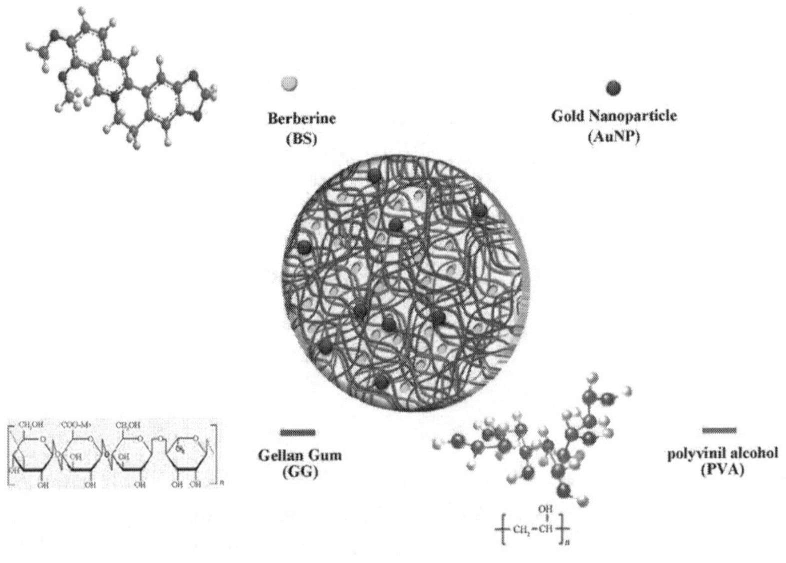

Figure 1. Schematic illustration of possible interaction of BS-GG2%PVA2%cys and AuNPs into the composite.

the ability of dissipating heat when absorbing light, being an interest approach for thermal therapy as a trigger to release of encapsulated drugs from thermosensitive hydrogels [9] [17] . Moreover, there are no reports on the use of BS for preparing crosslinked AuNPs hydrogel nanocomposites, despite its relevance in the context of new formulations for biocompatible hydrogels. The aim of this article was to study the different types of hydrogelbiodegradable composites and the inclusion of gold nanoparticles to this system, evaluating parameters of surface and it's correlation to release time course profile of BS improving the efficiency and versatility to be used in existing therapies.

2. EXPERIMENTAL

2.1. Materials

Berberine (BS from Sigma Chem. Co.); polyvinil alcohol (88% PVA from Sigma Chem. Co.); Deacylated gellan gum GG-LA (Kelcogel from Cpkelco); chloridric acid (from Akros Chem); Au(III) solution (1%) was prepared with HAuCl4, citric acid (from Sigma Chem. Co.); HEPES buffer (from Sigma Chem. Co.); Cisteyne (CYS). Double distilled water was used for the preparation of all solutions.

2.2. Preparation of Hydrogel Composites

PVA 10%. Hydrogel composites were prepared with 10% PVA solution, 7.5% citric acid solution in distilled water,BS was and used as received. All the reagents were of analytical grade. The preparation of the hydrogel has been described elsewhere [20] . In short, 100 mL of a 10% aqueous solution of PVA and citric acid solution and concentrated hydrochloric acid (HCl, 0.05 mL) were mixed, and the resulting dispersion was stirred (using a overhead stirrer at 100 ± 5 rpm) at 70°C for a half-hour to carry out the crosslinked PVA hydrogel production. The thick dispersion so obtained was converted into a membrane by the conventional solution casting method. The resultant membrane was washed thoroughly with distilled water to remove the HCl and was stored in a desiccators at 37°C. The empty hydrogel was designated PVA 10% and the berberine loaded was designated BS-PVA10%.

Gellan Gum. The gellan gum (GG) hydrogel composite were prepared using HEPES (50 mM, pH = 4), following proportion: 2% of GG-LA (89.1 COOH) [21] [22] . In short, 6 mL of HEPES and 20μM of solution CYS and GG-LA were mixed, and the resulting dispersion was stirred (using a overhead stirrer at 100 ± 5 rpm) at 80°C for ten minutes to carry out the crosslinked kelcogell hydrogel production. The thick dispersion so obtained was converted into a membrane by the conventional solution casting method. The empty hydrogel was designated GG2% and the berberine loaded was designated BS-GG2%.

PVA 2% plus Gellan Gum 2%. The hydrogel composite were prepared using HEPES 50 mM, pH = 4), following proportion: 2% of GG-LA (89.1 COOH) and 2% of PVA. In short, 6 mL of HEPES, 20 μM of solution CYS, GG-LA and PVA were mixed, and the resulting dispersion was stirred (using a overhead stirrer at 100 ± 5 rpm) at 80°C for ten minutes to carry out the crosslinked kelcogell hydrogel production. The thick dispersion so obtained was converted into a membrane by the conventional solution casting method. The empty hydrogel was designated GG2%PVA2%cys and the berberine loaded was designated BS-GG2%PVA2%cys.

The BS in micromolar range (10-6M) concentration was added before stirring time to produce hydrogels composites. AuNPs: Gold nanoparticle was prepared by modified sodium citrate reduction method by Turkevich [23] . Placed 0.70 mL of 0.1% HAuCl4 into a 150 mL beaker, then added water to near 60 mL and the solution was heated to boiling. An amount of 2 mL of 1% sodium citrate solution was added dropwhise into beaker solution. The solution became rubbish during stirring for 30 minutes. This solution is called naked AuNPs.

AuNPs-PVA 2% plus Gellan Gum 2%. The nanocomposites were prepared by blending the Au NPs with the PVA 2% plus Gellan Gum 2%. The colloidal AuNPs was added to the BS in micromolar range (10-6 M) concentration was added before stirring time to produce hydrogels composites in the same procedure as described before.

2.3. In Vitro Stability and Swelling of Biodegradable Hydrogel Composite

Approximately 30 mg of each type of composites loaded with 465 μM of BS was deposited into pre-weighed polypropylene sample tubes, weighed, warmed to 37 °C, and combined with 1000 μL of warm HEPES solution at pH 7.4. Samples were incubated at 37°C on a rotary shaker and the HEPPES buffer sampled with total replacement after: 30, 60,120 minutes and daily during 8 days showing stable weight. Recovered buffer was sonicated to disrupt residual hydrogel before Uv-vis analysis (Uv-vis Hitachi U-3900). The swelling ratio, DSw, of each type of composite were determined by accurately weighing each gel sample in the stability study after the HEPES buffer had been removed, correcting for residual buffer, and dividing by the original hydrated sample mass [24] -[26] . To compare the degradation of composites with different swelling characteristics, normalized DSw was defined as:

$$DSw = \frac{m}{m^*} \tag{1}$$

where m and m^* are the weight of hydrated composite and dried composite respectively.

2.4. Drug Release from Biodegradable Hydrogel Composite

Release profiles of each loaded drug composite were obtained by depositing approximately 30 mg of drug loaded dry hydrogel and adding 1000 μL of HEPES buffer into polypropylene sample tubes, warming to 37°C. Samples were incubated at 37°C on a rotary shaker and the HEPPES buffer sampled with total replacement after 15, 30, 60, 120, 360 minutes and 24 h and daily during 90 days for each preparation. All samples were fresh analyzed by UV-vis measurements were performed on Hitachi U-3900H UV-Vis Spectrophotometer in the wavelength interval of 200 - 800 nm at 37°C using HEPES buffer as blank. The measurements were accomplished when the values of absorbance of BS in the medium in each sample reached the plateau. These values were transformed to percentage of release (100%) and correlated to time (hours) and displayed as time course profile (TC). The fraction of retained water (Wf); hydration percentage (%H); Swelling (DSw) and time course profile (t = 100%) (TC) were evaluated for all hydrogel composite as described in Equations (2)-(4):

$$Wf = \frac{(m - m^*)}{1} \quad (2)$$

$$EWC = Wf \times 100 \quad (3)$$

$$H = 100 \times \frac{(m - m^*)}{m^*} \quad (4)$$

where m and m^* are the weight of hydrated composite and dried composite respectively.

2.5. Fourier Transform Infrared (FTIR) Spectroscopy of Biodegradable Hydrogel Composite

The chemical structure of the biodegradable composite materials and the composite itself (unloaded and loaded with BS) were analyzed by FTIR (IR Prestige-21 FRIT-8400S, Shimadzu, Japan) in transmission mode. For that, each composite (1.0 mg) were mixed with KBr (40.0 mg) and then formed into a disc in a manual press. Transmission spectra were recorded using at least 32 scans with 4.0 cm^{-1} resolution, in the spectral range 4000 - 400 cm^{-1}.

2.6. Differential Scanning Calorimetry (DSC) Biodegradable Hydrogel Composite

Thermal characterization of biodegradable hydrogel composite was performed with a Shimadzu DSC-60A. The equipment was calibrated with indium. The sample (approximately 3.0 mg) was heated twice from 25°C to 400°C at 20°C/min in a nitrogen atmosphere (flow rate 30 mL/min). The melting temperature (Tm) was determined from the endothermic peak of the DSC curve recorded in the first heating scan.

2.7. Thermogravimetric Analyses (TGA) of Biodegradable Hydrogel Composite

The composites and its material were analyses. The samples weighting about 3 mg heated from 25°C to 400°C in a platinum pan rate 20°C/min in the nitrogen atmosphere at flow of 30 mL/min. The glass transition temperature (Tg) was determined from the DSC curve recorded in the second heating scan.

2.8. Scanning Electron Microscopy Measurements (SEM)

Scanning electron microscopy (SEM) was used to evaluate the surface of biodegradable composite hydrogel. Samples containing biodegradable composite hydrogel were mounted on aluminum stubs and due to their lack of electrical conductivity they were coated with 50 nm gold coating under an argon atmosphere. The composites were examined and photographed by a Jeol 840 A (Tokyo, Japan) Scanning Electron Microscope operating at 20 kV in the traditional mode (SE1 detector). SEM images were obtained to characterize the surface of the BS-PVA 10%, BS-GG2%cys, BS-GG2%PVA2%cys and AuNPs-BS-GG2%PVA2%cys hydrogel films.

2.9. Statistical Analysis

Experiments were performed in triplicate using freshly prepared samples. Statistical analysis data are expressed as means± standard deviation unless otherwise noted. Comparisons of groups of means were determined by ANOVA and pairs of mean by Student's t-test where appropriate. Significance was assigned at $p < 0.05$.

3. RESULTS AND DISCUSSION

3.1. Hydrogel Composite Preparation

The first attempt to production of biodegradable hydrogel composite with PVA was done in 7.5% w/w however, this hydrogel type dissolved when added to solution of HEPES. As alternative, the concentration was increased to 10%. In this way, we characterized the hydrogel in terms of thickness as displayed for GG2% hydrogel preparation in dried and swollen form (Figure 2) by using a pachymeter. The values obtained for each preparation were GG2% 0.4 mm for dry and 3.1 mm for swollen. Also for PVA10% were 0.2 mm dry and 2.0 swollen; GG2%PVA2%cys were 0.2 dry and 4.0 mm swollen and AuNP-GG2%PVA2%cys were 0.2 dry and 3.5 mm swollen.

3.2. Water Content Parameters and Time Course Release of BS

Water content in the hydrogel composite was affected by the concentration of gellan gum and also by the cysteine content. The values of the fraction of retained water (Wf); hydration percentage (%H); Swelling (DSw) is displayed in Table 1. The hydrogel of BS-PVA10% showed the lower values for all the parameters assayed. Only the time course (TC) of the BS release was higher and this might be due the rigid network during gel formation. Also the others hydrogel composite BS-GG 2%, BS-GG2%PVA2%cys and AuNP-GG2%PVA2%cys preparations showed increased values of water parameters with a faster time course of BS release. This results could be explained by the characteristics of GG, a well known polysaccharide manufactured by microbial fermentation of the Sphingomonas paucimobilis microorganism, that can be dissolved in water, and when heated and mixed with mono or divalent cations its forms gels [13] ; this behavior associated with PVA and cross-linked with cysteine, turned it in the most suitable composite from the preparations evaluated. These features indicate that this preparation is suitable for a drug delivery system for BS in biological systems with high content of water. The citrate-capped AuNPs decreases the intensity of the crosslinking reaction between the PVA/GGcys and BS showing a less cross-linked hydrogels, with reduced swollen profile when compared with BS-GG2%PVA 2%cys.

Figure 2. The pachymeter for determining the thickness of GG2% hydrogel in dried and swollen form.

Table 1. Biodegradable hydrogel composite: swelling feature and time course release using BS [465 μM].

Hydrogel composite/ [BS] = 465 μM	Hydrogel properties			
	Wf	%H	DSw	TC (hs)
BS-PVA 10%	0.4962 ± 0.005	96.9 ± 2.06	1.82 ± 0.22	72 ± 3.90
BS-GG 2%	0.6747 ± 0.438	211.74 ± 40.0	3.1174 ± 0.40	22 ± 0.04
BS-GG 2%PVA2%cys	0.9184 ± 0.013	1150.87 ± 186.4	12.508 ± 1.84	22 ± 0.15
AuNP-BS-GG2% PVA2%cys	0.9623 ± 0.021	1450.87 ± 100.2	8.41 ± 1.105	14 ± 0.50

Wf (fraction of retained water); %H (hydration percentage); DSw (Swelling); TC (Time course).

3.3. SEM Measurements of Hydrogel Composites

In Figure 3 are displayed the SEM of the hydrogels surface of BS-PVA 10%, BS-GG2%, and BS-GG2%PVA2%cys hydrogel films respectively, of dried hydrogels films. The flat and featureless images indicate that the films have a condensed structure and clearly show the appearance of a smooth or fibrous structure. The soft and stiff surface of PVA 10% can explain the lower time course release of the BS from matrix. However, the porous and fibrous like surface of BS-GG2%, and BS-GG2%PVA2%cys could explain the faster BS release and the increased values of physical chemistry parameters analyzed. In addition, 10%PVA hydrogel are less able to hydrate than the types with GG2% explaining its lower time course profile. In these words, GG2% hydrogels com- posite and associated with PVA are capable of storage a large amount of BS and maintain longer time at thera- peutic line. Both images from Figures 3(B)-(C) displayed more fibrous in topographical surface than the PVA10%. The AuNP-BS-GG2%PVA2%cys displayed in Figure 3(D) shows less fibrous than BS-GG2%PVA2%cys due to less crosslink, but due to thermal characteristics of AuNPs embebed into hydrogel, under the experimental conditions the release was faster (36%) in relation to BS-GG2%PVA2%cys alone.

3.4. FTIR Analysis of Hydrogel Composites

FTIR spectra was obtained with absorption in the region of 400 - 4000 cm^{-1} at room temperature and used to confirm the hydrogels structures. In Figure 4(A) is displayed the PVA 10% hydrogel FTIR spectra with a characteristic peak at 3343 cm^{-1} due typical OH stretching [16] . Also, there is a 2914 cm^{-1} the asymmetric stretching vibrations of methylene group [24] . At region of 1730 cm^{-1} it is possible to observe C = O stretching [10] . The peak in 1431 cm^{-1} and 1135 cm^{-1} indicate the presence of CH-OH [27] also in 1381 cm^{-1} peak is due to CH asymmetrical bending [26] , and in the region of 1095 cm^{-1} correspond to CO linkage [24] . In Figure 4(B) is displayed the GG2% hydrogel spectra showing the following peaks in the region 3432 cm^{-1} that correspond the OH group as shown before in the PVA hydrogel; 1620 cm^{-1} due to bending vibration of hydroxyl group, 1462 due to CH asymmetrical bending [28] . In the region of absorption of 1219 cm^{-1} is possible see the sulfate ion. At the region 1219 cm^{-1} has a peak that correspond ester linkages [27] and the peak in the 1043 cm^{-1} indicate the presence of C-O-C group. Both hydrogels of PVA10% and GG2% showed preservation of major groups. These characteristics indicate that have no news groups formed in the blend of this hydrogels. It is possible observe, in Figure 4(C) that the GG2% peaks were preserved as 1620 cm^{-1} shifted to absorption in 1607 cm^{-1}, ester linkages were also preserved, as show in the 1192 cm^{-1}. This phenomenon is shown in the same graphic with PVA10% hydrogel peaks. The C = O stretching in 1730 cm^{-1}, characteristic peak of this hydrogel, shifted to 1721 cm^{-1}. The infrared spectra did not show absorption bands of the typical bridges resulted from the reaction of the hydrogel and BS [10] [29] . It seem just characteristic bands from BS.

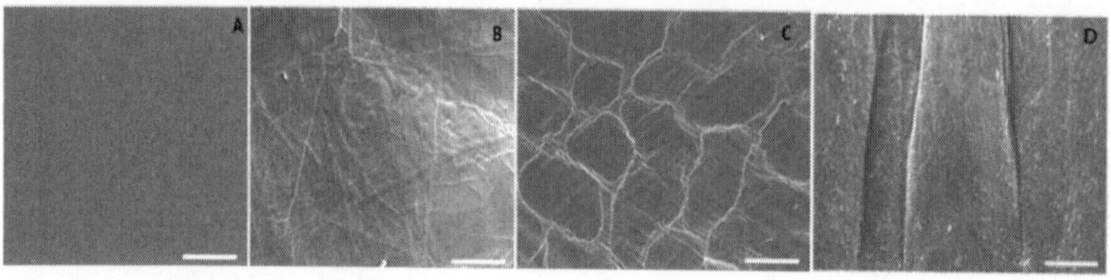

Figure 3. SEM images of hydrogel films. Surface of (a) BS-PVA10%, (b) BS-GG2%, (c) BS-GG2%PVA2%cys and (d) AuNP-BS-GG2%PVA2%cys hydrogel films (270× magnification).

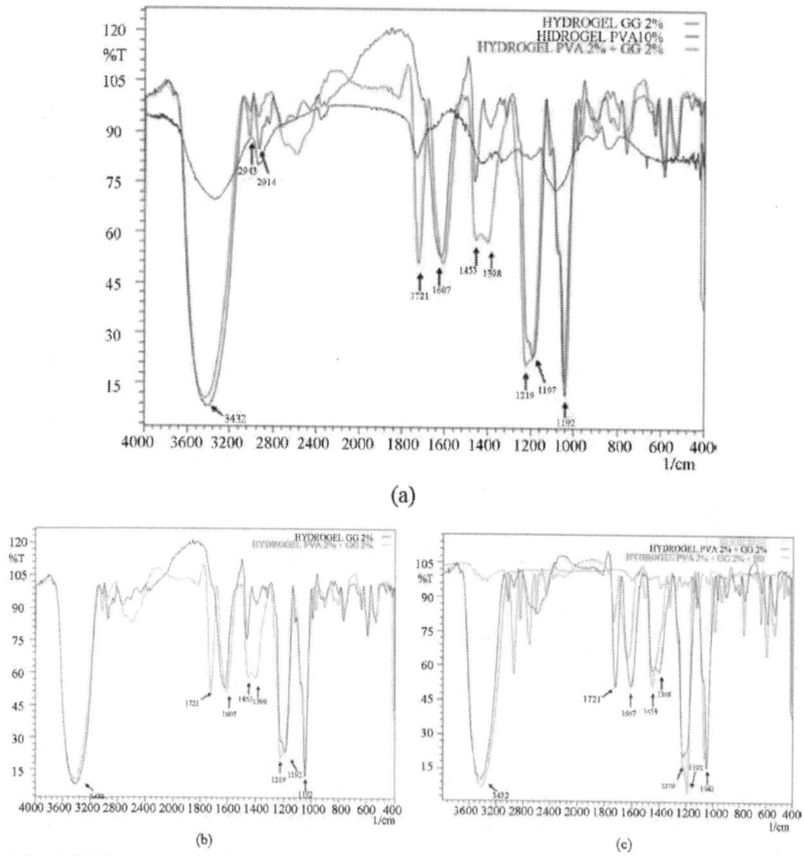

Figure 4. FTIR spectra of all hydrogel composite: (a) GG2%; PVA10%; GG2%PVA2%; (b) GG2% , GG2%PVA2%; and (c) BS; GG2%PVA2%; BS-GG2%PVA2%cys.

3.5. Thermal Analysis of Hydrogel Composites (DSC and TGA)

The DSC heating curves of all hydrogel samples displayed some endothermic peaks due to the melting of the various forms of water that freeze during the cooling stage (Figure 5). The endothermic peak around 40°C - 420°C displayed a peak, which have been ascribed previously to the freezing bound water (the sharp one) [30] -[34] . The thermogram is characterized by peaks crystallization (Tc), melting (Tm) and glass transition (Tg) of all hydrogel composite. The behavior observed for 10%PVA, GG2%, PVAGG2% were very similar showing its Tc in the range of 40°C - 42°C in the first peak and 190°C - 260°C in the following transitions. The biodegradable hydrogel composite BS-GG2%PVA2%cys displayed a similar behavior, however most of the process occurs in exothermic profile. When association occurs, the formation of more structured complex so formed exhibits different characteristics in comparison to components-alone. AuNPs-BS-GG2%PVA2%cys showed simi- lar behavior

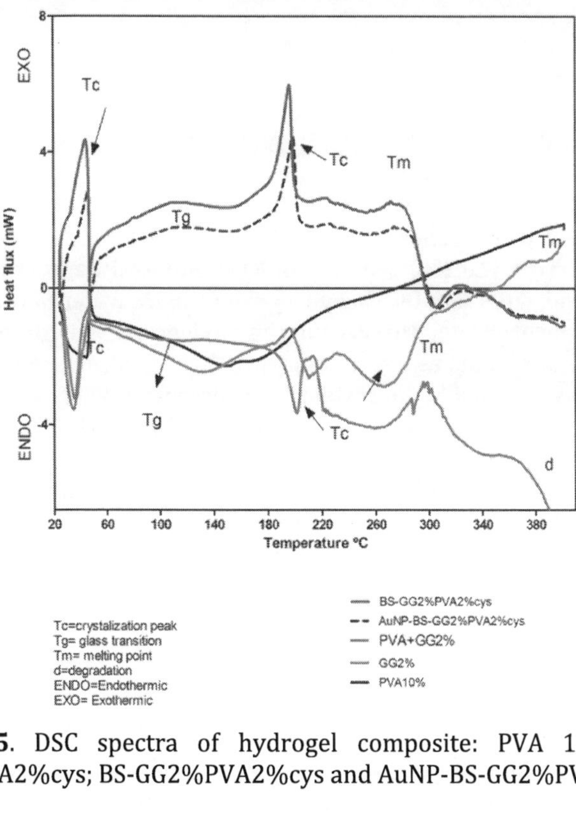

Figure 5. DSC spectra of hydrogel composite: PVA 10%; GG2 %; GG2%PVA2%cys; BS-GG2%PVA2%cys and AuNP-BS-GG2%PVA2%cys.

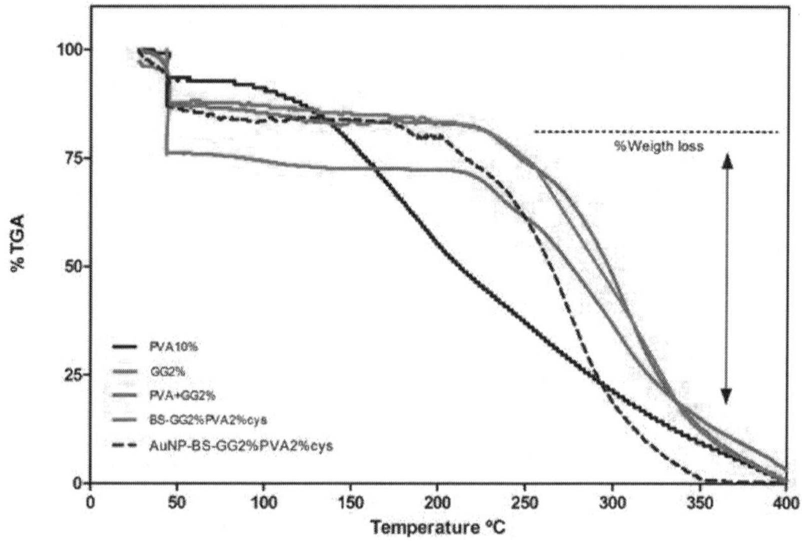

Figure 6. TGA spectra of hydrogel composite: PVA 10%; GG2 %; GG2% PVA2%cys; BS-GG2%PVA2%cys and AuNP-BS-GG2%PVA2%cys.

In Figure 6 is displayed the normalized TGA measurements of all components used in the synthesis biodegradable hydrogel composite. The evaporation process occurs more drastically in the GG2%-(23.0%) and the other component were respectively PVA10%-(5.02%); PVAGG2%-(12.1%); BS-GG2%PVA2%cys-(10.2%) and AuNP-BS-GG2%PVA2%cys-(12.1%). These results indicated that GG2% hydrogel lose more water than PVA10%, also when it is incorporated to PVA and the biodegradable hydrogel loaded with BS, this water lost is reduced indicating that the structure is modified entrapping the water inside the system. The percentage weight loss and the start temperature for all formulation were GG2% (71.5% at 214°C); PVA10%-(90% at 100°C); PVAGG2%-(81.4% at 225°C); BS-GG2%PVA2%cys-(81.2% at 225°C) and AuNP-BS-GG2%PVA2%cys-(80% at 215°C) were observed for all system These results indicate that the GG2% has less mass to lose since its structure is less porous than the network achieved with PVAGG2% BS-GG2%PVA2%cys and AuNP-BS-GG2%PVA2%cys.

4. CONCLUSIONS

Polymeric drug delivery system allows the development of innovative concepts that could create more efficient systems to achieve a prolonged and controlled drug delivery. In this study, by the use of the crosslink method, we have prepared several hydrogel composite by varying amount of GG and PVA loaded with BS and its association to gold nanoparticles, and then characterized its water holding capacities, time course release profile; physical-chemical behavior (FTIR, DSC; TGA) and surface analysis (SEM).

The results have demonstrated distinct surface profile of the hydrogel composites and two different time course profiles, which could explain by the feature of hydrogel matrix.

Among all biodegradable hydrogel composites evaluated it was clear that the addition of GG associated to PVA and association to gold nanoparticle, produced the most suitable drug delivery system for BS, due to all characteristics (higher entrapment efficiency) of the matrix structure, making a promising drug delivery system for delivery hydrophilic drugs. Future application of gold nanoparticles (AuNPs) added into the transparent loaded hydrogel can be a promising efficient drug delivery system in skin, by enabling the cell activating capacity once its size is only 1/2000 as large as pores, very easy to penetrate into the dermis. It can carry the drug such as BS deep into the skin to activate skin cells, increases vasodilatation and can enhance the compactness of the fibrous tissue of collagen to turn the skin compact and elastic against aging and other process.

ACKNOWLEDGEMENTS

Camila Rufino Souza, Henrique R. Oliveira, Wagner M. Pinheiro, Lubhandwa S. Biswaro, Ricardo B. Azevedo, Anderson J. Gomes, Claure N. Lunardi* This material is based upon work supported by the CNPq, FAPDF, FINATEC, DPP-UnB, CAPES.

REFERENCES

1. Huang, Z.J., Zeng, Y., Lan, P., Sun, P.H. and Chen, W.M. (2011) Advances in Structural Modifications and Biological Activities of Berberine: An Active Compound in Traditional Chinese Medicine. Mini-Reviews in Medicinal Chemistry, 11, 1122-1129.

2. Tillhon, M., Guamán Ortiz, L.M., Lombardi, P. and Scovassi, A.I. (2012) Berberine: New Perspectives for Old Remedies. Biochemical Pharmacologyhem Pharmacol, 84, 1260-1267.

3. Xie, W. and Du, L. (2011) Diabetes Is an Inflammatory Disease: Evidence from Traditional Chinese Medicines. Diabetes, Obesity and Metabolism, 13, 289-301.

4. Wang, X.H., Jiang, S.M. and Sun, Q.W. (2011) Effects of Berberine on Human Rheumatoid Arthritis Fibroblast-Like Synoviocytes. Experimental Biology and Medicine (Maywood), 236, 859-866.

5. Chen, X.W., Di, Y.M., Zhang, J., Zhou, Z.W., Li, C.G. and Zhou, S.F. (2012) Interaction of Herbal Compounds with Biological Targets: A Case Study with Berberine. Scientific World Journal, 2012, Article ID: 708292.

6. Chen, Y., Wang, Y., Zhang, J., Sun, C. and Lopez, A. (2011) Berberine Improves Glucose Homeostasis in Streptozotocin-Induced Diabetic Rats in Association with Multiple Factors of Insulin Resistance. ISRN Endocrinology, 2011, Article ID: 519371

7. Chidambara Murthy, K.N., Jayaprakasha, G.K. and Patil, B.S. (2012) The Natural Alkaloid Berberine Targets Multiple Pathways to Induce Cell Death in Cultured Human Colon Cancer Cells. European Journal of Pharmacology, 688, 14- 21

8. Tan, W., Li, Y., Chen, M. and Wang, Y. (2011) Berberine Hydrochloride: Anticancer Activity and Nanoparticulate Delivery System. International Journal of Nanomedicine, 6, 1773-1777.

9. Trimarco, V., Cimmino, C.S., Santoro, M., Pagnano, G., Manzi, M.V., Piglia, A., et al. (2012) Nutraceuticals for Blood Pressure Control in Patients with High-Normal or Grade 1 Hypertension. High Blood Pressure & Cardiovascular Prevention, 19, 117-122.

10. Wang, J., Liu, Q. and Yang, Q. (2012) Radiosensitization Effects of Berberine on Human Breast Cancer Cells. International Journal of Molecular Medicine, 30, 1166-1172.

11. Wang, Y., Huang, Y., Lam, K.S., Li, Y., Wong, W.T., Ye, H., et al. (2009) Berberine Prevents Hyperglycemia-Induced Endothelial Injury and Enhances Vasodilatation via Adenosine Monophosphate-Activated Protein Kinase and Endothelial Nitric Oxide Synthase. Cardiovascular Research, 82, 484-492.

12. Zhou, L., Wang, X., Yang, Y., Wu, L., Li, F., Zhang, R., et al. (2011) Berberine Attenuates cAMP-Induced Lipolysis via Reducing the Inhibition of Phosphodiesterase in 3T3-L1 Adipocytes. Biochimica et Biophysica Acta (BBA)—Mole- cular Basis of Disease, 1812, 527-535

13. Minhas, M.U., Ahmad, M., Ali, L. and Sohail, M. (2013) Synthesis of Chemically Cross-Linked Polyvinyl Alcohol-co- Poly (methacrylic acid) Hydrogels by Copolymerization; A Potential Graft-Polymeric Carrier for Oral Delivery of 5- Fluorouracil. DARU Journal of Pharmaceutical Sciences, 21.

14. Zhang, X.Z., Zhuo, R.X., Cui, J.Z. and Zhang, J.T. (2002) A Novel Thermo-Responsive Drug Delivery System with Positive Controlled Release. International Journal of Pharmaceutics, 235, 43-50.

15. Liu, L., Wang, B., Gao, Y. and Bai, T.C. (2013) Chitosan Fibers Enhanced Gellan Gum Hydrogels with Superior Mechanical Properties and Water-Holding Capacity. Carbohydrate Polymers, 97, 152-158.

16. Moura, M.J., Faneca, H., Lima, M.P., Gil, M.H. and Figueiredo, M.M. (2011) In Situ Forming Chitosan Hydrogels Prepared via Ionic/Covalent Co-Cross-Linking. Biomacromolecules, 12, 3275-3284.

17. Cencetti, C., Bellini, D., Pavesio, A., Senigaglia, D., Passariello, C., Virga, A., et al. (2012) Preparation and Characterization of Antimicrobial Wound Dressings Based on Silver, Gellan, PVA and Borax. Carbohydrate Polymers, 90, 1362-1370.

18. Cobley, C.M., Chen, J., Cho, E.C., Wang, L.V. and Xia, Y. (2011) Gold Nanostructures: A Class of Multifunctional Materials for Biomedical Applications. Chemical Society Reviews, 40, 44-56.

19. Bedford, E.E., Spadavecchia, J., Pradier, C.M. and Gu, F.X. (2012) Surface Plasmon Resonance Biosensors Incorporating Gold Nanoparticles. Macromolecular Bioscience, 12, 724-739.

20. Cristallini, C., Guerra, G.D., Barbani, N. and Bianchi, F. (2007) Biodegradable Bioartificial Materials Made with Chitosan and Poly(vinyl alcohol). Part I: Physicochemical Characterization. Journal of Applied Biomaterials & Biomechanics, 5, 184-191.

21. Cascone, M.G., Barbani, N., Cristallini, C., Giusti, P., Ciardelli, G. and Lazzeri, L. (2001) Bioartificial Polymeric Materials Based on Polysaccharides. Journal of Biomaterials Science, Polymer Edition, 12, 267-281.

22. Oliveira, J.T., Martins, L., Picciochi, R., Malafaya, P.B., Sousa, R.A., Neves, N.M., et al. (2010) Gellan Gum: A New Biomaterial for Cartilage Tissue Engineering Applications. Journal of Biomedical Materials Research Part A, 93, 852- 863.

23. Kimling, J., Maier, M., Okenve, B., Kotaidis, V., Ballot, H. and Plech, A. (2006) Turkevich Method for Gold Nanoparticle Synthesis Revisited. Journal of Physical Chemistry B, 110, 15700-15707.

24. Cai, X., Shao, W., Luan, Y., Pang, J., Li, F. and Li, Z. (2011) Metformin Hydrochloride-Loaded Poly(vinyl alcohol) Composites as Drug Delivery Systems. Journal of Nanoscience and Nanotechnology, 11, 8621-8627.

25. Chen, C.H., Wang, F.Y., Mao, C.F., Liao, W.T. and Hsieh, C.D. (2008) Studies of Chitosan: II. Preparation and Characterization of Chitosan/Poly(vinyl alcohol)/Gelatin Ternary Blend Films. International Journal of Biological Macromolecules, 43, 37-42

26. Lima, A.C., Sher, P. and Mano, J.F. (2012) Production Methodologies of Polymeric and Hydrogel Particles for Drug Delivery Applications. Expert Opinion on Drug Delivery, 9, 231-248.

27. Hu, Y., Wang, Q. and Tang, M. (2013) Preparation and Properties of Starch-g-PLA/Poly(vinyl alcohol) Composite Film. Carbohydrate Polymers, 96, 384-388.

28. Krauland, A.H., Leitner, V.M. and Bernkop-Schnürch, A. (2003) Improvement in the in Situ Gelling Properties of Deacetylated Gellan Gum by the Immobilization of Thiol Groups. Journal of Pharmaceutical Sciences, 92, 1234-1241.

29. Bashmakova, N., Kutovyy, S., Kornienko, M., Yashchuk, V., Hovorun, D. and Zhurakivsky, R. (2010) Experimental and Calculated by the DFT Method Vibration Spectra of Berberine. Xxii International Conference on Raman Spectroscopy, 1267, 426-427.

30. Ahuja, M., Singh, S. and Kumar, A. (2013) Evaluation of Carboxymethyl Gellan Gum as a Mucoadhesive Polymer. International Journal of Biological Macromolecules, 53, 114-121.

31. Coutinho, D.F., Sant, S.V., Shin, H., Oliveira, J.T., Gomes, M.E., Neves, N.M., et al. (2010) Modified Gellan Gum Hydrogels with Tunable Physical and Mechanical Properties. Biomaterials, 31, 7494-7502.

32. Liu, Y., Geever, L.M., Kennedy, J.E., Higginbotham, C.L., Cahill, P.A. and McGuinness, G.B. (2010) Thermal Behavior and Mechanical Properties of Physically Crosslinked PVA/Gelatin Hydrogels. Journal of the Mechanical Behavior of Biomedical Materials, 3, 203-209.

33. Peng, Z., Kong, L.X., Li, S.D. and Spiridonov, P. (2006) Poly(vinyl alcohol)/Silica Nanocomposites: Morphology and Thermal Degradation Kinetics. Journal of Nanoscience and Nanotechnology, 6, 3934-3938.

34. Wang, H.Y., Lu, S.S. and Lun, Z.R. (2009) Glass Transition Behavior of the Vitrification Solutions Containing Propanediol, Dimethyl Sulfoxide and Polyvinyl Alcohol. Cryobiology, 58, 115-117.

12

A CORRELATIVE IMAGING BASED METHODOLOGY FOR ACCURATE QUANTITATIVE ASSESSMENT OF BONE FORMATION IN ADDITIVE MANUFACTURED IMPLANTS

Hua Geng[1] • Naomi M. Todd[2] • Aine Devlin-Mullin[2] • Gowsihan Poologasundarampillai[1] • Taek Bo Kim[3] • Kamel Madi[1] • Sarah Cartmell[1] • Christopher A. Mitchell[2] • Julian R. Jones[3] • Peter D. Lee[1]

[1]School of Materials, The University of Manchester, Oxford Road, Manchester M13 9PL, UK

[2]School of Biomedical Sciences, Centre for Molecular Biosciences (CMB), Ulster University, Coleraine, Northern Ireland, UK

[3]Department of Materials, Imperial College London, London SW7 2AZ, UK

ABSTRACT

A correlative imaging methodology was developed to accurately quantify bone formation in the complex lattice structure of additive manufactured implants. Micro computed tomography (μCT) and histomorphometry were combined, integrating the best features from both, while demonstrating the limitations of each imaging modality. This semi-automatic methodology registered each modality using a coarse graining technique to speed the registration of 2D histology sections to high resolution 3D µCT datasets. Once registered, histomorphometric qualitative and quantitative bone descriptors were directly correlated to 3D quantitative bone descriptors, such as bone ingrowth and bone contact. The correlative imaging allowed the significant volumetric shrinkage of histology sections to be quantified for the first time (~15 %). This technique demonstrated the importance of location of the histological section, demonstrating that up to a 30 % offset can be introduced. The results were used to quantitatively demonstrate the effectiveness of 3D printed titanium lattice implants.

1 INTRODUCTION

Histology remains the gold standard for assessing bone formation within implants because it provides information with a high level of specificity in terms of cellular makeup coupled with excellent image resolution. Additionally, quantitative results, such as area-to-volume ratio, perimeter-to-area ratio [1] and trabecular width (Tb Wi) [2], can be acquired through histomorphometric analysis. Though histology can provide detailed and effective results, it has several drawbacks. Firstly, the preparation of bony sections containing metal implants is difficult and time-consuming [3] and requires special equipment [4]. Secondly, significant deformation and shrinkage of the sample may occur during processing [5], both altering the result and making it difficult to carry out further experiments. Thirdly, a single 2D image from a large volume sample may not represent bone formation over the entire implant. For a typical defect (3–5 mm diameter) in a small animal model, histology only looks at 3–5 % of the tissue when a 100 µm thickness bone implant section is prepared. Although serial sectioning may be an alternative, current methods are inadequate for serial sectioning of tissue in a metal implant without significant tissue loss. Additionally, histomorphometric evaluation of an image poses other difficulties, such as multiple tissue types and variations in staining uniformity within histological images making it difficult to form an objective judgment of overall regeneration in bone lesions [6].

More recently, µCT has been employed to image and characterise the three-dimensional (3D) structure of porous scaffolds [7, 8] and hard tissue [9, 10] in a non-invasive manner. Compared with histology, the substantial advantages of µCT lie in its simple sample preparation, ability to image whole 3D structures and high contrast (in the case of mineralised tissue) [11, 12]. Thus, it is an attractive technique to visualise and quantify bone regeneration within the defect site as well as around the implant [13, 14]. Additional information such as volume fraction on both the implant [7, 15] and newly formed bone has been derived.

Correlative imaging is the combination of multiple techniques, such as μCT with histology to provide complementary information [16] and it is a frequently used technique in medical imaging; for example, combining MRI with PET [17]. For bone formation, few authors have used μCT with histology [18–21] at the resolution required to achieve direct comparison of bone formation. Bernhardt et al. [18] were among the first to quantitatively analyse bone formation by using μCT and histology, but the implant had a simple geometric structure and the newly formed bone was quantified in predefined gaps only. Stalder et al. [22] reported the regenerative capacity of ceramic bone grafting materials with the dual application of μCT and histology. Their use of synchrotron tomography to image ceramic bone graft yields a much cleaner 3D dataset which is free of artefacts such as beam hardening and shadowing when bone is present in its vicinity [18, 22]. Synchrotron sources are, at present, scarce and do not completely eliminate the shadowing artefacts when titanium-implants are imaged. Therefore, there is a need for a new methodology to correlatively image and quantify bone formation in titanium-implants using histomorphometry and laboratory x-ray sources.

A direct comparison can be obtained simply by repeating a μCT scan of the histological sections [19], but it is limited to non-decalcified specimens. From the above, a robust technique that enables a direct comparison, taking advantage of the strengths of both laboratory μCT and histology is highly desirable. In order to match the histological section with its corresponding μCT region, a 2D–3D multimodal (different intensity level) registration is required.

2D–3D registration has been adopted in clinical applications such as image-guided interventions [23]. However, the specific problem of registering histology images to μCT registration has gained less attention and has never been validated, which may be due to the challenges associated with this process. Biological samples are subject to complex morphological deformations, staining artefacts, and missing tissue in individual preparation processing [5]. To address this issue, Museyko et al. [24] applied affine and elastic registration to tibiae and vertebra to estimate the effect of deformation. A main finding of Museyko's study was that segmentation-based registration could achieve comparable accuracy as intensity-based registration.

Further, the issue of tissue shrinkage during histology section preparation is an important one when considering morphometry in fields such as biomechanics and implant surgery. However, to our knowledge, only one study [25] has attempted to quantify the volumetric shrinkage following histology procedures. This may be because of the difficulty in identifying the correspondence between different images.

Here, we report the first application of correlative imaging to quantify bone formation into a novel additive manufactured porous titanium scaffold. A 2D 50–80 µm thick haematoxylin and multiple staining solution stained histological section was registered into a 3D µCT dataset of implanted rat tibia from which the section was prepared. The presence of the non-deformed titanium scaffold allowed rigid registration of µCT and histology sections independently of bone and surrounding tissue. The quality and reliability of the registration was assessed to demonstrate a well-defined correlation between µCT and histology features. The rigid transformation was then used, for the first time, to achieve a comparable study of the bone ingrowth (BI) into porous titanium implant between µCT and histology. After that, a non-rigid registration was performed between the cortical bone and the histological image and its corresponding 2D µCT to quantify the volumetric shrinkage of bone in the preparation of the histological section.

2 MATERIALS AND METHODS

2.1 Production of the titanium scaffolds

The porous titanium scaffolds [26] were designed as a cylinder (diameter Ø = 3 mm, height h = 1.8 mm) consisting of orthogonal struts (Ø = 180 µm) and an overall porosity of 65 % [15].

The structures were fabricated from grade 1 commercially pure titanium (Sumitomo, Japan) using a selective laser melting technique [26]. The laser and scanner were computer-controlled using the Stereo Lithography (STL) data converted from the CAD data. The titanium powder was produced by gas-atomisation with a median particle size of 28.5 µm. All implants were sterilised by immersion in 200 µl of 70 % ethanol for 2 h. The ethanol was removed, then the samples were further sterilised with dry heat (200 °C for 2 h) before implantation.

2.2 Animal experiment

A total number of 12 male Wistar rats (12 weeks old; body mass 250–300 g) were used in this study. Animal protocols were conducted in accordance with the institutional (Ulster University, United Kingdom (UK)) and national guidelines for animal care and welfare. All rats were anaesthetised by isoflourane gas inhalation followed by an intraperitoneal injection of 0.5 ml/100 g bodyweight of a mixture of 2 ml Ketaset (100 mg/ml), 1 ml Xylapan (20 mg/ml) and 5 ml phosphate buffered saline (PBS, pH 7.4). A deep plane of anaesthesia was confirmed by the loss of pedal reflex. The animals were premedicated with 1.5 ml of subcutaneous injection of Metacam (Boehringer Ingelheim, Germany) prepared in water (1:10 dilution) to reduce postoperative pain. The lower right leg was shaved with mechanical hair clippers. The area was cleared of all hair using depilatory cream (Veet®, Rickitt Benckiser Group Plc, UK), cleaned and sterilized using three consecutive washes of pre-warmed chlorhexidin (Hibisrub®, Regent Medical Ltd, UK), followed by 70 % isopropanol. Throughout the duration of the surgical procedure, animals were kept warm on a heating mat (37 °C). A 15 mm full-thickness longitudinal skin incision was made above the middle third of the medial aspect of the tibia. The skin flap was opened using fine spreaders (InterFocus Ltd, Cambridge, UK) exposing the underlying tibia. The tibial surface was cleared of connective tissue and periosteum. A 3 mm circular defect was created using a trephine burr, which extended into the level of the marrow cavity. The defect site was continuously cleared of blood and bone fragments with the aid of suction and the tissue kept moist by saline irrigation. During the surgical procedure, a suture was used to stabilise the porous titanium scaffold. After insertion of the scaffold, the overlying muscle was closed using with three sutures using 4-0 Ethicon Ethilon polyamide (Johnson & Johnson, U.S), one directly over the implant and one above and below the defect area. The skin was closed using a running suture. Post-operation, the animals received topical application of 0.2 % chloramphenicol solution to the sutured skin to prevent infection of the operative site followed by an intra-peritoneal administration of 5 ml of 5 % dextrose-saline solution. The rats were housed separately with 12 h light/dark cycles and were given unrestricted access to food and water. The right tibiae were harvested after either 2, 3, 4 or 6 weeks' implantation (n = 3) for further evaluation.

2.3 μCT image acquisition

The specimens were placed in an ABS plastic sample holder and scanned in a laboratory source μCT machine (nano-focus, Phoenix|x-ray General Electric Company, Measurement and Control, Wunstorf, Germany) with an isotropic voxel size of 9 μm (spatial resolution 19–24 μm). The scanning parameters were set to 85 kV and 111 μA. A 0.5 mm copper filter was placed in the x-ray path to reduce beam hardening. Each scan consisted of 1000 projections over 360°, with the sample rotated in equiangular steps along its longitudinal axis. For each projection, the exposure time was set to 2000 ms. Ring artefacts were reduced using detector jitter [27].

The projections were imported into the reconstruction software (Phoenix dato s| × 2 reconstruction) to generate 3D images of dimensions 990 × 990 × 1000 voxels. All specimens were scanned and reconstructed with identical settings.

2.4 μCT image processing

After reconstruction, scan datasets were normalised to a predefined histogram. To remove digital noise and artefacts, the μCT volume was smoothed with anisotropic diffusion and edge-preserving filters, as described by Chao et al. [28]. The filtered volumes were then segmented into bone tissue and implant using an in-house algorithm. The implant was a high-signal low-noise region and was segmented by global thresholding. However, the bone region was more difficult to distinguish due to metal artefacts. An iterative resampling algorithm correction was applied to compensate for the effect of metal artefacts.

For all scans, the segmented bone and titanium phases were then registered with an average leg model and a cylindrical mask (whose dimensions were identical to those of the implant), respectively. 3D morphometric parameters were calculated in the volume of interest (VOI), which was defined by the overlap of the average leg model and the cylindrical mask. Bone ingrowth and bone contact (BC) were then measured for the new bone.

2.5 Histology

Following fixation in 10 % neutral buffered formalin (Sigma-Aldrich, UK), all bone implant specimens were decalcified in 14 % EDTA (Sigma-Aldrich, UK), dehydrated through a series of increasing concentrations of ethanol, and embedded in LR white resin (TAAB Laboratories Equipment Ltd, UK) for sectioning.

The tissue blocks were then trimmed and a 100–200 μm section was cut using an EXAKT 310 Macro Band System with a diamond blade (EXAKT, USA), ground on an EXAKT 400CS grinding system (EXAKT, USA), using K800 and K1200 Grinding paper and P2500 and P4000 polishing paper (EXAKT, USA) down to a thickness of 50–80 μm. One transverse section was obtained from each specimen. All sections from the tibia were transverse, and the chosen section was at the location of the first cut exposing the implant. This meant the histological section was offset towards the implant edge.

All sections were stained with Gill's Haematoxylin III (Fisher Scientific, Loughborough, UK) and multiple staining solution according to the following protocol. Briefly, slides with adhered sects. (50–80 μm) were placed in successive solutions of 1 % formic acid (2 min), 50 % ethanol (5 min), de-ionized water (5 min) followed by dropwise addition of Gill's Haematoxylin to the slide for 30 min. Slides were subsequently rinsed well in water, stained with multiple staining solution (Polysciences Inc., Warrington, USA) for 20 min, further rinsed in distilled water, air dried and directly imaged via light microscopy using an Axio Scope 1 (Carl

Zeiss, Germany) microscope at a range of objective magnifications (a magnification of 1.25 for a 5.23 μm/pixel and a magnification of 5 for a 1.26 μm/pixel).

2.6 Registration

The registration framework consisted of a segmentation stage for the porous titanium implant followed by two registration steps, as illustrated in Fig. 1. The segmented titanium scaffolds were used in the rigid registration to define the correspondence between the multimodal images independent of the surrounding bone and soft tissues. A more detailed description of the three steps of the registration is presented below.

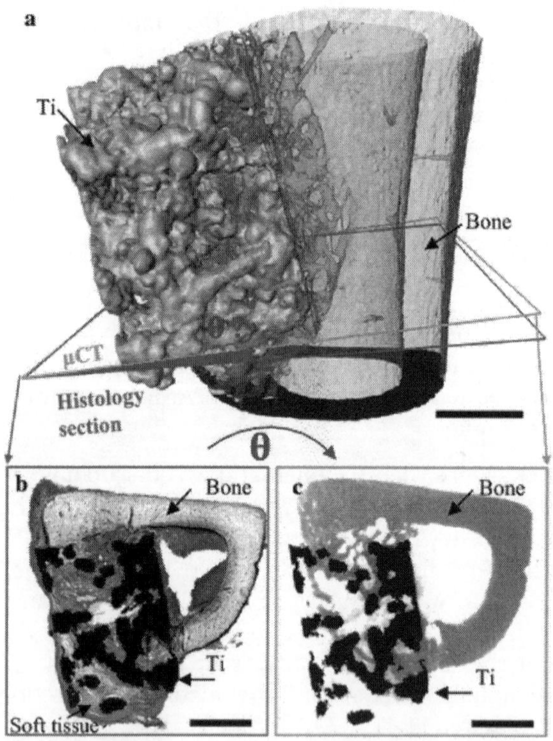

Figure 1.3D rendering of a 3 mm titanium scaffold within a rat tibiae defect illustrating a mismatch between the histology slice (*blue*) and the 2D slices of the μCT data (*green*). A multi-step registration was performed and final alignment was achieved between histology (**b**) and its corresponding 2D μCT images (**c**). *θ?* Indicates the unknown displacements and rotations required to register. *Scale bar* in all figures is 1000 μm (Color figure online)

Step 1 Segmentation of the titanium scaffold from the histological image and μCT volume.

The haematoxylin and multiple staining solution stained histology RGB (red, green and blue) images were split into the three RGB channels and the segmentation was performed on the red channel due to its superior brightness and contrast, which enabled easier segmentation of the histological image. An anisotropic diffusion filter was applied first to smooth the image while preserving the edge [29]. After that, global thresholding was applied to segment the titanium from the bone, allowing the modes of the histogram to be clearly distinguished in the 8-bit grayscale image. The thresholding value was acquired from the histogram by taking an average of the peak value of the titanium from three specimens.

Step 2 Coarse registration between the slice and volume datasets.

A search tool based on the correlation coefficient was used to identify the coarse correspondence between titanium segmented from histology and μCT (Fig. 1). The similarity between the porous titanium scaffold segmented from the μCT and histological results was defined by the correlation coefficient [30–32], implemented in Avizo (Visualisation Sciences Group, Merignac Cedex, France) and calculated as follows:

$$C = \frac{\sum_{i=1}^{n}(T_i - T_m) * (R_i - R_m)}{\sqrt{\sum_{i=1}^{n}(T_i - T_m)^2} * \sqrt{\sum_{i=1}^{n}(R_i - R_m)^2}}$$

where T_i is the intensity of the ith pixel in the target image and R_i is the ith intensity of the i^{th}pixel in the reference image. T_m and R_m are the mean intensity of the target image and reference image, respectively. A C value of 1 implies a positively identical registration between the two images.

Step 3 Refined registration of the target image.

Prior to the analysis, a parametric study was conducted to investigate the effect of the histological section thickness on the registration. If the thickness was equal to single voxel size, only a 2D transformation was allowed during the registration. Above 5 pixels, 6 degrees of freedom were allowed and the degree of similarity was stable. All the transformation metrics were recorded before validation.

2.7 Accuracy of the registration method

A cylindrical titanium implant was scanned twice in conditions similar to those used for the implanted tibia. The porous implant was then cut using a diamond saw at 80° to the axis of the cylinder.

The section was scanned at three different resolutions (2.5 μm/voxel, 4.5 μm/voxel, 9.0 μm/voxel). All data were processed and segmented based on the same protocol (Sect. 2.4).

The accuracy of the registration method was assessed by measuring the effect of image noise, scanning resolution and cutting (distortion) on the degree of similarity. To quantify the effect of image noise, a 2D μCT section was numerically extracted from the repeat scan and registered to the first scanned volume (reference). The cut section was then scanned at a different scanning resolution and registered to the reference volume. Finally, the CT sections before/after cutting were inspected visually to investigate any potential damage (material loss) and registered to assess any cutting induced distortion.

2.8 Histological and 2D μCT quantification of bone formation

Once the transformation metrics were applied to the 3D μCT data, the histology correspondence 2D μCT could then be acquired. In addition, a defect equivalent region of interest (ROI) was acquired by applying the same method (in 2.4) to the transformed average leg model and cylinder mask.

Bone ingrowth and BC inside the 2D ROI and volumetric shrinkage were calculated for each histological image and its corresponding 2D μCT. During the histomorphometric analysis, each specimen was quantified over five random orientations using ImageJ [33] (*XY* grid). Values of BI and BC were calculated as an average of the five values for each specimen. All the measurements were performed for three specimens per time point and the results were based on the average of these measurements. The histology corresponding to the 2D μCT image was analysed using the same procedure as the 3D μCT.

The following parameters were investigated histologically and by 2D μCT: (1) BI: bone ingrowth [(bone area/ROI) × 100 %]; (2) BC: bone contact [(BC area/(total scaffold area) × 100 %]; (3) Volumetric shrinkage of bone: (a) [(cortical bone area histology/cortical bone area in 2D μCT) × 100 %] (b) [scale factor (*X*-axis) × scale factor (*Y*-axis)].

2.9 Statistical analysis

All data were presented as a mean ± standard deviation. The Mann–Whitney U test was used for comparison of 3D μCT and 2D μCT/histological analysis across time points. A p value of < 0.05 was considered significant. After that, corresponding BI/BC values were used to show the regression line between histological and 2D μCT results. A 95 % confidence band for the regression line was applied (XLSTAT, 2014, Addinsoft, Inc., Brooklyn, NY, USA).

3 RESULTS

3.1 Accuracy of registration

Satisfactory correlation was achieved after 39 fine iterations corresponding to a correlation ratio (C) of 0.85 (Supplementary Fig. 1) where a correlation ratio of 1 corresponds to a perfect match between the histology section and μCT data. Error in registration, i.e., lower correlation coefficient could result from noise in μCT data, difference in resolution of the two type of images and from sample deformation on histology section preparation. Among these factors, the sample deformation was found to be the major source of the mismatch (data not shown).

3.2 Bone ingrowth (BI)

In total, twelve samples (n = 3 at 2, 3, 4 and 6 weeks' post-implantation) were imaged and reconstructed. Figure 2 shows the corresponding 2D μCT slice and histology section for 2, 4 and 6 weeks following the semiautomated registration method. The defect equivalent ROI has good consistency between different specimens and time points. Results of quantitative data measured from 2D μCT/histology and 3D μCT have been summarised in Table 1 and plotted in Fig. 3. The BI increased with time as measured from both μCT data and histology images. Both 2D μCT and histology had similar BI for all time points. In comparison, the mean of BI measured from 3D μCT was lower than that of histology for all the time points, this was statistically significant at 6 weeks.

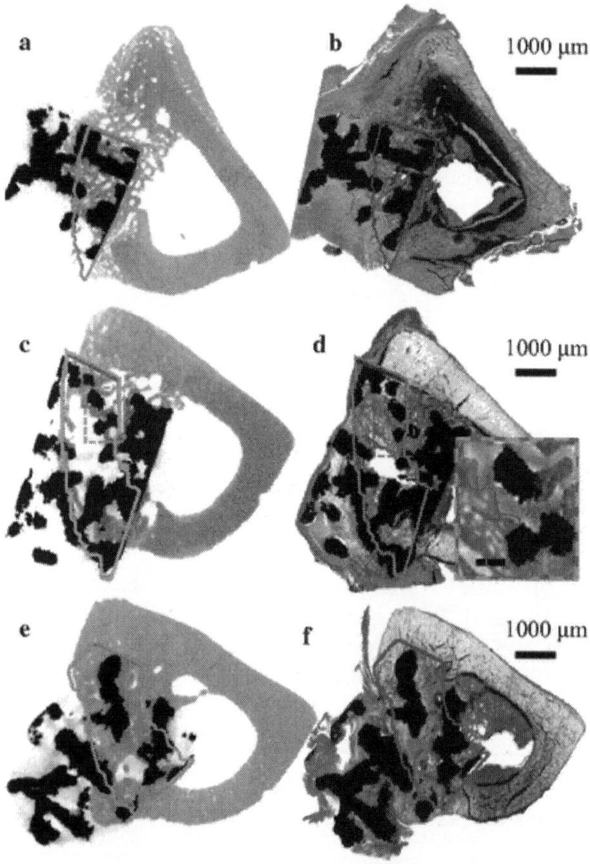

Figure 2. Examples of bone growth into the defect area at 2 (**a**, **b**), 3 (**c**, **d**) and 6 (**e**, **f**) weeks post-surgery. Bone formation and contact area were quantified in haematoxylin and multiple staining solution stained histological slices (ROI marked by *blue solid line*), which correlated well with its corresponding 2D μCT (ROI marked by *green solid line*). *Inset* of **d** shows zoomed in areas highlighted (*dashed line*) in **c** and **d**overlaid to show region of bone not detected (*blue arrow*) on μCT (Color figure online)

Table 1. Bone ingrowth into the porous titanium scaffold as a function of time measured from 2D μCT, histology and 3D μCT

Time (weeks)	BI (%)			BC (%)		
	2D μCT	Histology	3D μCT	2D μCT	Histology	3D μCT
2	25 ± 18	26 ± 17	23 ± 15	6 ± 9	7 ± 10	19 ± 14
3	52 ± 23	58 ± 22	40 ± 10	35 ± 27	26 ± 21	39 ± 19
4	55 ± 19	62 ± 19	44 ± 4	48 ± 21	48 ± 17	47 ± 2
6	84 ± 7	86 ± 7	61 ± 13	67 ± 11	72 ± 6	65 ± 9

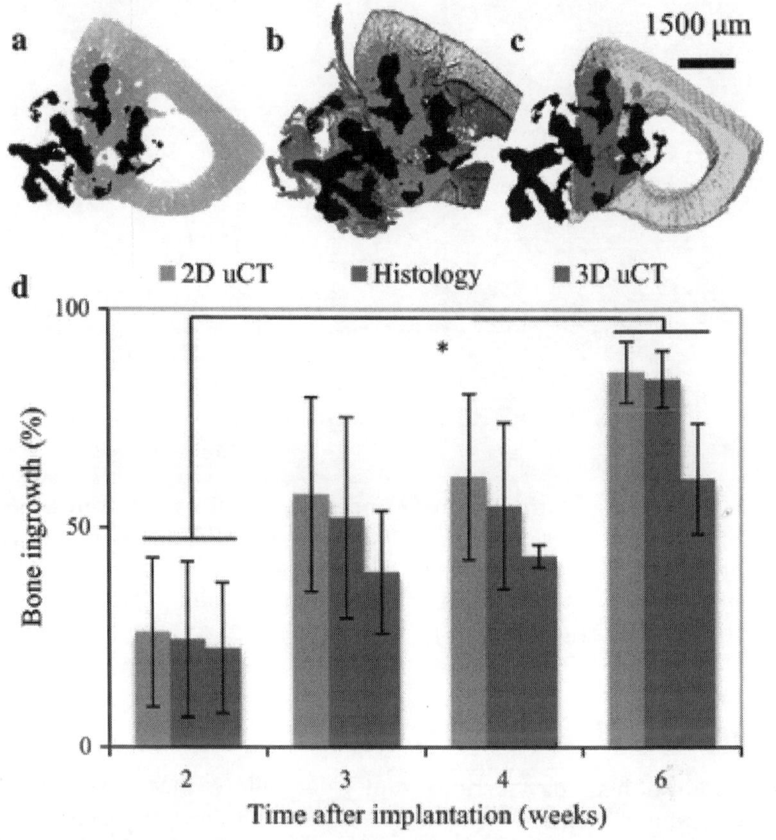

Figure 3. Bone ingrowth in corresponding **a** 2D μCT **b** histology and **c** 3D μCT image were measured as a function of time for 2, 3, 4 and 6 weeks (**d**). The quantity of newly formed bone increases significantly between 2 and 3 weeks and then after 4 weeks post-implantation

Figure 4 shows a plot of the bone formation within titanium implants from one end of the defect to the other after 2 and 6 weeks implantation. The exemplary plot demonstrates the increase in the amount of BI from 2 to 6 weeks. Further, after 6 weeks implantation the bone formation at the periphery was higher than at the core of the implant.

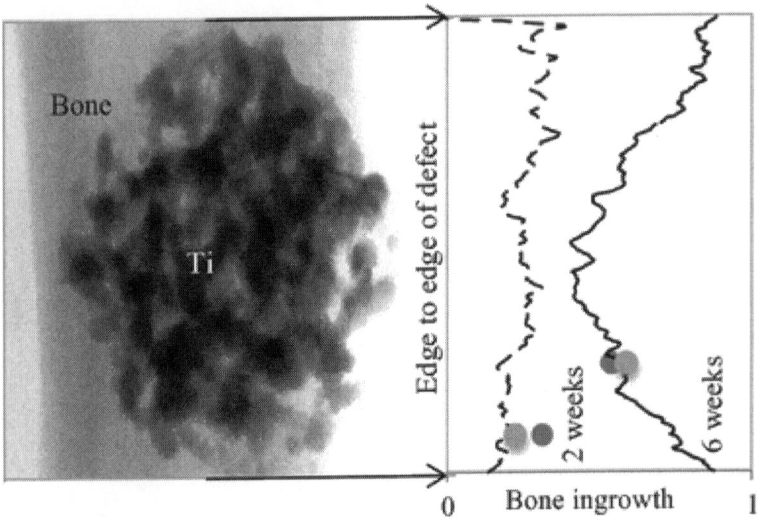

Figure 4. Bone ingrowth measured from 2D μCT slices and from edge to edge of a defect after 2 (*dashed line*) and 6 (*solid line*) weeks post-implantation. *Blue* and *greendots* on the *lines* mark the amount of bone ingrowth and location of histology sections and 2D μCT slices, respectively (Color figure online)

3.3 Bone implant contact (BC)

Registration of histology sections to μCT data allows measurement of BC from the same region and comparison of data obtained via each method. Quantification of BC measured on histology sections, 2D μCT slices and 3D μCT data for the samples at different time points are shown in Table 1 and Fig. 5. All three measurements show that the BC increases with time where more than 65 % of the titanium surface had been covered with the newly formed bone after 6 weeks of implantation. Figure 6 shows registered images of histology sections and 2D μCT slice after 2 and 6 weeks implantation. Tissue separation was observed at 2 weeks implantation while extensive tissue shrinkage (~15 %) was after 6 weeks implantation.

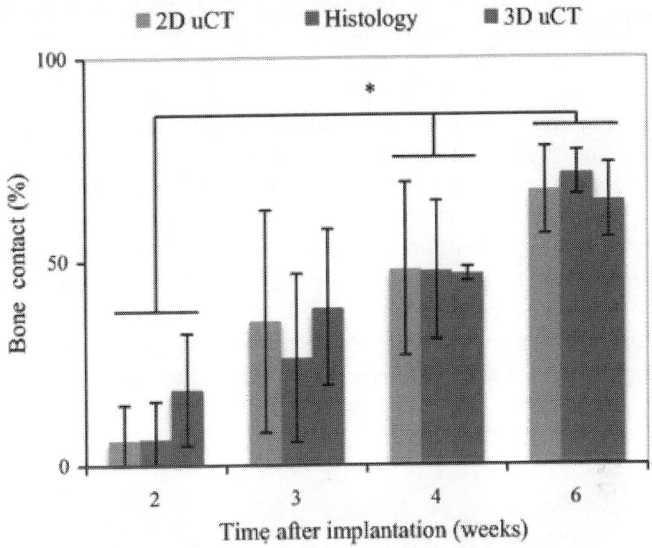

Figure 5. Bone implant contact was measured by 3D μCT, histology and its corresponding 2D μCT image. The amount of the bone implant contact shows a significant increase between 2 and 4 weeks, 2 and 6 weeks post-implantation

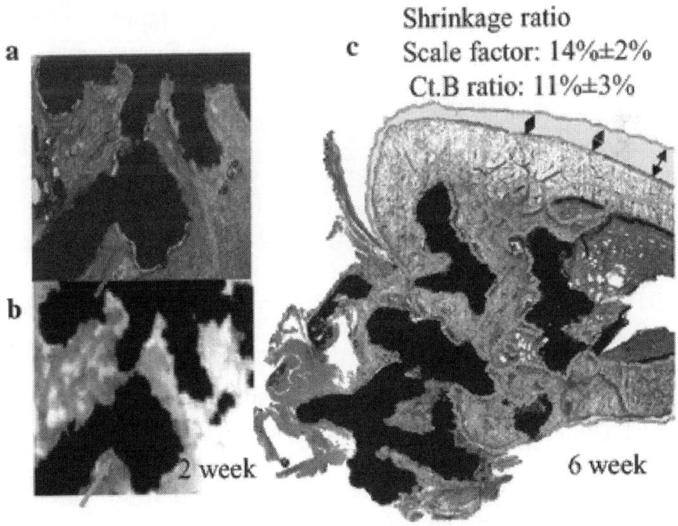

Figure 6. Tissue separation was observed on histology **a** in comparison to its corresponding 2D μCT slice **b** at 2 weeks post-implantation. Direct comparison of bone segmented from histology slices (haematoxylin and multiple staining solution stained tibiae) and its corresponding 2D μCT (marked as *green*) reveal a shrinkage (*black arrow*) of the bone tissue. Volumetric shrinkage was calculated based on scaling factor and cortical bone area ratio (**c**) (Color figure online)

4 DISCUSSION

Both μCT and histology methods have inherent advantages and limitations: with μCT a high number of slices can be generated per sample leading to volumetric information. A drawback is the relatively low resolution compared with histology. However, using a combination of μCT and histology, it is possible to perform a quantitative study in the same region. By using the non-deformed titanium (Supplementary Fig. 2) as the feature that was common in both μCT data and histology images registration was successfully performed. Importantly the accuracy of the registration was not affected by staining artefacts or soft tissue deformation. Figure 2 shows the corresponding 2D μCT slice and histology section for 2, 4 and 6 weeks following the semiautomated registration method.

4.1 Bone ingrowth (BI)

The BI increased with time as measured from both μCT data and histology images. In the case of 2D μCT and histology, no significant differences in BI were observed at any of the time points. The data obtained showed a high correlation ($r = 0.99$) of BI between 2D μCT and histology, suggesting that that the semiautomated registration process adopted in this study is robust and 2D μCT has promise for achieving comparable results as histology in terms of quantifying BI. However, a difference in BI (mean) between 3D μCT and histology analysis was revealed at 2, 3, 4 and 6 weeks (Table 1; Fig. 3) this was significant at 4 and 6 weeks. This is due to a combination of preferential BI in porous implants and histological sectioning method that has a bias towards the edge of the implant hence the BI measured from histology sections is always an overestimation of the average BI in the whole implant. Figure 4 shows a representative plot of the BI from edge to edge of the implants after 2 and 6 weeks post-implantation measured from 2D μCT slices. It shows that BI is consistent throughout the implant 2 weeks after implantation. After 6 weeks, significant increase in bone growth is observed; however, the edge to edge profile (Fig. 4) of BI is different to that after 2 weeks. At 6 weeks, preferential BI is observed where, at the edges >70 % of the pores are filled with new bone while at the core only ~50 % of the void space had bone. This fits well with the observation by Muschler et al. [34]. who show that once an osteoconductive porous implant is fitted into a defect, the porous structure provides channels for cell attachment and migration. The newly formed bone grows in from the edge, continuing to grow towards the core of the implant while the bone at the edge remodels. In the case of 2 weeks, the low attenuation coefficient of the non-mineralised bone at the early time points (2 and 3 weeks) cannot be resolved on μCT (Fig. 2d inset), this may explain the flat profile seen on Fig. 4. However, after 6 weeks' implantation, the majority of the bone is remodelled to a mineralised form which could be detected with confidence using μCT. This result suggests that 3D μCT is complimentary to histomorphometry analysis where the BI measured using 3D μCT is an average of BI of the whole implant, both edge and centre, to give a more representative result for bone formation, which is inherently volumetric.

4.2 Bone contact (BC)

In this study, the mean of BC-histology is smaller than that of 3D μCT at 2 and 3 weeks. This is because the BC is more sensitive to the deformation of the newly formed bone within the implant surface on sample preparation for histology [35]. This situation changes at weeks 4 and 6, where, the BC-histology is larger than that of 3D μCT, suggesting that the bone had remodelled and adhered well to the implant. Figure 6 shows examples of tissue deformation on histology section preparation, in particular after 2 weeks (Fig. 6a); the tissue adjacent to the implant separated away from the surface of the implant leading to a lower measure of the BC from histology images (Table 1). Figure 6b shows that the volumetric shrinkage of cortical bone through histology section preparation 6 weeks after implantation could be as high as 15 %. Although, there was a large amount of tissue shrinkage and deformation, the tissue separation from the implant was negligible. This could be due to large amount of mature bone occupying majority of the void space that gives rise to stable bone-implant interface, which enables calculation of BC with confidence from histology sections. However, Fig. 6 undoubtedly reaffirms the need for a non-destructive means of measuring BC.

5 CONCLUSIONS

A methodology was developed to perform semi-automatic correlative imaging using both μCT and quantitative histology (histomorphometry), allowing quantitative analysis of BI into titanium additive manufactured lattice implants. The evolution of bone density and quality over a 6 week period was tracked using BI and BC. It was shown that μCT provides full volumetric information, a higher degree of statistical significance, and is a simple and fast experimental technique. Histomorphometry is shown to complement μCT, providing higher resolution local information, particularly at early stages of BI.

The benefits of using correlative μCT and histomorphometric assessment over the gold standard technique of histology alone were demonstrated. Two potential areas where bias can be introduced when using histology alone are identified: errors due to the histology section measuring only a single plane within an complex 3D implant; and shrinkage during preparation of the histology section. The latter volumetric shrinkage error was quantified, and for the histology preparation used was ~15 %.

The correlative imaging methodology was then applied to quantify BI into an additive manufactured titanium lattice implant, quantifying the excellent BI and attachment achieved.

ACKNOWLEDGMENTS

This work was made possible by the facilities and support provided by the Diamond-Manchester Collaboration and the Research Complex at Harwell, funded in part by the EPSRC (EP/I02249X/1). We also thank professor David Hukins for valuable comments on the manuscript.

REFERENCES

1. Dempster DW, Compston JE, Drezner MK, Glorieux FH, Kanis JA, Malluche H, Meunier PJ, Ott SM, Recker RR, Parfitt AM. Standardized nomenclature, symbols, and units for bone histomorphometry: a 2012 update of the report of the ASBMR Histomorphometry Nomenclature Committee. J Bone Miner Res. 2013;28:2–17.
2. Parfitt AM. Bone histomorphometry: standardization of nomenclature, symbols and units (summary of proposed system). Bone. 1988;9:67–9.
3. Gielkens PFM, Schortinghuis J, de Jong JR, Huysmans MCDNJM, van Leeuwen MBM, Raghoebar GM, Bos RR, Stegenga B. A comparison of micro-CT, microradiography and histomorphometry in bone research. Arch Oral Biol. 2008;53:558–66.
4. Donath K, Breuner G. A method for the study of undecalcified bones and teeth with attached soft tissues. J Oral Pathol Med. 1982;11:318–26.
5. Lane J, Ráliš ZA. Changes in dimensions of large cancellous bone specimens during histological preparation as measured on slabs from human femoral heads. Calcif Tissue Int. 1983;35:1–4.
6. Compston JE, Croucher PI. Histomorphometric assessment of trabecular bone remodelling in osteoporosis. Bone Miner. 1991;14:91–102.
7. Jones JR, Poologasundarampillai G, Atwood RC, Bernard D, Lee PD. Non-destructive quantitative 3D analysis for the optimisation of tissue scaffolds. Biomaterials. 2007;28:1404–13.

8. Yue S, Lee PD, Poologasundarampillai G, Yao Z, Rockett P, Devlin AH, Mitchell CA, Konerding MA, Jones JR. Synchrotron X-ray microtomography for assessment of bone tissue scaffolds. J Mater Sci Mater Med. 2010;21:847–53.

9. David V, Laroche N, Boudignon B, Lafage-Proust MH, Alexandre C, Ruegsegger P, Vico L. Noninvasive in vivo monitoring of bone architecture alterations in hindlimb-unloaded female rats using novel three-dimensional microcomputed tomography. J Bone Miner Res. 2003;18:1622–31.

10. Midha S, Kim TB, van den Bergh W, Lee PD, Jones JR, Mitchell CA. Preconditioned 70S30C bioactive glass foams promote osteogenesis in vivo. Acta Biomater. 2013;9:9169–82.

11. Clark DP, Badea CT. Micro-CT of rodents: state-of-the-art and future perspectives. Phys Med. 2014;30:619–34.

12. Postnov A, Zarowski A, De Clerck N, Vanpoucke F, Offeciers F, Van Dyck D, Peeters S. High resolution micro-CT scanning as an innovatory tool for evaluation of the surgical positioning of cochlear implant electrodes. Acta Otolaryngol. 2006;126:467–74.

13. Buie HR, Bosma NA, Downey CM, Jirik FR, Boyd SK. Micro-CT evaluation of bone defects: applications to osteolytic bone metastases, bone cysts, and fracture. Med Eng Phys. 2013;35:1645–50.

14. Jones AC, Arns CH, Sheppard AP, Hutmacher DW, Milthorpe BK, Knackstedt MA. Assessment of bone ingrowth into porous biomaterials using MICRO-CT. Biomaterials. 2007;28:2491–504.

15. Kim TB, Yue S, Zhang Z, Jones E, Jones JR, Lee PD. Additive manufactured porous titanium structures: through-process quantification of pore and strut networks. J Mater Process Technol. 2014;214:2706–15.

16. Hedberg EL, Kroese-Deutman HC, Shih CK, Lemoine JJ, Liebschner MA, Miller MJ, Yasko AW, Crowther RS, Carney DH, Mikos AG. Methods: a comparative analysis of radiography, microcomputed tomography, and histology for bone tissue engineering. Tissue Eng. 2005;11:1356–67.

17. Hutton BF, Thomas BA, Erlandsson K, Bousse A, Reilhac-Laborde A, Kazantsev D, Pedemonte S, Vunckx K, Arridge S, Ourselin S. What approach to brain partial volume correction is best for PET/MRI? Nucl Instr Meth Phys Res A. 2013;702:29–33.

18. Bernhardt R, van den Dolder J, Bierbaum S, Beutner R, Scharnweber D, Jansen J, Beckmann F, Worch H. Osteoconductive modifications of Ti-implants in a goat defect model:

characterization of bone growth with SR μCT and histology. Biomaterials. 2005;26:3009–19.

19. Baril E, Lefebvre L, Hacking S. Direct visualization and quantification of bone growth into porous titanium implants using micro computed tomography. J Mater Sci Mater Med. 2011;22:1321–32.

20. Jimbo R, Coelho PG, Vandeweghe S, Schwartz-Filho HO, Hayashi M, Ono D, Andersson M, Wennerberg A. Histological and three-dimensional evaluation of osseointegration to nanostructured calcium phosphate-coated implants. Acta Biomater. 2011;7:4229–34.

21. Park YS, Yi KY, Lee IS, Jung YC. Correlation between microtomography and histomorphometry for assessment of implant osseointegration. Clin Oral Implants Res. 2005;16:156–60.

22. Stalder AK, Ilgenstein B, Chicherova N, Deyhle H, Beckmann F, Müller B, Hieber SE. Combined use of micro computed tomography and histology to evaluate the regenerative capacity of bone grafting materials. Int J Mater Res. 2014;105:679–91.

23. Markelj P, Tomaževič D, Likar B, Pernuš F. A review of 3D/2D registration methods for image-guided interventions. Med Image Anal. 2012;16:642–61.

24. Museyko O, Marshall RP, Lu J, Hess A, Schett G, Amling M, Kalender WA, Engelke K. Registration of 2D histological sections with 3D micro-CT datasets from small animal vertebrae and tibiae. Comput Methods Biomech Biomed Eng. 2015;18:1658–73.

25. Buytaert J, Goyens J, De Greef D, Aerts P, Dirckx J. Volume shrinkage of bone, brain and muscle tissue in sample preparation for micro-CT and light sheet fluorescence microscopy (LSFM). Microsc Microanal. 2014;20:1208–17.

26. Mullen L, Stamp RC, Brooks WK, Jones E, Sutcliffe CJ. Selective laser melting: a regular unit cell approach for the manufacture of porous, titanium, bone in-growth constructs, suitable for orthopedic applications. J Biomed Mater Res B. 2009;89:325–34.

27. Zhu Y, Zhao M, Li H, Zhang P. Micro-CT artifacts reduction based on detector random shifting and fast data inpainting. Med Phys. 2013;40:031114.

28. Chao SM, Tsai DM, Chiu WY, Li WC, editors. Anisotropic diffusion-based detail-preserving smoothing for image restoration. Image processing (ICIP), 2010 17th IEEE international conference on; 2010: IEEE.

29. Perona P, Malik J. Scale-space and edge detection using anisotropic diffusion. Pattern analysis and machine intelligence. IEEE Trans On. 1990;12:629–39.

30. Pearson K. Mathematical contributions to the theory of evolution. III. Regression, heredity, and panmixia. Philos Trans Royal Soc London A. 1896;187:253–318.

31. Goshtasby AA. Similarity and dissimilarity measures. Image registration principles, tools and methods. London: Springer; 2012. p. 7–66.

32. Roche A, Malandain G, Pennec X, Ayache N. The correlation ratio as a new similarity measure for multimodal image registration. Medical image computing and computer-assisted interventation—MICCAI'98. Berlin: Springer; 1998. p. 1115–24.

33. Schneider CA, Rasband WS, Eliceiri KW. NIH image to imageJ: 25 years of image analysis. Nat Methods. 2012;9:671–5.

34. Muschler GF, Nakamoto C, Griffith LG. Engineering principles of clinical cell-based tissue engineering. J Bone Joint Surg. 2004;86:1541–58.

35. Johansson CB, Morberg P. Importance of ground section thickness for reliable histomorphometrical results. Biomaterials. 1995;16:91–5.

INDEX

O

R

S

T

Z